T0134889

Springer Theses

Recognizing Outstanding Ph.D. Research

Aims and Scope

The series "Springer Theses" brings together a selection of the very best Ph.D. theses from around the world and across the physical sciences. Nominated and endorsed by two recognized specialists, each published volume has been selected for its scientific excellence and the high impact of its contents for the pertinent field of research. For greater accessibility to non-specialists, the published versions include an extended introduction, as well as a foreword by the student's supervisor explaining the special relevance of the work for the field. As a whole, the series will provide a valuable resource both for newcomers to the research fields described, and for other scientists seeking detailed background information on special questions. Finally, it provides an accredited documentation of the valuable contributions made by today's younger generation of scientists.

Theses are accepted into the series by invited nomination only and must fulfill all of the following criteria

- They must be written in good English.
- The topic should fall within the confines of Chemistry, Physics, Earth Sciences, Engineering and related interdisciplinary fields such as Materials, Nanoscience, Chemical Engineering, Complex Systems and Biophysics.
- The work reported in the thesis must represent a significant scientific advance.
- If the thesis includes previously published material, permission to reproduce this must be gained from the respective copyright holder.
- They must have been examined and passed during the 12 months prior to nomination.
- Each thesis should include a foreword by the supervisor outlining the significance of its content.
- The theses should have a clearly defined structure including an introduction accessible to scientists not expert in that particular field.

More information about this series at http://www.springer.com/series/8790

Elodie Resseguie

Electroweak Physics at the Large Hadron Collider with the ATLAS Detector

Standard Model Measurement, Supersymmetry Searches, Excesses, and Upgrade Electronics

Doctoral Thesis accepted by
University of Pennsylvania, Philadelphia, USA

Author
Dr. Elodie Resseguie
Physics Department
Lawrence Berkeley National Laboratory
Berkeley, CA, USA

Supervisor
Prof. Elliot Lipeles
Department of Physics and Astronomy
University of Pennsylvania
Philadelphia, PA, USA

ISSN 2190-5053 ISSN 2190-5061 (electronic)
Springer Theses
ISBN 978-3-030-57018-7 ISBN 978-3-030-57016-3 (eBook)
https://doi.org/10.1007/978-3-030-57016-3

© The Editor(s) (if applicable) and The Author(s), under exclusive license to Springer Nature Switzerland AG 2020
This work is subject to copyright. All rights are solely and exclusively licensed by the Publisher, whether the whole or part of the material is concerned, specifically the rights of translation, reprinting, reuse of illustrations, recitation, broadcasting, reproduction on microfilms or in any other physical way, and transmission or information storage and retrieval, electronic adaptation, computer software, or by similar or dissimilar methodology now known or hereafter developed.
The use of general descriptive names, registered names, trademarks, service marks, etc. in this publication does not imply, even in the absence of a specific statement, that such names are exempt from the relevant protective laws and regulations and therefore free for general use.
The publisher, the authors and the editors are safe to assume that the advice and information in this book are believed to be true and accurate at the date of publication. Neither the publisher nor the authors or the editors give a warranty, expressed or implied, with respect to the material contained herein or for any errors or omissions that may have been made. The publisher remains neutral with regard to jurisdictional claims in published maps and institutional affiliations.

This Springer imprint is published by the registered company Springer Nature Switzerland AG
The registered company address is: Gewerbestrasse 11, 6330 Cham, Switzerland

Supervisor's Foreword

It is an honor to introduce the context of this thesis and highlight its contents. It has been a pleasure to work with and mentor Elodie Resseguie in her graduate career. This thesis covers a significant set of scientific results and technical developments related to testing the standard model (SM) of particle physics and searching for new as of yet unknown particles.

In 2012, the ATLAS and CMS collaborations at the Large Hadron Collider, LHC, succeeded in finding the predicted and long-sought Higgs boson. This discovery also confirmed the existence of a major problem with the SM, a lack of explanation for the light Higgs mass. This problem, known as the Higgs naturalness problem, manifests is self as a fine-tuning required to get the right observed Higgs mass. A candidate solution to this problem is Supersymmetry (SUSY), which introduces a symmetry between fermions and bosons such that, for every particle in the SM, there is a new particle whose spin differs from the original by a half-integer. The standard model, its deficiencies, and a SUSY are reviewed in Chap. 2.

The superpartner particles that couple to the strong force already have tight constraints on them from recent LHC data. Because the remaining partners only couple through electroweak processes and are much less copiously produced in proton-proton collisions, these are much less constrained. The current experimental situation and phenomenological context are reviewed in Chap. 6.

This thesis presents contributions to three searches as part of the ATLAS collaboration in three different relevant regions of the SUSY model parameter space, described in Chaps. 7, 10, and 11 as well a study of a dominant *WZ* background to these searches, described in Chap. 5. Each of theses searches required the development of selection to optimize sensitivity, complex modeling of the background processes which could be mistaken for the signal, modeling of the detector response, and detailed validation of that model. Contributions to all of these aspects of the analysis work are described.

Within the ATLAS collaboration, a separate analysis in one of these SUSY search regions found an intriguing excess that could be evidence of a signal. Chapter 8 describes detailed studies of this result including reproducing the result with a simplified method. Chapter 9 then describes adding approximately three times more data. The excess did not persist in the additional data suggesting that the original excess was a statistical fluctuation.

In addition to this broad set of analysis projects, Elodie Resseguie also contributed significantly to both the operation of the existing detector and the development of the next-generation detector. Operating a complex detector requires constant effort including visits to the control room at all hours. Elodie worked on such operations as an expert on the Transition Radiation Tracker (TRT) data acquisition system. Elodie also used her extensive electronics knowledge to help test the custom electronics chips to be used in the ATLAS tracking detector upgrade.

Taken as a whole, this thesis represents a very productive graduate career and I'm sure we can expect many more accomplishments in the future.

Philadelphia, USA
September 2020

Elliot Lipeles

Abstract

This thesis presents searches for the production of charginos and neutralinos, supersymmetric partners of the gauge bosons, decaying via W and Z bosons. To understand the dominant background in these searches, WZ, the WZ cross section measurement is also performed. The measurement and searches were performed with both 36.1 fb^{-1} and 139 fb^{-1} of data collected at p $\sqrt{s} = 13\,\text{TeV}$ with the ATLAS detector. The WZ cross section is measured to be 50.6 ± 3.6 fb (statistical and systematic), which is consistent with the theoretical prediction calculated at next-to-leading order in QCD. The supersymmetric searches are motivated by two models: Higgsino and wino production. The Higgsino search makes use of low-transverse-momentum leptons and leads to the first result of Higgsino production at a collider experiment since the Large Electron-Positron Collider. The wino production searches discussed in this thesis show a tension in the exclusion limit with one search finding an excess of data above the background prediction and another search targeting the same model and phase space which sees data consistent with the background prediction. The excess is followed up with 139 fb^{-1} of data and the excess is found to be no longer significant. The signatures of electroweak SUSY will continue to be very important measurements throughout the life of the LHC but are difficult to perform because they require a lot of luminosity; as a result, these searches will remain interesting during the High-Luminosity LHC (HL-LHC) program, which will deliver 3000 fb^{-1} of data. The increase in radiation during HL-LHC will require a new inner detector. To test the performance of this new detector, the thesis also describes the testing of one of the read-out chips, the HCC130, through calibration tests and irradiation.

Elodie Resseguie

Preface

This thesis presents my work on the ATLAS experiment from 2014 to 2019. As a graduate student at the University of Pennsylvania, I worked on the ATLAS experiment focusing on detector operations, detector upgrades, *WZ* cross section measurement, and electroweak supersymmetry (SUSY) searches.

Hardware: HCC Testing and Irradiation

I began my research on ATLAS by working on the electronics for an upgraded tracker. In order to gain more sensitivity to processes with small production cross sections, the rate of proton-proton collisions must increase by a factor of ten; this configuration of the LHC is known as High-Luminosity LHC (HL-LHC). The particles emerging from the collisions are measured by the ATLAS detector. In the higher rate conditions, the inner part of the current ATLAS detector will no longer be able to resolve charged particles due to the large number of collisions per crossing of proton bunches. As a result, a new inner detector, made of silicon will take its place. I worked on testing one of the application-specific integrated circuits (ASICs), the hybrid controller chip or HCC130, used read-out data from the silicon-strip sensors. I tested the HCC130, the prototype chip for the upgrade to the inner detector before HL-LHC. The work is summarized in Chap. 4. I designed two boards for the testing: one that would hold the HCC130 chip with protection diode, called the passive board, and another which would have active components and the field-programmable gate array (FPGA) and system-on-chip manufactured by XILINX, the MicroZed, called the active board. I selected Digital-to-Analog converter (DAC) and Analog-to-Digital converter (ADC) integrated circuits to be used on the active board to generate voltages for components on the board and read back voltages from the HCC130. I designed the layout of the board, figuring out how all the signals would be routed from the HCC130 and from each pin of each chip on the two boards.

During this time, I also worked on the FPGA code for the DAC and ADC. By looking at the chip's specification sheets, I determined which signals were needed to control the DAC and ADC and read back their values, as well as the clocks needed for both. I also wrote code in C that allows the user to select the DAC and ADC and, for the DAC, which voltage to generate. This code interfaces between the user and the FPGA code that generates and reads back signals from the DAC/ADC.

I wrote additional code to develop testing protocols to test the HCC130, which allowed us to uncover some bugs in the chip design as well as provide a protocol for collaborators at Lawrence Berkeley National Laboratory to test all HCC130 given to the collaboration for further studies. The tests I performed on one block of the HCC130, the Analog Monitor (AM), were important in the development of the AMAC, a monitoring chip based on the HCC130's AM. This work can be found in the upgrade Technical Design Report [1].

The detector will be exposed to a high level of radiation during operation of the HL-LHC, so it is important to test the radiation tolerance of the chips; furthermore, early tests of a different readout chip had shown that there was a rise in the current at low radiation doses. The HCC130 was irradiated to test for this current rise. For the HCC130 irradiation at Brookhaven National Laboratory with cobalt-60, I was responsible for writing the data acquisition (DAQ) code. I installed Linux on the microcontroller to be able to save the data collected. I wrote code to be able to perform different tests at different frequencies. While we were waiting for irradiation, I took months of calibration data to determine that the HCC130 and the DAQ system was stable in room temperature conditions. I was also responsible for the analysis of the data from the irradiation and from the HCC130 testing. The results from irradiation have been helpful in determining that the HCC130 did not suffer from a large current bump at low total ionizing dose.

Operation: TRT DAQ

After working 2 years on the HCC130 testing, I moved to CERN where I started working as part of the Transition Radiation Tracker (TRT) DAQ team. The TRT is part of the current inner detector (ID), and we were responsible for ensuring that the TRT was functional and the TRT data was of good quality during data-taking and during machine development times. I was a DAQ expert so part of my responsibilities included being on-call for a week at a time and responding to calls from the personnel monitoring the TRT in the ATLAS control room, called ID shifters, at all hours. The 2018 data taking would be taken at higher pileup conditions, meaning that the TRT would not be able to sustain the increase in data. As a result, we upgraded some of the electronics on our boards. I participated in testing the new electronics and in the installation of electronics. After the electronics upgrade, I calibrated the TRT to ensure uniform levels of occupancy. I also checked the TRT for radiation damage by analyzing calibration runs; however, in 2016, only one calibration run was completed. In 2017, I wanted the TRT to have many calibrations to see the impact of radiation damage on the TRT as a function of luminosity. In order to collect these runs, the ID shifters need to take calibration runs when there is no data taking. I wrote a training manual so they would be able to take these calibrations and the experts would only need to analyze them. During this time, there was a migration to make and all our software needed to change to work with this new method of compilation, so I was involved in software changes as well.

After a year of working as part of the TRT DAQ team, I became a TRT DAQ coordinator and was responsible for training new students and postdocs working on the TRT as well as organizing the TRT tests conducted during machine development weeks. I led the effort in investigating software resynchronizations that the TRT experienced during the high-μ part of the heavy ion run. We believe that the resyncs are due to current increasing in the read-out electronics when the TRT receives many trigger requests after a period of not receiving triggers, which cause the buffers to fill up rapidly. I also wrote code so we would be able to change the settings of our trigger during cosmic running to match the conditions during a special LHC run.

Standard Model Measurement: WZ Cross Section

The first physics analysis I was involved with was the *WZ* cross section measurement at the beginning of Run 2. I got involved in this measurement because there was not enough data to make a significant new SUSY limit using 2015 data. This also allowed me to understand the modeling of *WZ*, the dominant background in the electroweak SUSY searches I performed. I studied the *ZZ* background, the second-most dominant background, and how it enters the *WZ* signal region since *ZZ* produces four leptons, while the *WZ* search requires events with three leptons. The fourth lepton can enter the signal region either because it is outside detector acceptance, the lepton has a large η or small p_{T}, or it is not identified. The leptons that are not identified have anti-ID scale factors applied to them, which corrects the *ZZ* prediction. After applying anti-ID scale factors to the unidentified leptons, the *ZZ* yield increases by 14% in the signal region. The *WZ* cross section measurement is further described in Chap. 5 and in our 2016 publication [2].

Electroweak Supersymmetry: Searches and Excess Follow-Up

Most of my thesis work has been spent on Supersymmetry searches. Physics beyond the Standard Model (SM) is an important search program because the SM does not explain the observation of a light Higgs mass, despite the quadratic divergence of the quantum corrections to the Higgs mass, or provide a dark matter candidate. R-parity conserving SUSY is a compelling model of new physics addressing these issues, and discovering it or excluding the simplest scenarios is well motivated. Given the strong LHC constraints on squark and gluino production, electroweak (EWK) production is an interesting mode to search for SUSY and will benefit from the increase in luminosity during Run 3 and during HL-LHC. I have focused most of my efforts on EWK SUSY searches on the ATLAS experiment, guided by two different simplified models: Bino lightest SUSY particle (LSP) with Wino next-to-lightest SUSY particle (NLSP), motivated by dark matter, and Higgsino LSP, motivated by naturalness since the mass parameter μ, governing the mass of Higgsinos, is on the order of the weak scale. Higgsino LSPs result in small ("compressed") mass splittings between the neutralinos resulting in soft leptons, missing transverse momentum, and a dilepton mass endpoint. Compressed scenarios are also worth pursuing for Bino LSP models because these contain uncovered regions of phase space, and SUSY particles need to be light to satisfy the dark matter relic density constraint.

In the Bino LSP search, the SUSY particles decay via on-shell W and Z bosons with a dominant WZ background. The signal is differentiated from the background by the presence of two additional invisible particles (LSPs). This results in extra missing energy ($E_{\mathrm{T}}^{\mathrm{miss}}$) quantified as m_T, the invariant mass of the leptons and $E_{\mathrm{T}}^{\mathrm{miss}}$ using only quantities transverse to the beamline. This search is performed using 2015 and 2016 data. The traditional variable has a problem that WZ background can accidentally be reconstructed with large m_T if the wrong lepton is chosen as the lepton from the W. I developed a new variable to mitigate this problem called $m_{\mathrm{T}}^{\mathrm{min}}$ which allowed for greater background rejection than the nominal assignment. I worked on the signal region optimization, separating the region into a jet-veto region (to target larger mass splittings between SUSY particles) and an initial-state-radiation (ISR) region (to target smaller mass splittings). Control regions are used to normalize the dominant backgrounds to data, while validation regions check the modeling of the normalized background. Both regions are kinematically similar but orthogonal to the signal regions. I also developed control regions for the WZ background and validation regions that also contain a jet veto and an ISR region to be close kinematically to the signal region. This search is further described in Chap. 7 and in our published work [3].

After verifying that there is a good background modeling in the validation regions, we look at the data in the signal regions, a process called "unblinding". After unblinding, we saw good agreement between the observed data and the background expectation; however, another search also using 2015–16 data but using the Recursive Jigsaw Reconstruction (RJR) technique saw an excess of data above the background expectation in a similar region of phase space [4]. The RJR technique is a complex method to reconstruct an event, boosting to different frames of references and calculating kinematic variables in those frames. I developed a simplified selection for the same phase space and reproduced the excess, a technique called "emulated RJR" (eRJR). The eRJR technique is described in Chap. 8.

My study was instrumental in understanding the excess seen in this search and in explaining more intuitively the RJR technique.

I studied whether the excess of data above background remains with the addition of 2017 and 2018 data to determine if the excess grows with the additional data or is due to statistical fluctuation or mismodeling of the background. The excess decreased with the full Run 2 data set and the observed candidates in the signal region are consistent with the expected backgrounds. This work is detailed in Chap. 9. For the excess follow-up, I have studied the migration to the new reconstruction algorithm of the 2015–16 data that was unblinded by the RJR search. I studied the efficiencies as a function of pile-up for the selection for 2015–16 data and MC, taken at lower pileup conditions, and for 2017 data and MC, taken at higher pileup conditions. Due to changes of efficiencies in a variable used to define a validation region (VR) in the selection with an ISR jet, VR-ISR, I developed two additional ISR-VR to further check the background modeling. Most of the selection was kept identical to the RJR search; however, in the control (CR) and validation regions with a jet veto, CR-low, and VR-low definitions were changed to make the regions orthogonal to the fake lepton measurement region and to the ISR regions. These changes had a negligible impact on the *WZ* normalization factor extracted derived in the control region. I also estimated the fake lepton background using the fake factor data-driven estimate. The fake background originates from photon conversion, heavy flavor decay, or from jets faking a lepton and cannot be properly estimated using MC. I also measured the normalization factor for backgrounds from top quark production and decay. I calculated the fake lepton systematics and the *WZ* theory systematics. Finally, I worked on the statistical interpretations of the results using the HISTFITTER package [5] producing an exclusion limit, model-independent limits, and final yields with full systematics. This work was published as a conference note [6, 7].

I have also been working on two compressed searches, searches with small p_T leptons: Higgsino and off-shell Wino production with Bino LSP. The higgsino model produces very small p_T leptons, called soft leptons. Because these leptons are so soft, we cannot use lepton triggers. Instead, we use an ISR jet to boost the system, which allows us to use $E_{\mathrm{T}}^{\mathrm{miss}}$ triggers. I studied the efficiency of a trigger that has been developed for this search which contains jets, leptons, and $E_{\mathrm{T}}^{\mathrm{miss}}$, as well as different $E_{\mathrm{T}}^{\mathrm{miss}}$ triggers. I also did other studies related to the $Z \rightarrow \tau\tau$ background, acceptances of $m_{\ell\ell}$ reweighting of higgsino samples as opposed to using wino-bino samples, and truth studies to validate the signal MC. Finally, I worked on the HistFitter implementation for the statistical interpretations of results. This search led to the first limits set on Higgsino production since LEP. This work can be found in Chap. 10 and in our publication [8].

I have also been involved in developing a signal region optimization for an off-shell wino-bino search decaying to three leptons. I have used many lessons learned from participating in the compressed search. The compressed search makes use of the invariant mass of the two leptons, $m_{\ell\ell}$, for higgsino models and the transverse mass, m_{T2}, for slepton models to perform shape fits, since those models have kinematic edges at their splittings in their respective discriminating variable. I have used the three lepton kinematics in the calculation of m_{T2}, and this variable has a signal kinematic edge at its mass splitting. This signal region also takes advantage of the kinematic edge in the $m_{\ell\ell}^{\mathrm{min}}$ distribution by performing a shape fit. A cut on $m_{\ell\ell}^{\mathrm{max}}$ reduces contamination from on-shell *WZ* background. The signal region is separated into a jet veto region, which makes use of lepton triggers, and an ISR region, triggered with $E_{\mathrm{T}}^{\mathrm{miss}}$ triggers. I have also added additional variables, m_{T} for the jet veto, and $\frac{p_{\mathrm{T}}^{3\ell}}{E_{\mathrm{T}}^{\mathrm{miss}}}$ for the ISR region to further reject backgrounds. This search is currently ongoing but my contribution can be found in Chap. 11.

I have been involved in many areas of ATLAS during my time at the University of Pennsylvania. As we move toward Run 3 and beyond, I would love the opportunity to collaborate on projects that would improve how we use the data we have already collected, have an impact on the data that will be collected, and finally, work on testing chips and boards for ITK so that we can continue taking data in the future. If light SUSY particles are accessible at the LHC, we could have the opportunity to not only discover them but be able to measure their properties. Exclusions for light SUSY masses could limit the most appealing solution to the open questions of the Standard Model and the Higgs naturalness problem.

Philadelphia, USA
June 2019

Elodie Resseguie

References

1. ATLAS Collaboration (2017) Technical design report for the ATLAS inner tracker strip detector. Tech Rep CERN-LHCC-2017-005. ATLAS-TDR-025, CERN, Geneva, https://cds.cern.ch/record/2257755. (document), 4.1, 4.1, 4.2, 4.5, 4.6, 4.23
2. ATLAS Collaboration (2016) Measurement of the W±Z boson pair-production cross section in pp collisions at $\sqrt{s} = 13$ TeV with the ATLAS detector. Phys Lett **B**762 pp 1–22, arXiv: 1606.04017 [hep-ex]. (document), 5
3. ATLAS Collaboration (2018) Search for electroweak production of supersymmetric particles in final states with two or three leptons at $\sqrt{s} = 13$ TeV with the ATLAS detector. Eur Phys J **C**78(12):995, arXiv:1803.02762 [hep-ex]. (document), 7, 7.9.1, 8.1, 8, 8.5
4. ATLAS Collaboration (2018) Search for chargino-neutralino production using recursive jigsaw reconstruction in final states with two or three charged leptons in proton-proton collisions at $\sqrt{s} = 13$ TeV with the ATLAS detector. Phys Rev **D**98 9:092012, arXiv:1806.02293 [hep-ex]. (document), 8.1, 8, 8.3.1, 8.5, 9, 9.5.4.3, 9.6
5. Baak M, Besjes G J, Cï£·te D, Koutsman A, Lorenz J, Short D (2015) HistFitter software framework for statistical data analysis. Eur Phys J **C7**5, 153, arXiv:1410.1280 [hep-ex]. (document), 7.9, 9.5, 9.5.4.3, 10.8
6. ATLAS Collaboration (2019) Search for chargino-neutralino production with mass splittings near the electroweak scale in three-lepton final states in $\sqrt{s} = 13$ TeV pp collisions with the ATLAS detector. Tech Rep ATLAS-CONF-2019-020, CERN, Geneva, http://cds.cern.ch/record/2676597. (document), 9

7. ATLAS Collaboration, Aad G et al. (2020) Search for chargino-neutralino production with mass splittings near the electroweak scale in three-lepton final states in $\sqrt{s}$ = 13 TeV pp collisions with the ATLAS detector. Phys Rev **D**101 7:072001, arXiv:1912.08479 [hep-ex]. (document), 9
8. ATLAS Collaboration (2018) Search for electroweak production of supersymmetric states in scenarios with compressed mass spectra at $\sqrt{s}$ = 13 TeV with the ATLAS detector. Phys Rev **D**97, 5:052010, arXiv:1712.08119 [hep-ex]. (document), 10

Acknowledgements

I would like to thank my mentors and advisors who have encouraged me and cultivated my passion for physics. They include my undergraduate advisors Tony Tyson and Max Chertok. They introduced me to physics research and instrumentation and taught me to love both. I would like to thank my graduate advisor Elliot Lipeles. You taught me to look at physics critically and to think outside of the box. I really appreciate the support and advice I have received over the years. Thank you to the other Penn faculty: Joe Kroll, Brig Williams, and Evelyn Thomson. Your advice and mentorship has been invaluable during these years. The Penn group fosters collaboration across advisors and this has been my favorite part of graduate school. I really benefited by being able to learn from the other faculty members and also share ideas with my fellow graduate students.

When I first joined the Penn group, I started working on instrumentation. I would like to thank the Penn Instrumentation Group for taking me in and teaching me so much about board design, chip testing, and just chatting in general about electronics. Thank you to Mitch Newcomer, Paul Keener, Adrian Nikolica, Godwin Mayers, and Mike Reilly. I would also like to thank my collaborators at Lawrence Berkeley National Laboratory for working with us to develop the firmware for our chip testing: Carl Haber, Niklaus Lehmann, and Timon Heim. Finally, I would like to thank our collaborators at Brookhaven National Laboratory who helped with the chip irradiation, including Dave Lynn, Stefania Stucci, and Jim Kierstead.

At CERN, I joined the TRT DAQ team. Working on a part of the ATLAS detector and having to change electronic boards when the magnetic field was on was one of my favorite moments in graduate school. I would like to thank Anatoli Romaniouk, Dominik Derendarz, and Andrea Bocci for their leadership and guidance. I learned that the successful operation of the TRT is only possible due to collaboration between many groups of experts, so I would also like to thank the DCS team: Jolanta Olszowska, Elziebeta Banas, and Zbyszek Hajduk. Finally, thanks to the DAQ team. It was a lot of fun working with you all: Chris Meyer, Khilesh Mistry, Bijan Haney, Keisuke Yoshihara, Shion Chen, Daniil Ponomarenko, Vincent Wong, Ian Dyckes, and Joe Mullin.

A large part of my experience on ATLAS was working on physics analyses. I want to especially thank Joana Machado Miguens. You were the first person I worked with on physics analysis and you have been a tremendous help to me; I definitely would not have been successful without you. I'm also glad we are now friends and that you showed me that life is better with a scooter. Another collaborator I want to thank is Joey Reichert. Joey, Joana, and I worked on many analyses together and it was so great working with both of them. I will truly miss working with you both! In addition, I would like to thank my other collaborators on the *WZ* cross section measurement, Kurt Brendlinger, and Will Di Clemente and those I've spent countless hours working on SUSY searches including Sarah Williams, Nicola Abraham, Ben Hooberman, Mike Hance, Stefano Zambito, Moritz Backes, Jesse Liu, Lorenzo Rossini, Sheena Schier, Julia Gonski, Jeff Shahinian, Iacopo Vivarelli, Jeff Dandoy, Ian Dyckes, Sara Alderweireldt, Shion Chen, Broos Vermeulen, Eli Baeverfjord Rye, and Lucia Pedraza.

There are other Penn ATLAS members not yet mentioned that I'd like to thank, including Christian Herwig, Ben Rosser, Lucas Flores, James Heinlein, and Rachael Creager. I especially want to thank two of my friends Leigh Schaefer and Bill Balunas. I will miss our conversations and our pranks.

Surviving grad school would not be have been possible without my friends. I would like to thank my roommates Christian Herwig, Ashley Baker, and Saul Kohn. I will never forget my graduate classmates; we were a really special group to have remained close during the years. I would like to thank especially Eric Horsley ("neighbs"). I will miss sushi dinners, walking to DRL in the morning, your wit, and sense of humor. I would also like to thank Ashley Baker, Charlotte Pfeifer, Tatyana Gavrilchenko, and Lia Papadopoulos. I will always be grateful for our brunches, crossword puzzles, animated shorts, and gym times. I would also like to thank the Penn soft matter group who adopted me as an honorary member, including Randy Kamien, Asja Radja, Jason Rocks, Desi Todorova, Tristan Sharp, Francesca Serra, Kevin Chiou, and Daniel Sussman. I would like to thank my friends at CERN: Magda Chesltowska, Aaron Armbruster, Alex Tuna, Jonathan Long, Doug Schaefer, Stany Sevova, Stefan Gadatsch, Tina Ojeda, Reyer Band, Riju Dasgupta, and Christine McLean. Also, I would like to thank my friends Sasha Fedotova and Alyssa Schluter. Thanks for always visiting me wherever in the world I might be.

Finally, I would like to thank my family, especially my parents Florence and Patrick, and my brother Richard. I would not be here without their love and support throughout the years. Your example of hard work and dedication has inspired me to be the best I can be. I would not be where I am today without you. I would also like to thank Max Lavrentovich. You have been a constant support during grad school and you are my best friend. I can't imagine what grad school would have been like without you; you always believed in me when I doubted myself. I'll fondly remember the many dinners we shared, our love of trivia, our many travels, and your tolerance for my love of all things autumn. We have shared so many adventures around the world and I can't wait to see what's next for us.

Contents

Chapter 1
Introduction

A particle consistent with the Higgs boson hypothesized in the Standard Model was discovered in the first data-taking period (Run 1) of the Large Hadron Collider (LHC). With this discovery, all the particles of the Standard Model have been found and the model can be said to be complete. The Standard Model, described in Chap. 2 has been successful, resulting in theory predictions that have good agreement with experimental observations. Also, improvements in theoretical calculations have resolved tensions between theoretical predictions and observation, as in the WZ cross section measurement, discussed in Chap. 5. However, the Standard Model does not provide a satisfactory explanation for the relatively small Higgs mass (called the Higgs naturalness problem) and does not provide a dark matter candidate. Supersymmetric (SUSY) models have solutions to these and many other problems, and it is crucial to test these theories at the LHC during Run II.

R-parity conserving SUSY is a compelling model of new physics addressing these issues, and discovering it or excluding the simplest scenarios is well-motivated. Given the strong LHC constraints on squark and gluino production, electroweak (EWK) production is an interesting mode to search for SUSY. The motivation for EWK SUSY and the SUSY parameter mixing is discussed in Chap. 6. A few searches for EWK SUSY on the ATLAS experiment are discussed, guided by two different simplified models: a bino lightest SUSY particle (LSP) with wino next-to-lightest SUSY particle (NLSP) (motivated by dark matter and discussed in Chap. 7) and Higgsino LSP in Chap. 10 (motivated by naturalness, since the mass parameter μ, governing the mass of Higgsinos, is on the order of the weak scale). Higgsino LSPs result in small ("compressed") mass splittings between the neutralinos resulting in soft leptons, missing transverse momentum, and a dilepton mass endpoint. Compressed scenarios are also worth pursuing for bino LSP models because these contain unexplored regions of phase space. Also, SUSY particles need to be light to satisfy the dark matter relic density constraint. These issues are discussed in Chap. 11.

© The Editor(s) (if applicable) and The Author(s), under exclusive license
to Springer Nature Switzerland AG 2020

E. Resseguie, *Electroweak Physics at the Large Hadron Collider with the ATLAS Detector*, Springer Theses,
https://doi.org/10.1007/978-3-030-57016-3_1

While the two searches discussed above did not observe any excesses above the background prediction, a search using the Recursive Jigsaw Reconstruction (RJR) technique targeting the wino-bino model had excesses with local significances of 2.1 σ and 3.0 σ in two orthogonal regions. A new technique, called emulated RJR (eRJR), explores the intersection between the conventional and RJR approaches to better understand the tension in the exclusion limits produced by the two analyses. It emulates the variables used by the RJR technique with conventional laboratory frame discriminating variables, providing a simple set of variables that are easily reproducible. This technique reproduces the excesses seen by the RJR analysis. The excess is then followed-up using a larger dataset corresponding to 139 fb^{-1} of pp collision data collected between 2015 and 2018. The technique and search are described in Chaps. 8–9.

EWK production of SUSY will benefit from the increase in luminosity during Run 3 and during HL-LHC (high luminosity LHC) since those SUSY models have small production cross sections. During HL-LHC, there were will be 200 interactions per beam crossing, known as pile-up, while the current run conditions have only 60 interactions per beam crossing. The inner detector of the ATLAS detector, described in Chap. 3, will not be able to sustain those higher radiation conditions. As a result, the inner detector will be replaced by a silicon detector called the Inner Tracker (ITk), which is composed of two subdetectors: Pixels and Strips. Prototypes and production chips for the readout system of ITk are being tested for resilience against radiation damage, as well as high performance in high pile-up conditions. Chapter 4 discusses the testing of one of the readout chips, the HCC130 and its performance during irradiation. These tests are crucial to the ATLAS experiment and ensure good data taking in the future. This data can be used to deepen our understanding of what lies beyond the Standard Model and to narrow down possible solutions to the Higgs naturalness problem.

Chapter 2
Theoretical Framework

This chapter introduces the theoretical framework for the work presented in this thesis. The Standard Model (SM) particles will be introduced before building the SM from symmetry arguments. Deficiencies of the SM will be described before introducing supersymmetry as a proposed solution to the open questions of the SM. The formalism presented summarizes the following books by Thomson [1], Langacker [2], Schwartz [3], Halzen [4], and Georgi [5].

2.1 Introduction to the Standard Model

The Standard Model of particle physics is a theory that provides a unified description of three of the four forces [strong, electromagnetic (EM), and weak] that govern all known particles. The SM particles, summarized in Fig. 2.1, can be separated into groups: quarks, leptons, the force carriers, and the Higgs boson:

- Charged leptons: electron, muon, tau
- Neutral leptons (neutrinos): ν_e, ν_μ, ν_τ
- Up-type quarks with spin $2/3$: up, charm, top
- Down-type quarks with spin $-1/3$: down, strange, bottom
- Force carriers: photon, W boson, Z boson, gluon
- Scalar boson: Higgs

The SM particles are divided further into three generations summarized in Table 2.1. The electron, its associated neutrino, the up quark, and the down quark are known as the first generation of particles. For each of the four first-generation particles, there are two other particles (one for each other generation) which differ

© The Editor(s) (if applicable) and The Author(s), under exclusive license
to Springer Nature Switzerland AG 2020

E. Resseguie, *Electroweak Physics at the Large Hadron Collider with the ATLAS Detector*, Springer Theses,
https://doi.org/10.1007/978-3-030-57016-3_2

Table 2.1 Three generations of quarks and leptons with their masses and spins

	Leptons			Quarks		
	Particle	Spin	Mass/GeV	Particle	Spin	Mass/GeV
First generation	Electron (e^-)	−1	0.0005	down (d)	−1/3	0.003
	Neutrino (ν_e)	0	$< 10^{-9}$	up (u)	+2/3	0.005
Second generation	Muon (μ^-)	−1	0.106	strange (s)	−1/3	0.1
	Neutrino (ν_μ)	0	$< 10^{-9}$	charm (c)	+2/3	1.3
Third generation	Tau (τ^-)	−1	1.78	bottom (b)	−1/3	4.5
	Neutrino (ν_τ)	0	$< 10^{-9}$	top (t)	+2/3	174

only by their mass. These particles form the fundamental particles; all other particles are composite.

Protons and neutrons are formed from quarks and are composite. These particles are known collectively as hadrons. Particles composed of three quarks, called baryons, have half-integer spin. Protons and neutrons are baryons because the proton is made of two up-quarks and a down-quark, while the neutron consists of two down-quarks and one up-quark. Mesons are composite particles composed of two quarks. These include the pions (bound state of up and down quarks), kaons (formed with strange quarks), and D mesons (formed with charm quarks). Mesons have spin 1 or 0, so they are bosons. Of course, baryons, having half-integer spin, are fermions.

The particles interact with each other through the four fundamental forces: gravity, electromagnetism, the strong force and the weak force. The gravitational interaction between the particles is very small compared to the other forces and is not incorporated in the SM. The quarks interact through the strong, EM, and weak force. The charged leptons interact through the weak and EM forces only. Finally, the neutral leptons only interact via the weak force.

Each of the three forces in the SM is described by a Quantum Field Theory (QFT) corresponding to the exchange of a spin 1 boson, known as a gauge boson. All particle interactions involve the exchange of a gauge boson. The photon is the gauge boson of Quantum Electrodynamics (QED), the theory that explains the EM force. The photon is a massless particle. The charged W boson and the neutral Z bosons are the gauge boson of the weak force, responsible for the nuclear beta decay. The gluon, the gauge boson of the strong force, described by Quantum Chromodynamics (QCD), is massless. The forces with their masses and strengths are summarized in Table 2.2.

Electrically charged particles, such as the quarks and charged leptons, can interact via the EM force. Because the photon has no mass, the EM interaction has an infinite range, but with a strength proportional to $1/r^2$. The strength of the interaction

Table 2.2 Three forces of the Standard Model and their relative strength and mass

Force	Strength	Boson	Spin	Mass/GeV
Strong	1	Gluon (g)	1	0
EM	10^{-3}	Photon (γ)	1	0
Weak	10^{-8}	W boson ($W^{\pm}$)	1	80.4
		Z boson (Z)	1	91.2

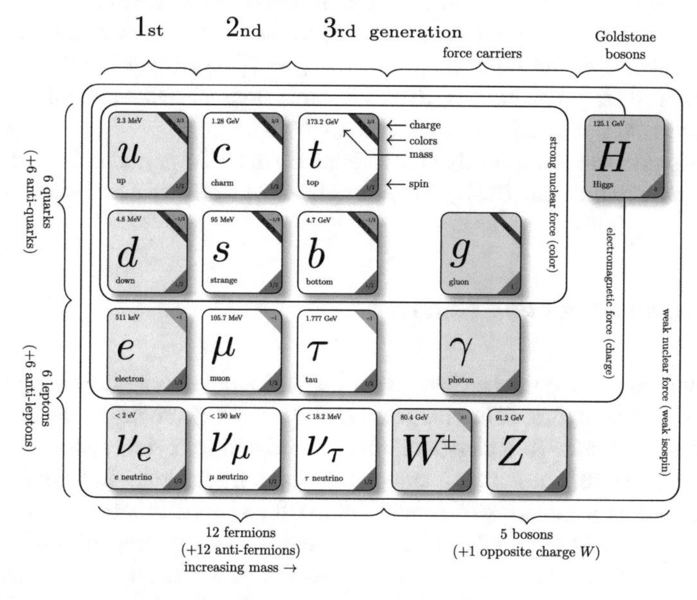

Fig. 2.1 The Standard Model of particle physics includes the quarks, leptons, their force mediators (photon, gluon, and W/Z bosons), and the Higgs boson. The mass, charge, spin, and color is shown when appropriate [6]

between the gauge boson and the fermion is determined by a coupling constant g (which for EM is just the electric charge e). Rather than working with the coupling constant, a dimensionless constant, α, is used. In EM, the constant is the fine structure constant, $\alpha = \frac{e^2}{4\pi\epsilon_0\hbar c} = \frac{1}{137}$.

The particles that are able to interact with each force carry the charge of that force. Thus the quarks carry the color charge of QCD. Each flavor of quark comes in three possible "colors": red (R), green (G), and blue (B), which triples the number of quarks shown in Fig. 2.1. Anti-quarks carry the anti-color charge ($\bar{\mathrm{R}}$, $\bar{\mathrm{G}}$, $\bar{\mathrm{B}}$). In the SM, all particles are color singlets and no states with free quarks are observed, a phenomenon known as color confinement. Gluons also have color and, unlike photons, can directly interact with each other, which has consequences for long range interactions. Each gluon carries one unit of color and one unit of anti-color, in one of a set of mixed-color states known as the "color octet". Hence, there are a total of eight different gluons. The QCD interaction is more strongly-coupled than EM, with $\alpha_s \approx 1$. This makes the QCD theory notoriously difficult to analyze perturbatively.

The weak force interaction is governed by weak isospin (an analog of charge and color of EM and QCD, respectively). Leptons that participate in the weak force interaction differ by one unit of electric charge. Studies of β-decays, such as $n \rightarrow pe^{-}\nu_e$ have shown that only left-handed particles participate in charged-current interactions (interactions mediated by the charged W boson). Neutral-current interactions are mediated by the Z boson. Also, W and Z bosons have self-interactions, similar to the gluons. The W boson can also couple to the photon since it is charged. The intrinsic strength of the weak interaction, $\alpha_W \approx \frac{1}{30}$, is stronger than that of EM; however, since the gauge bosons, W and Z, have large masses, the weak interaction has a sharp cutoff at distances larger than 10^{-18} m.

The Higgs boson is responsible for giving masses to SM particles through spontaneous symmetry breaking (SSB). This particle and the SSB will be discussed below.

2.1.1 Symmetries and Currents

As mentioned earlier, particles that can participate in force interactions carry a charge associated with that force. The charge is associated to a symmetry of the Lagrangian describing that force. To understand this, it is helpful to review symmetries in the context of QFTs and the action S associated with a Lagrangian. In a QFT, particles are described as excitations of a field, which for simplicity we can take to be a complex scalar field $\phi(x)$. Lagrangians are used to describe properties of that field. Lagrangians are exclusively used in a QFT because they are Lorentz invariant and, as a result, are composed only of the field and derivatives of that field, $\mathcal{L} = \mathcal{L}(\phi, \partial_\mu \phi)$. The action S is then defined as the integral over time of the Lagrangian:

$$S = \int \mathrm{d}^4x \, \mathcal{L}(x) \tag{2.1}$$

If the field, ϕ, is varied by some arbitrary perturbing field $\delta\phi$, $\phi \rightarrow \phi + \delta\phi$, then the action transforms to

$$\delta S = \int d^4x \left[\frac{\partial \mathcal{L}}{\partial \phi} \delta\phi + \frac{\partial \mathcal{L}}{\partial(\partial_\mu \phi)} \delta(\partial_\mu \phi) \right] \tag{2.2}$$

Integrating by parts (allowing for $A\partial_\mu B = -(\partial_\mu A)B$ inside the integral), Eq. (2.2) becomes,

$$\delta S = \int \mathrm{d}^4x \left[\left[\frac{\partial \mathcal{L}}{\partial \phi} - \partial_\mu \frac{\partial \mathcal{L}}{\partial(\partial_\mu \phi)} \right] \delta\phi + \partial_\mu \left[\frac{\partial \mathcal{L}}{\partial(\partial_\mu \phi)} \delta\phi \right] \right], \tag{2.3}$$

where we have ignored boundary terms because we take the boundary to be at infinity where we assume all the fields vanish. Similarly, the last term in the integral in Eq. 2.3 is a total derivative and can thus be reduced to a vanishing boundary term, also. Suppose now that the action S is stationary under all perturbations $\delta\phi$ of the field ϕ. In this case, we must have a vanishing variation δS:

$$\frac{\delta S}{\delta \phi} = \frac{\partial \mathcal{L}}{\partial \phi} - \partial_\mu \frac{\partial \mathcal{L}}{\partial(\partial_\mu \phi)} = 0. \tag{2.4}$$

These conditions are the Euler-Lagrange equations, or the equations of motion for the field ϕ. In other words, variations of the action S under arbitrary perturbations $\delta\phi$ vanish if ϕ obeys the equations of motion. This is the celebrated principle of stationary action, which we may now exploit to understand symmetries.

If a Lagrangian $\mathcal{L}$ is invariant under a *specific* type of variation, $\phi \to \phi + \delta\phi$, then the condition $\delta\mathcal{L} = 0$ under such variations gives us a conserved quantity and the corresponding transformation is called a symmetry of the Lagrangian. The fact that each symmetry is associated with a conserved quantity is known as Noether's theorem. There are two types of symmetries considered in the SM. Global symmetries change the field over all of space-time in the same manner while local symmetries vary the field differently at each point in space-time. Let us examine how this works. The vanishing variation condition yields

$$\begin{aligned} 0 &= \delta\mathcal{L} \\ &= \frac{\partial \mathcal{L}}{\partial \phi}\delta\phi + \frac{\partial \mathcal{L}}{\partial(\partial_\mu \phi)}\delta(\partial_\mu\phi) \\ &= \left[\frac{\partial \mathcal{L}}{\partial \phi} - \partial_\mu \frac{\partial \mathcal{L}}{\partial(\partial_\mu \phi)}\right]\delta\phi + \partial_\mu\left[\frac{\partial \mathcal{L}}{\partial(\partial_\mu \phi)}\delta\phi\right] \end{aligned} \tag{2.5}$$

If the equations of motion are satisfied, Eq. (2.5) simplifies to $\partial_\mu J_\mu = 0$, where

$$J_\mu = \frac{\partial \mathcal{L}}{\partial(\partial_\mu \phi)}\delta\phi \tag{2.6}$$

J_μ is the conserved current. The total charge, Q, defined as,

$$Q = \int \mathrm{d}^3x\, J_0 \tag{2.7}$$

satisfies

$$\partial_t Q = \int \mathrm{d}^3x\, \partial_t J_0 = 0 \tag{2.8}$$

meaning that the total charge does not change with time and is conserved. This is the essence of Noether's theorem in the QFT context.

The Standard Model is a gauge theory built using symmetry considerations. The SM is a locally-gauged symmetry, making it invariant under local gauge transformations. It is invariant under the $SU(3)_C \times SU(2)_L \times U(1)_Y$ symmetry. The $SU(3)$ symmetry describes QCD where the color charge (C) is conserved. The electroweak (EW) sector, after the unification of EM and weak forces, is invariant under $SU(2)_L \times U(1)_Y$ where Y refers to the weak hypercharge and L refers to the fact that the weak force acts only on left-handed particles. These symmetry considerations are used to specify the form of the SM Lagrangian, which may then be analyzed in the way described in this section.

2.1.2 Electromagnetic Force: Quantum Electrodynamics

The Dirac Lagrangian describes a fermion with mass m,

$$\mathcal{L}_{\text{Dirac}} = \bar{\psi}(i\gamma^\mu \partial_\mu - m)\psi, \tag{2.9}$$

with the γ-matrices, or Dirac matrices, satisfying the anticommutator relationship, $\{\gamma^\mu, \gamma^\nu\} = 2g^{\mu\nu}$. The Euler Lagrange equation (2.5) associated with this Lagrangian (called the Dirac equation of motion) is

$$(i\gamma^\mu \partial_\mu - m)\psi = 0. \tag{2.10}$$

Quantum electrodynamics, QED, which describes the electromagnetic force, is invariant under a local $U(1)$ symmetry, meaning that the field transforms as,

$$\psi(x) \rightarrow e^{-i\alpha(x)}\psi(x), \tag{2.11}$$

where $\alpha(x)$ is an arbitrary scalar function. This transformation commutes, meaning that $e^{-i\alpha^{(1)}(x)}e^{-i\alpha^{(2)}(x)} = e^{-i\alpha^{(2)}(x)}e^{-i\alpha^{(1)}(x)}$ for two arbitrary functions $\alpha^{(1,2)}(x)$. A symmetry with a commuting transformation is called an abelian gauge symmetry. The Noether current associated with this $U(1)$ symmetry is given by

$$J_\mu = \bar{\psi}\gamma_\mu\psi \tag{2.12}$$

and the conserved charge,

$$Q = \int \mathrm{d}^3x J_0 = \int \mathrm{d}^3x \psi^\dagger\psi, \tag{2.13}$$

is the electron number, which is the number of electrons minus the number of positrons.

The Lagrangian, $\mathcal{L}_{\text{Dirac}}$, however, is not invariant under this transformation. It transforms as,

$$\partial_\mu \psi(x) \to e^{-i\alpha(x)} \partial_\mu \psi(x) - i e^{-i\alpha(x)} \psi(x) \partial_\mu(\alpha(x)) \tag{2.14}$$

To make a proper Lagrangian for QED, a generalization of the derivative is defined, called the covariant derivative, which transforms under the $U(1)$ transformation as

$$D_\mu \psi(x) \to e^{-i\alpha(x)} D_\mu \psi(x). \tag{2.15}$$

Then, an auxiliary field, A_μ (the vector potential), is introduced to cancel the additional term in Eq. (2.14). The covariant derivative is chosen to be

$$D_\mu \equiv \partial_\mu + ieA_\mu, \tag{2.16}$$

where e is a constant (the electric charge). The auxiliary field transforms as

$$A_\mu \to A_\mu - \frac{1}{e} \partial_\mu \alpha(x). \tag{2.17}$$

All of these rules ensure that local gauge invariance is preserved. The invariance under these transformations can be checked,

$$\begin{aligned} \mathcal{D}'_\mu e^{-i\alpha(x)} \psi &= (\partial_\mu + ieqA'_\mu) e^{-i\alpha} \psi \\ &= e^{-i\alpha(x)} (\partial_\mu \psi + i(\partial_\mu \alpha)\psi + ieA_\mu - i\frac{e}{e}(\partial_\mu \alpha)\psi) \\ &= e^{-i\alpha(x)} \mathcal{D}_\mu \psi \end{aligned} \tag{2.18}$$

Thus, to make $\mathcal{L}_{\text{Dirac}}$ invariant under the local $U(1)$ symmetry, ∂_μ is replaced with the covariant derivative $\mathcal{D}_\mu$ in Eq. (2.9)

$$\begin{aligned} \mathcal{L} &= \bar{\psi}(i\gamma^\mu D_\mu - m)\psi \\ &= \mathcal{L}_{\text{Dirac}} - e\bar{\psi}\gamma^\mu A_\mu \psi. \end{aligned} \tag{2.19}$$

Note that requiring local invariance has generated a new term in the Lagrangian, $ie\bar{\psi}\gamma^\mu A_\mu \psi$, which represents the interaction between a fermion and a photon, as illustrated in Fig. 2.2.

To complete the construction of the QED Lagrangian, the dynamics for the field A_μ are introduced as a commutator of the covariant derivative. The commutator is gauge invariant and is proportional to the curvature (field strength) $F_{\mu\nu} = \partial_\mu A_\nu - \partial_\nu A_\mu$. It is an effect of parallel transport of the vector field A_μ around closed paths. If it is 0, vector returns to the point of origin pointing in original direction. For an arbitrary function f, we have

$$\frac{i}{e}[\mathcal{D}_\mu, \mathcal{D}_\nu] f = \frac{e}{i}[\partial_\mu, \partial_\nu] f + [\partial_\mu, A_\nu] f + [\partial_\nu, A_\mu] f - ieq[A_\mu, A_\nu] f \tag{2.20}$$

$$= (\partial_\mu A_\nu - \partial_\nu A_\mu) f = F_{\mu\nu} f. \tag{2.21}$$

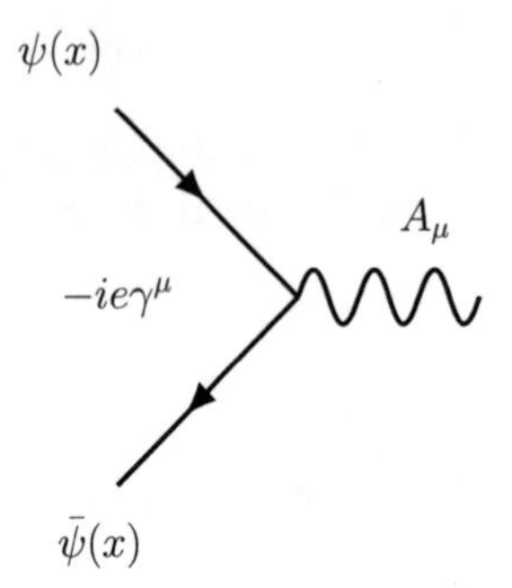

Fig. 2.2 Feynman diagram of an interaction between fermions and a photon

So, the kinetic term for the photon is described by the Lagrangian

$$\mathcal{L}_{\text{kin}} = -\frac{1}{4} F_{\mu\nu} F^{\mu\nu}. \tag{2.22}$$

This Lagrangian is invariant under the $U(1)$ described in Eqs. (2.11) and (2.17). A photon mass term of the form $\frac{1}{2}m^2 A_\mu A^\mu$ is forbidden by gauge invariance and, thus, the photon field is massless.

In summary, the Lagrangian of QED is

$$\mathcal{L}_{\text{QED}} = \bar{\psi}(i\gamma^\mu \partial_\mu - m)\psi - e\bar{\psi}\gamma^\mu A_\mu \psi - \frac{1}{4} F_{\mu\nu} F^{\mu\nu}, \tag{2.23}$$

and describes the kinetic energy and mass of the fermion, the kinetic energy of the photon field, and the interaction between the two.

2.1.3 *Strong Force: Quantum Chromodynamics*

Quantum Chromodynamics, QCD, is the quantum field theory that describes the strong force. The QCD Lagrangian has the same structure as the QED Lagrangian but the invariance is with respect to the non-abelian group $SU(3)$ instead of the abelian $U(1)$ group in QED. The $SU(3)$ symmetry group has 8 parameters, $\frac{\lambda^a}{2}$, the Gell-Mann matrices, which are the eight generators of the gluon fields. The matrices satisfy the Lie algebra,

$$[\lambda^a, \lambda^b] = if^{abc}\lambda^c, \tag{2.24}$$

where f^{abc} is the structure constant of the group. We can define $T^a = \frac{\lambda^a}{2}$. Similar to the abelian case, the local transformation of any field, $q(x)$, can be written as the exponentiation of the generator times a parameter

$$q(x) \to e^{-i\alpha(x)^a T^a} q(x) \tag{2.25}$$

Unlike in the abelian cases, two gauge transformations do not commute. To make the Lagrangian invariant under local SU(3) transformation, new auxiliary fields are introduced just like in QED. Since there are 8 parameters, 8 auxiliary fields are needed. These gluon fields are denoted $G_\mu^a(x)$. The covariant derivative is defined as,

$$\partial_\mu q \to (\partial_\mu - i g_s T^a G_\mu^a(x))\psi \equiv \mathcal{D}_\mu q \tag{2.26}$$

where g_s is the strong coupling constant. The gluon fields transform under $SU(3)$ as

$$G_\mu^a \to G_\mu^a - \frac{1}{g_s}\partial_\mu \alpha_a - f_{abc}\alpha_b G_\mu^c \tag{2.27}$$

which includes the structure constant of the group since the fields do not commute. Unlike the transformation of the photon field, the transformations of the gluon fields include self-interaction terms. Just like for the photon, the kinetic energy of the gluon fields is derived from the commutation relation,

$$\begin{aligned} \mathcal{D}_\mu, \mathcal{D}_\nu t &= [\partial_\mu - i g_s T^a G_\mu^a, \partial_\nu - i g_s T^b G_\nu^b] \\ &= -i g_s [\partial_\mu G_\nu^a - \partial_\nu G_\mu^a + g_s f^{abc} G_\mu^b G_\nu^c] T^a \end{aligned} \tag{2.28}$$

Thus, the kinetic term for the gluon fields is

$$\begin{aligned} G_{\mu\nu}^a &= \frac{i}{g_s T^a}[\mathcal{D}_\mu, \mathcal{D}_\nu] \\ &= [\partial_\mu G_\nu^a - \partial_\nu G_\mu^a + g_s f^{abc} G_\mu^b G_\nu^c] \end{aligned} \tag{2.29}$$

This has a similar structure as the $F_{\mu\nu}$ of the photon field but with the addition of the group structure constant. The Lagrangian for QCD is

$$\mathcal{L}_{\text{QCD}} = \sum_c \bar{q}_c (i\gamma^\mu \partial_\mu - m_c) q_c - g_s \sum_c (\bar{q}_c \gamma^\mu \frac{\lambda_a}{2} q_c) G_\mu^a - \frac{1}{4} G_{\mu\nu}^a G_a^{\mu\nu}, \tag{2.30}$$

The first term describes the kinetic energy and mass of the quark fields q_c. The second term describes the interaction between quarks and gluons, shown in the Feynman diagram Fig. 2.3. The third term describes the self-interaction of the gluons,

$$G_{\mu\nu}^a G_a^{\mu\nu} = \ldots + (\partial_\mu G_\nu^a - \partial_\nu G_\mu^a) g_s f^{abc} G_\mu^b G_\nu^c + g_s^2 f^{abc} f^{aef} G_\mu^b G_\nu^c G^{c\mu} G^{f\nu} \tag{2.31}$$

which gives a 3-gluon and 4-gluon interaction vertices shown in Fig. 2.4. The ellipses correspond to non-interacting quadratic terms.

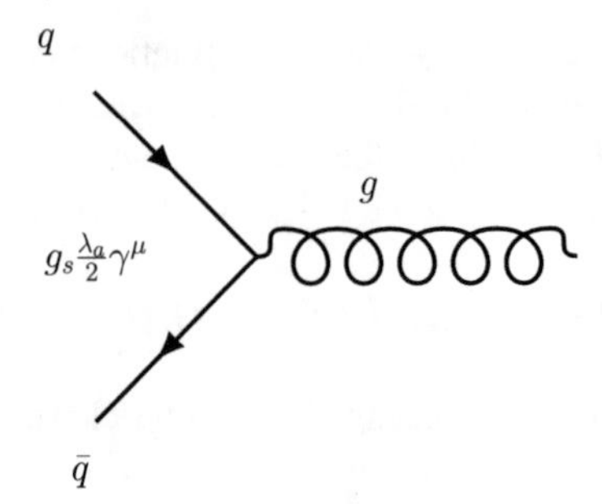

Fig. 2.3 Feynman diagram of an interaction between quarks and a gluon

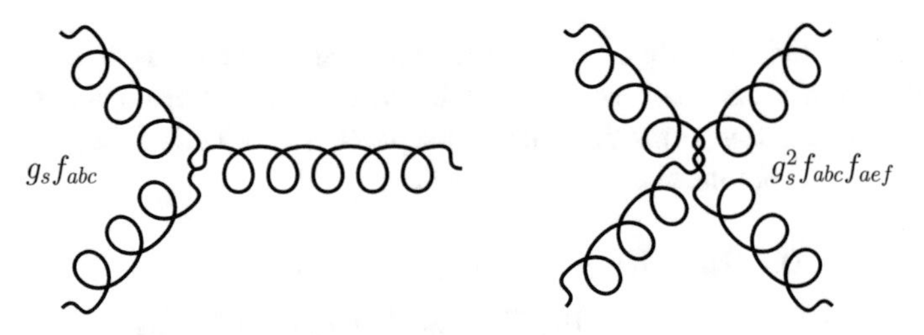

Fig. 2.4 Feynman diagram of gluon self-interactions

2.1.4 *Weak Force and Electroweak Unification*

There have been many experiments that have uncovered the nature of the weak force. Two weak force interactions were found in experiments: the charged current and the neutral current. Detailed studies of energies and angular distributions β, such as those for the decays $\mu^- \rightarrow e^- \bar{\nu}_e$ and $n \rightarrow pe^- \bar{\nu} e$, have shown that only left-handed fermions and right-handed anti-fermions participate in charged current weak interactions. This means that charged current interactions do not conserve parity, P, $(x \rightarrow -x)$ or charge conjugation, C, (particle $\rightarrow$ antiparticle) but they do conserve the combined parity and charged conservation.

The other two interactions relevant for the weak force is the interaction with a photon and the neutral weak current interactions. The neutral weak current is flavor-conserving, while the charged currents are flavor-changing. Also, unlike photons that can couple equally to both fermion chiralities, the neutral current couples only to left-handed particles.

Using symmetry arguments, the structures of QED and QCD have been determined. To do the same here, the model for the combined electroweak interactions should contain a doublet for the left-handed particles, described by $SU(2)$, and the electromagnetic interactions, which from Sect. 2.1.2, is described by a $U(1)$ group. We may guess, then, that the symmetry group to consider is $SU(2)_L \times U(1)_Y$. This is indeed accurate, but subtle, as we for now assume that all the gauge bosons are massless and there is no direct correspondence between the larger gauge group and the $U(1)$ of electromagnetism, for example. Indeed, the $U(1)_Y$ here is a group corresponding to hypercharge. We shall see this in the following.

Both quarks and leptons couple to these forces so they both need to be represented by fields ψ_i, with $i = 1, 2, 3$. The quarks are described as

$$\psi_1(x) = \begin{pmatrix} u \\ d \end{pmatrix}_L, \ \psi_2(x) = u_R, \ \psi_3(x) = d_R. \tag{2.32}$$

Even though the derivation considers only the quarks, this is also valid for the lepton sector, described by

$$\psi_1(x) = \begin{pmatrix} \nu_e \\ e^- \end{pmatrix}_L, \ \psi_2(x) = \nu_{eR}, \ \psi_3(x) = e^-_R. \tag{2.33}$$

The Lagrangian for the weak interactions with no kinetic terms can be written as

$$\begin{aligned} \mathcal{L} &= i\bar{u}(x)\gamma^\mu \partial_\mu u(x) + i\bar{d}(x)\gamma^\mu \partial_\mu d(x) \\ &= \sum_{j=1}^{3} i\bar{\psi}_j(x)\gamma^\mu \partial_\mu \psi_j(x). \end{aligned} \tag{2.34}$$

This Lagrangian is invariant under global transformations of $SU(2)_L \times U(1)_Y$. The $U(1)_Y$ transformations are analogous to the transformation in QED,

$$\begin{aligned} \psi_1(x) &\rightarrow e^{iy_1\beta}\psi_1(x) \\ \psi_2(x) &\rightarrow e^{iy_2\beta}\psi_2(x) \\ \psi_3(x) &\rightarrow e^{iy_3\beta}\psi_3(x), \end{aligned} \tag{2.35}$$

where the parameters y_i are called hypercharges and β is an arbitrary parameter. Just like in QED, this transformation is abelian. The $\psi_1(x)$ field, which is a left-handed doublet, also transforms under $SU(2)$

$$\psi_1(x) \rightarrow e^{iy_1\beta} U_L \psi_1(x), \tag{2.36}$$

where the 2×2 unitary matrix, U_L, is defined as

$$U_L = e^{i\frac{\sigma_i}{2}\alpha^i} \tag{2.37}$$

where $\sigma_i/2$ (for $i = 1, 2, 3$) are the generators of $SU(2)$, with σ_i the Pauli matrices. The α^i are arbitrary parameters. The Pauli matrices are traceless and have the following commutation relation

$$[\sigma_a, \sigma_b] = 2i\epsilon_{abc}\sigma_c. \tag{2.38}$$

Thus the U_L matrix is non-abelian and $SU(2)$ has a similar group structure as $SU(3)$.

To require local gauge invariance, where the arbitrary parameters become arbitrary functions of space-time, $\alpha^i = \alpha^i(x)$ and $\beta = \beta(x)$, four auxiliary fields are defined (one for each arbitrary function) and covariant derivatives,

$$\begin{aligned} D_\mu\psi_1(x) &= [\partial_\mu - ig\tilde{W}_\mu(x) - ig'y_1 B_\mu(x)]\psi_1(x) \\ D_\mu\psi_2(x) &= [\partial_\mu - ig'y_2 B_\mu(x)]\psi_2(x) \\ D_\mu\psi_3(x) &= [\partial_\mu - ig'y_3 B_\mu(x)]\psi_3(x), \end{aligned} \tag{2.39}$$

where the g and g' are the couplings of the $SU(2)_L$ and $U(1)_Y$. The auxiliary fields associated with the $SU(2)_L$ transformation, $\tilde{W}_\mu(x)$, are defined as

$$\tilde{W}_\mu(x) = \frac{\sigma_i}{2}\, W^i_\mu(x). \tag{2.40}$$

The B_μ field transforms similarly to the QED A_μ field while the $\tilde{W}_\mu$ fields transform like the QCD fields, G_μ,

$$\begin{aligned} B_\mu(x) &\rightarrow B_\mu(x) + \frac{1}{g'}\partial_\mu\beta(x) \\ \tilde{W}_\mu(x) &\rightarrow U_L(x)\tilde{W}_\mu(x)U_L^\dagger(x) - \frac{1}{g}\partial_\mu U_L(x)U_L^\dagger(x). \end{aligned} \tag{2.41}$$

The field strengths for the auxiliary fields are introduced by again taking the commutation of the covariant derivatives. The kinetic term for the B_μ field has similar structure as in QED, with

$$B_{\mu\nu} = \partial_\mu B_\nu - \partial_\nu B_\mu \tag{2.42}$$

while the field strength for the $\tilde{W}_\mu$ field has similar structure as in QCD,

$$\begin{aligned} \tilde{W}_{\mu\nu} &= \frac{i}{g}[(\partial_\mu - ig\tilde{W}_\mu), (\partial_\nu - ig\tilde{W}_\nu)] \\ &= \partial_\mu\tilde{W}_\nu - \partial_\nu\tilde{W}_\mu - ig[\tilde{W}_\mu, \tilde{W}_\nu] \end{aligned} \tag{2.43}$$

Since $\tilde{W}_{\mu\nu} = \frac{\sigma_i}{2}W^i_{\mu\nu}$, Eq. (2.43) becomes,

$$W^i_{\mu\nu} = \partial_\mu W^i_\nu - \partial_\nu W^i_\mu + g\epsilon^{ijk}W^j_\mu W^k_\nu \tag{2.44}$$

The $W^i_{\mu\nu}$ terms have the the same structure as the QCD $G_{\mu\nu}$.

The Lagrangian for $SU(2)_Y \times U(1)_Y$ group is

$$\mathcal{L}_{\text{EWK}} = \sum_{j=1}^{3} i\bar{\psi}_j(x)\gamma^\mu D_\mu\psi_j(x) - \frac{1}{4}B_{\mu\nu}B^{\mu\nu} - \frac{1}{4}W^i_{\mu\nu}W_i^{\mu\nu} \tag{2.45}$$

Expanding Eq. (2.45) with the covariant derivative from Eq. (2.39), the terms of the interactions between the fermions and the gauge bosons are

$$g\bar{\psi}_1\gamma^\mu \tilde{W}_\mu \psi_1 + g' B_\mu \sum_{j=1}^{3} y_j \bar{\psi}_j \gamma^\mu B_\mu \psi_j. \tag{2.46}$$

The terms containing the $SU(2)_L$ matrix,

$$\begin{aligned}\tilde{W}_\mu &= \frac{\sigma^i}{2} W^i_\mu \\ &= \frac{1}{\sqrt{2}} \begin{pmatrix} \sqrt{2}W^3_\mu & W^\dagger_\mu \\ W_\mu & -\sqrt{2}W^3_\mu \end{pmatrix},\end{aligned} \tag{2.47}$$

gives rise to the charged current interactions with the boson field. Here we have defined

$$W_\mu = \frac{(W^1_\mu + iW^2_\mu)}{2} \text{ and } W^\dagger_\mu = \frac{(W^1_\mu - iW^2_\mu)}{2}, \tag{2.48}$$

where W_μ is called a creation operator and $W^\dagger_\mu$ is called an annihilation operator of the W boson. Thus the charged-current (CC) interaction Lagrangian is,

$$\mathcal{L}_{\text{CC}} = \frac{g}{2\sqrt{2}} \left\{ W^\dagger_\mu [\bar{u}\gamma^\mu (1-\gamma_5) d + \bar{\nu}_e \gamma^\mu (1-\gamma_5) e] + h.c. \right\} \tag{2.49}$$

where u and d are the spinor fields for the up- and down-type quarks, and the charged lepton fields (for the electron e^- and the neutrino ν_e) have been added. The structure of the charged current of the weak interaction, $\bar{\psi}\gamma^\mu(1-\gamma^5)\psi$ is referred to as the vector minus axial (V-A) structure of the weak interaction. The operators, $\frac{1}{2}(1 \pm \gamma^5)$, are the left- and right-handed chiral projection operators that reveal the parity violating nature of the weak interaction. The interaction vertex for a charged current interaction is shown in Fig. 2.5. In Eq. (2.49), there are no mass terms, so the charged current gauge bosons are massless. The process of how the weak interaction gauge bosons get their masses will be described later.

Only the charged current interactions of Eq. (2.49) have been discussed; however, there also is the interaction due to W^3_μ and B_μ, which represent the neutral current interactions. The Z boson and the photon from QED must be combinations of the two auxiliary fields and be can represented as

$$\begin{pmatrix} W^3_\mu \\ B_\mu \end{pmatrix} = \begin{pmatrix} \cos\theta_W & \sin\theta_W \\ -\sin\theta_W & \cos\theta_W \end{pmatrix} \begin{pmatrix} Z_\mu \\ A_\mu \end{pmatrix}, \tag{2.50}$$

where A_μ is the photon field from QED and Z_μ is the Z gauge boson field. The neutral current (NC) Lagrangian in terms of the photon and Z boson fields is

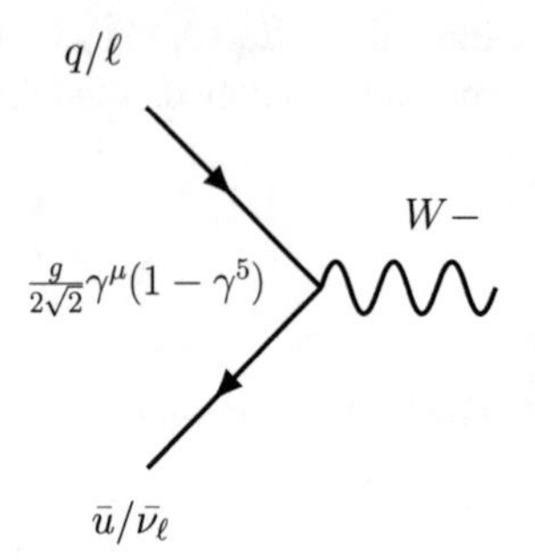

Fig. 2.5 Feynman diagram of a vertex for a charged-current weak interaction

$$\mathcal{L}_{\rm NC} = \sum_j \bar{\psi}_j \gamma^\mu \left\{ A_\mu \left[g\frac{\sigma_3}{2}\sin\theta_W + g' y_j \cos\theta_W \right] + Z_\mu \left[g\frac{\sigma_3}{2}\cos\theta_W - g' y_j \sin\theta_W \right] \right\} \psi_j \tag{2.51}$$

In order to get the QED charge, e, from the A_μ term in Eq. (2.51), the following conditions are imposed

$$g\sin\theta_W = g'\cos\theta_W = e, \text{ and } \frac{Y}{2} = Q - T_3. \tag{2.52}$$

Here, Y is the hypercharge, $T_3 = \frac{\sigma_3}{2}$, and Q is the electromagnetic charge operator, where

$$\begin{aligned} Q_1 &= \begin{pmatrix} Q_{u/\nu} & 0 \\ 0 & Q_{d/e} \end{pmatrix} \\ Q_2 &= Q_{u/\nu} \\ Q_3 &= Q_{d/e}. \end{aligned} \tag{2.53}$$

The first identity in Eq. (2.52) relates the $SU2_L$ and $U(1)_Y$ couplings to the electromagnetic coupling, providing the unification of the electroweak interactions. The second identity relates the hypercharge in terms of the electric charge and coupling, summarized in Table 2.3.

Table 2.3 Summary of neutral current couplings

		Q	T_3	Y
Quarks	$u_{\rm L}$	$\frac{2}{3}$	$\frac{1}{2}$	$\frac{1}{3}$
	$d_{\rm L}$	$-\frac{1}{3}$	$-\frac{1}{2}$	$\frac{1}{3}$
	$u_{\rm R}$	$\frac{2}{3}$	0	$\frac{4}{3}$
	$d_{\rm R}$	$-\frac{1}{3}$	0	$-\frac{2}{3}$
Leptons	$\ell_{\rm L}$	-1	$-\frac{1}{2}$	-1
	$\ell_{\rm R}$	-1	0	-2
	$\nu_{\rm L}$	0	$\frac{1}{2}$	-1

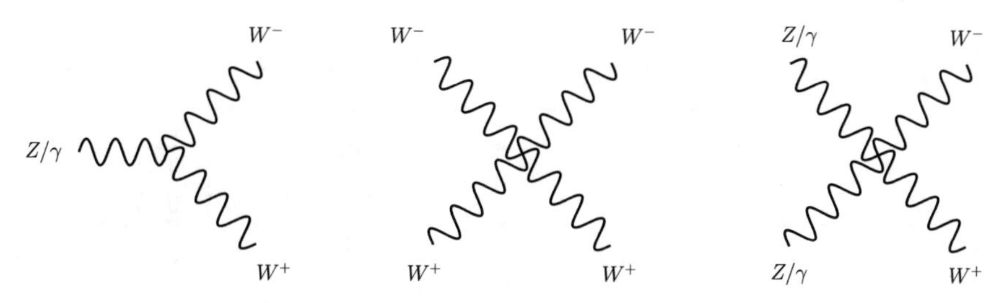

Fig. 2.6 Feynman diagram of cubic and quartic self-interactions of gauge bosons

Substituting the relations in Eq. (2.52) into Eq. (2.51), the Lagrangian for neutral current interactions becomes,

$$\begin{aligned}\mathcal{L}_{\mathrm{NC}} &= \mathcal{L}_{\mathrm{QED}} + \mathcal{L}^{Z}_{\mathrm{NC}} \\ &= eA_\mu \sum_j \bar{\psi}_j \gamma^\mu Q_j \psi_j + \frac{e}{2\sin\theta_W \cos\theta_W} Z_\mu \sum_j \bar{f}\gamma^\mu (v_f - a_f\gamma^5) f\end{aligned} \tag{2.54}$$

where $a_f = T_3^f$ and $v_f = T_3^f(1 - 4|Qf|\sin^2\theta_W)$, f indicates the fermionic field. Just as with the charged current, the neutral current interaction associated with the Z boson has the parity operator since this field couples only to left-handed fermions.

Substituting the same relations in Eq. (2.52) into the kinetic term for the W and Z gauge bosons from Eq. (2.52), the cubic and quartic self-interactions can be derived,

$$\begin{aligned}\mathcal{L}_3 =& - ie\cot\theta_W \left\{(\partial^\mu W^\nu - \partial^\nu W^\mu) W^\dagger_\mu Z_\nu - \left(\partial^\nu W^{\mu\dagger}\right) W_\mu Z_\nu + W_\mu W^\dagger_\nu (\partial^\mu Z^\nu - \partial^\nu A^\mu)\right\} \\ & - ie\left\{(\partial^\mu W^\nu - \partial^\nu W^\mu) W^\dagger_\mu A_\nu - \left(\partial^\mu W^{\nu\dagger} - \partial^\nu W^{\mu\dagger}\right) W_\mu A_\nu + W_\mu W^\dagger_\nu (\partial^\mu A^\nu - \partial^\nu A^\mu)\right\} \\ \mathcal{L}_4 =& - \frac{e^2}{2\sin^2\theta_W}\left\{\left(W^\dagger_\mu W^\mu\right)^2 - W^\dagger_\mu W^{\mu\dagger} W_\nu W^\nu\right\} - e^2\cot^2\theta_W \left\{W^\dagger_\mu W^\mu Z_\nu Z^\nu - W^\dagger_\mu Z^\mu W_\nu Z^\nu\right\} \\ & - e^2\cot\theta_W \left\{2W^\dagger_\mu W^\mu Z_\nu A^\nu - W^\dagger_\mu Z^\mu W_\nu A^\nu - W^\dagger_\mu A^\mu W_\nu Z^\nu\right\} \\ & - e^2\left\{W^\dagger_\mu W^\mu A_\nu A^\nu - W^\dagger_\mu A^\mu W_\nu A^\nu\right\}.\end{aligned} \tag{2.55}$$

The self-interactions terms are illustrated in Fig. 2.6. In each self-interaction vertex, there is at least one pair of W bosons because the $SU(2)$ algebra does not generate any neutral vertex with only photons or Z bosons. Again, just as for the charged current interaction, the gauge bosons are massless.

2.1.4.1 *W* and *Z* Boson Decays

After the electroweak symmetry breaking, the W and Z bosons are no longer massless, which means that there are many decays possible. The W boson decays hadroni-

Table 2.4 Summary of Z coupling to fermions assuming $\sin^2\theta_W = 0.23$

Fermion	Q_f	$I_W^{(3)}$	c_L	c_R
ν_e, ν_μ, ν_τ	0	$+\frac{1}{2}$	0.5	0
e^-, μ^-, τ^-	-1	$-\frac{1}{2}$	0.27	0.23
u, c, t	$+\frac{2}{3}$	$+\frac{1}{2}$	0.35	-0.15
d, s, b	$-\frac{1}{3}$	$-\frac{1}{2}$	-0.42	0.08

cally through weak interactions with amplitudes proportional to the Cabibbo-Kobayashi-Maskawa (CKM) matrix elements shown below.

$$V_{\text{CKM}} = \begin{pmatrix} |V_{ud}| & |V_{us}| & |V_{ub}| \\ |V_{cd}| & |V_{cs}| & |V_{cb}| \\ |V_{td}| & |V_{ts}| & |V_{tb}| \end{pmatrix} = \begin{pmatrix} 0.97 \pm 0.0001 & 0.22 \pm 0.001 & 0.0039 \pm 0.0004 \\ 0.23 \pm 0.01 & 1.02 \pm 0.04 & 0.0041 \pm 0.001 \\ 0.0084 \pm 0.0006 & 0.039 \pm 0.002 & 0.88 \pm 0.07 \end{pmatrix} \tag{2.56}$$

Decays to the top quark are forbidden since the W mass is smaller than the top quark mass; therefore, there are only two significant hadronic decay modes, decays to up and down quarks, and decays to charge and strange quarks. Each hadronic decay also has three colors. The branching ratio for the process of a W boson decaying to jets is 70%. The W boson can also decay leptonically to a lepton and a neutrino. There are three leptons and the branching ratio of the W boson decaying leptonically is 30%.

The Z boson couples to both left- and right- handed chiral states but not equally because the current associated with the Z boson has contributions from the weak interaction, which couples to left-handed particles, and from the field associated with the $U(1)$ symmetry which couples equally to both states. Thus, the Z boson couplings to fermions is proportional to $c_L = I_W^{(3)} - Q_f \sin^2\theta_W$ for left-handed particles, and $c_R = -Q_f \sin^2\theta_W$ for right handed particles. Taking $\sin^2\theta_W = 0.23$, Table 2.4 summarizes the coupling of fermions to the Z boson. The branching ratio is proportional to $(c_L + c_R)^2$ over all modes (except top since $m_t > m_Z$). The branching ratio of the Z boson to neutrinos is 20%, the branching ratio to jets is 70%, and the branching ratio to leptons is 10%.

2.1.5 Spontaneous Symmetry Breaking and the Higgs Mechanism

In the previous section, the Lagrangian for the electroweak interactions is derived; however, all bosons are massless in this derivation. From experiment, it is known that the W and Z bosons have mass. One would expect, then, that something is wrong with our formulation. However, as we will see, this is is not the case. We are only missing an extra ingredient: the scalar Higgs field which gives the gauge bosons their masses via the Higgs mechanism. Before we describe the mechanism, we first describe how ground states of a system may break certain global symmetries that remain present in the Lagrangian, a process known as Spontaneous Symmetry Breaking (SSB). This is the first step in understanding how the gauge bosons associated with the local $SU(2)_L \times U(1)_Y$ symmetry get their mass.

2.1.5.1 Spontaneous Symmetry Breaking

To understand how SSB occurs, the complex scalar field Lagrangian is considered, where $\phi = \frac{\phi_1 + i\phi_2}{\sqrt{2}}$.

$$\mathcal{L} = \partial_\mu \phi^* \partial^\mu \phi - V(\phi), \tag{2.57}$$

where the potential, $V(\phi)$ is defined as,

$$V(\phi) = -m^2 \phi\phi^* - \lambda \phi^2 \phi^{*2} \tag{2.58}$$

The Lagrangian in Eq. (2.57) is invariant under the global transformation

$$\phi(x) \to e^{i\alpha}\phi(x) \tag{2.59}$$

If $m^2 > 0$, Eq. (2.58) only has a minimum at $\phi = 0$. When $m^2 < 0$, the minimum at $\phi = 0$ is unstable. In this case, the potential $V(\phi) = m^2|\phi|^2 + \lambda|\phi|^4$ is minimized when $|\phi|^2 = -\frac{m^2}{\lambda}$. Thus, there are an infinite number of equivalent vacua $|\Omega_\alpha\rangle$ with $\langle\Omega_\alpha|\phi|\Omega_\alpha\rangle = -\frac{m^2}{\lambda}e^{i\alpha}$. Indeed, the system will choose one of these vacua which breaks rotational invariance. In other words, the ground state of this potential breaks the symmetry of the Lagrangian, while the Lagrangian itself remains symmetric! This is the essence of spontaneous symmetry breaking. This is now described in more detail.

The Lagrangian in Eq. (2.57) can be rewritten in terms of the two real fields, ϕ_1 and ϕ_2,

$$\mathcal{L} = \frac{1}{2}(\partial_\mu \phi_1)^2 + \frac{1}{2}(\partial_\mu \phi_2)^2 - \frac{1}{2}(\phi_1^2 + \phi_2^2) - \lambda(\phi_1^2 + \phi_2^2)^2. \tag{2.60}$$

There is a circle of minima of the potential $V(\phi)$ in the ϕ_1, ϕ_2 plane, with radius v, where

$$v^2 = \phi_1^2 + \phi_2^2 \text{ and } v^2 = -\frac{m^2}{\lambda}. \tag{2.61}$$

The potential is illustrated in Fig. 2.7.

Let us pick a particular vacuum without loss of generality, say $\phi_1 = v$ and $\phi_2 = 0$. The Lagrangian can be expanded about this vacuum with the following substitution for $\phi(x)$,

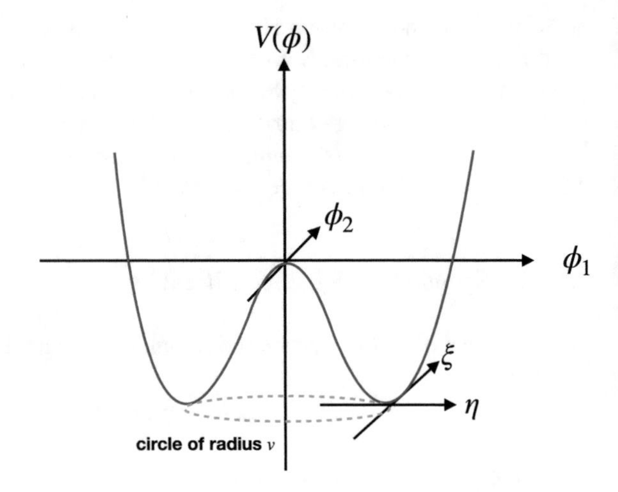

Fig. 2.7 Potential $V(\phi)$ for a complex scalar field with $m^2 < 0$ and $\lambda > 0$

$$\phi(x) = \sqrt{\frac{1}{2}}[v + \eta(x) + i\xi(x)], \tag{2.62}$$

where $\eta(x)$ and $\xi(x)$ are excitations around the vacuum v. The Lagrangian in terms of these new fields becomes

$$\mathcal{L} = \frac{1}{2}(\partial_\mu \xi_1)^2 + \frac{1}{2}(\partial_\mu \eta_2)^2 - \frac{\mu^2}{2}\eta^2 - \frac{\lambda}{2}\left[(v+\eta)^2 + \xi^2\right]^2 - \mu^2 v\eta - \frac{\mu^2}{2}\xi^2 - \frac{1}{2}\mu^2 v^2. \tag{2.63}$$

The field η acquires a mass, $m_\eta^2 = \mu^2$, while the ξ field is massless because it corresponds to zero-energy excitations. This is known as Goldstone's theorem, where SSB of global symmetries implies the existence of massless particles. These massless ξ particles are known as Goldstone bosons. These bosons correspond to the excitation of the system within the space of equivalent vacua.

It is now possible to combine the spontaneous breaking of this global symmetry with a Lagrangian that contains a gauge symmetry. This brings us to the Higgs mechanism. Let us consider the simplest case where our gauge symmetry is a simple $U(1)$.

2.1.5.2 Higgs Mechanism

Using the previous example of the complex scalar field and the $U(1)$ gauge symmetry from QED, the Lagrangian is

$$\mathcal{L} = -\frac{1}{4}F_{\mu\nu}^2 + (\partial_\mu \phi^* - ieA_\mu \phi^*)(\partial_\mu \phi - ieA_\mu \phi) - \mu^2|\phi|^2 - \lambda|\phi|^4 \tag{2.64}$$

which is known as the abelian Higgs model. If $\mu^2 > 0$, this is the Lagrangian from QED in Eq. (2.23); however, to generate masses from SSB, $\mu^2 < 0$ is considered. Just as in the previous section, ϕ is parametrized in terms of fields η and ξ. Then, substituting Eq. (2.62) into Eq. (2.64) yields

$$\mathcal{L} = \frac{1}{2}(\partial_\mu \xi)^2 + \frac{1}{2}(\partial_\mu \eta)^2 - v^2 \lambda \eta^2 + \frac{1}{2} e^2 v^2 A_\mu A^\mu - ev A_\mu \partial^\mu \xi - \frac{1}{4} F_{\mu\nu} F^{\mu\nu} + \text{interaction terms.} \tag{2.65}$$

This Lagrangian describes the interactions of a massless Goldstone boson, ξ, a massive scalar η with $m_\eta = \sqrt{2\lambda v^2}$, and a massive field A_μ with mass $m_A = ev$. In other words, the vacuum value of the scalar field contributes terms to the Lagrangian which look like *masses* for the gauge bosons. To complete our discussion, we have to now just deal with the massless Goldstone boson ξ.

As it turns out, we may use our gauge freedom to remove the massless Goldstone boson from the Lagrangian entirely. Consider that

$$\begin{aligned} \phi &= \sqrt{\frac{1}{2}}(v + \eta + i\xi) \\ &\approx \sqrt{\frac{1}{2}}(v + \eta) e^{i\xi/v}. \end{aligned} \tag{2.66}$$

Thus, the following gauge transformations are used

$$\begin{aligned} \phi &\to \sqrt{\frac{1}{2}}(v + h(x)) e^{i\theta(x)/v} \\ A_\mu &\to A_\mu + \frac{1}{ev}\partial_\mu \theta, \end{aligned} \tag{2.67}$$

where h, θ, and A_μ are real fields. With this transformation, the Lagrangian in Eq. (2.64) becomes

$$\mathcal{L} = \frac{1}{2}(\partial_\mu h)^2 - \lambda v^2 h^2 + \frac{1}{2} e^2 v^2 A_\mu^2 - \lambda v h^3 - \frac{\lambda}{4} h^4 + \frac{1}{2} e^2 A_\mu^2 h^2 + v e^2 A_\mu^2 h - \frac{1}{4} F_{\mu\nu} F^{\mu\nu} \tag{2.68}$$

The Goldstone boson does not appear in the theory. Instead, there are two massive particles. First, there is a vector gauge boson A_μ. Second, there is a massive scalar boson, h, called the Higgs boson. The massless Goldstone boson has been turned into a massive gauge boson, a process called the "Higgs Mechanism". In this example, the photon field acquires mass while the photon in QED is massless. When the electroweak (EWK) symmetry is broken using the Higgs Mechanism, only the W and Z bosons become massive while the photon remains massless. This shall now be discussed in more detail.

2.1.6 EWK Symmetry Breaking and the Higgs Mechanism

The Higgs mechanism is used to give masses to gauge bosons so now we want to apply this to our electroweak model to give masses to the W and Z bosons while keeping the photon massless. A $SU(2)_L$ doublet of scalar fields is defined,

$$\phi(x) = \begin{pmatrix} \phi^{(+)}(x) \\ \phi^{(0)}(x) \end{pmatrix} \tag{2.69}$$

where $\phi^{(+)}(x)$ is charged, and $\phi^{(0)}(x)$ is neutral, and both have hypercharge of 1. A scalar Lagrangian can be defined with similar structure to the Lagrangian in Eq. (2.57),

$$\mathcal{L}_S = \left(D_\mu \phi\right)^\dagger D^\mu \phi - \mu^2 \phi^\dagger \phi - \lambda \left(\phi^\dagger \phi\right)^2, \tag{2.70}$$

where $\lambda > 0$ and $\mu^2 < 0$. This Lagrangian is invariant under $SU(2)_L \times U(1)_Y$ with the covariant derivative

$$D^\mu \phi = \left[\partial^\mu - ig\widetilde{W}^\mu - ig' y_\phi B^\mu\right] \phi, \tag{2.71}$$

where y_ϕ is the hypercharge and $y_\phi = Q_\phi - T_3 = \frac{1}{2}$. The hypercharge is set by the conditions that the photon does not couple to $\phi^{(0)}$ and has a charge of 1 for $\phi^{(+)}$. The potential in Eq. (2.70) is the same as in Eq. (2.58) so the vacuum expectation value is the same as before,

$$\langle \Omega_0 | \phi | \Omega_0 \rangle = -\frac{\mu^2}{\lambda} \equiv \frac{v}{\sqrt{2}}. \tag{2.72}$$

The scalar doublet can be parametrized in the form

$$\phi(x) = e^{i\frac{\sigma_i}{2}\theta^i(x)} \frac{1}{\sqrt{2}} \begin{pmatrix} 0 \\ v + H(x) \end{pmatrix}, \tag{2.73}$$

where $\theta^i(x)$ and $H(x)$ are four real fields. Since the Lagrangian is invariant under $SU(2)_L$, the dependence on $\theta^i(x)$ is rotated away. Thus the three fields corresponding to $\theta^i(x)$, which would have been the Goldstone bosons, are removed. When taking the unitary gauge where $\theta^i(x) = 0$, the transformation of the field ϕ becomes

$$\phi(x) = \frac{1}{\sqrt{2}} \begin{pmatrix} 0 \\ v + H(x) \end{pmatrix}. \tag{2.74}$$

The covariant derivative from Eq. (2.71) can be rewritten as

$$D_\mu = \begin{pmatrix} \partial_\mu - \frac{i}{2}\left(gW_\mu^3 + g'B_\mu\right) & -\frac{ig}{2}\left(W_\mu^1 - iW_\mu^2\right) \\ -\frac{ig}{2}\left(W_\mu^1 + iW_\mu^2\right) & \partial_\mu + \frac{i}{2}\left(gW_\mu^3 - g'B_\mu\right) \end{pmatrix}. \tag{2.75}$$

With the substitution for $\phi(x)$ in Eq. (2.74), the kinetic piece of Eq. (2.70) becomes

$$\left(D_\mu\phi\right)^\dagger D^\mu\phi = \frac{1}{2}\partial_\mu H\partial^\mu H + \frac{1}{8}g^2(v+H)^2\left|W_\mu^1 + iW_\mu^2\right|^2 + \frac{1}{8}(v+H)^2\left|gW_\mu^3 - g'B_\mu\right|^2. \quad (2.76)$$

From Eq. (2.52), we can derive

$$\cos\theta_W = \frac{g}{\sqrt{g^2+g'^2}} \quad \text{and} \quad \sin\theta_W = \frac{g'}{\sqrt{g^2+g'^2}}. \quad (2.77)$$

From Eqs. (2.50) and (2.77), the Z_μ and A_μ fields can be rewritten as

$$\begin{aligned} Z_\mu &= \cos\theta_W W_\mu^3 - \sin\theta_W B_\mu = \frac{g' W_\mu^3 - gB_\mu}{\sqrt{g^2+g'^2}} \\ A_\mu &= \sin\theta_W W_\mu^3 + \cos\theta_W B_\mu = \frac{g' W_\mu^3 + gB_\mu}{\sqrt{g^2+g'^2}}. \end{aligned} \quad (2.78)$$

Using the relations in Eqs. (2.47) and (2.78), the kinetic term for the scalar field can be rewritten as

$$\left(D_\mu\phi\right)^\dagger D^\mu\phi = \frac{1}{2}\partial_\mu H\partial^\mu H + (v+H)^2\left\{\frac{g^2}{4}W_\mu^\dagger W^\mu + \frac{g^2}{8\cos^2\theta_W}Z_\mu Z^\mu\right\}. \quad (2.79)$$

The vacuum expectation value of the neutral scalar field has generated masses for the W and Z bosons (and the photon remains massless) that are related to each other via the weak angle, $\cos\theta_W$, where

$$m_Z\cos\theta_W = m_W = \frac{1}{2}vg. \quad (2.80)$$

With all these considerations, the Higgs Lagrangian now takes the form

$$\begin{aligned} \mathcal{L}_{\text{Higgs}} &= \frac{1}{2}\partial_\mu H\partial^\mu H + (v+H)^2\left(\frac{g^2}{4}W_\mu^\dagger W^\mu + \frac{g^2}{8\cos^2\theta_W}Z_\mu Z^\mu\right) - \lambda v^2H^2 - \lambda vH^3 - \frac{\lambda}{4}H^4 + \frac{1}{4}hv^2 \\ &= \frac{1}{2}\partial_\mu H\partial^\mu H - \frac{1}{2}M_H^2H^2 - \frac{M_H^2}{2v}H^3 - \frac{M_H^2}{8v^2}H^4 + m_W^2W_\mu^\dagger W^\mu\left\{1 + \frac{2}{v}H + \frac{H^2}{v^2}\right\} \\ &+ \frac{1}{2}m_Z^2Z_\mu Z^\mu\left\{1 + \frac{2}{v}H + \frac{H^2}{v^2}\right\} + \frac{1}{4}\lambda v^4, \end{aligned} \quad (2.81)$$

in which we can identify some terms that form the Higgs potential $V(H)$. These are

$$V(H) = \frac{1}{2}M_H^2H^2 + \frac{M_H^2}{2v}H^3 + \frac{M_H^2}{8v^2}H^4 - \frac{1}{4}\lambda v^4. \quad (2.82)$$

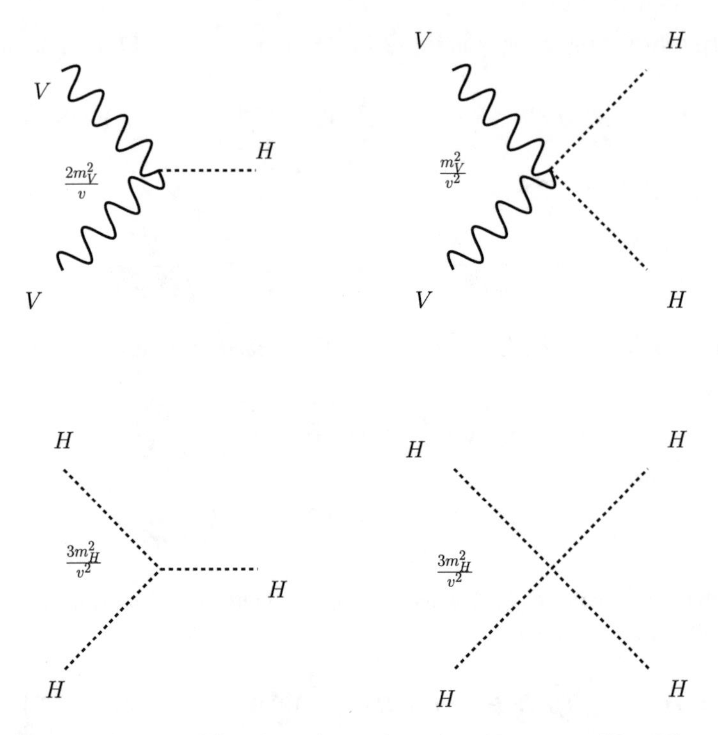

Fig. 2.8 Feynman diagram of Higgs boson interactions where V represents W or Z bosons

So, giving masses to the gauge bosons has generated an additional massive particle, the Higgs boson (H). This spin-0 particle is its own antiparticle and has mass $m_H = \sqrt{-2\mu^2} = v\sqrt{2\lambda}$. From experimental values, $v = 246$ Gev and λ can be obtained from the measured Higgs boson mass of 125 GeV [7, 8]. The structure of the interaction of the gauge bosons and the Higgs boson as well as the self-interaction of the Higgs boson becomes apparent. The coupling is proportional to the bosons coupled to the Higgs and is illustrated in Fig. 2.8.

Through the Higgs mechanism, the gauge bosons have acquired mass; however, the fermions are still massless. To fix this, the scalar field ϕ is now used to construct a new Lagrangian called the Yukawa Lagrangian. Here, the scalar field and the fermions are coupled with cubic terms,

$$\mathcal{L}_{\text{Yukawa}} = c_1(\overline{u}, \overline{d})_L \begin{pmatrix} \phi^{(+)} \\ \phi^{(0)} \end{pmatrix} d_R + c_2(\overline{u}, \overline{d})_L \begin{pmatrix} \phi^{(0)*} \\ -\phi^{(-)} \end{pmatrix} u_R + c_3\,(\overline{\nu}_e, \overline{e})_L \begin{pmatrix} \phi^{(+)} \\ \phi^{(0)} \end{pmatrix} e_R + \text{h.c.}, \tag{2.83}$$

where the second term involves the charge-conjugate scalar field, $\phi^c = i\sigma_2\phi^*$, that has opposite charge and hypercharge as the $\phi(x)$ doublet. In the unitary gauge, after SSB, Eq. (2.83) is simplified to

Fig. 2.9 Feynman diagram of Higgs boson coupling to fermions

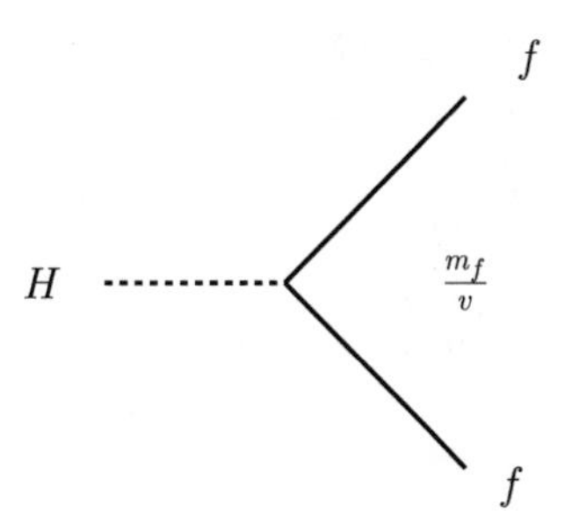

$$\mathcal{L}_{\text{Yukawa}} = \frac{1}{\sqrt{2}}(v + H)\left\{c_1\overline{d}d + c_2\overline{u}u + c_3\overline{e}e\right\}. \tag{2.84}$$

The fermions have acquired masses, $m_d = -c_1\frac{v}{\sqrt{2}}$, $m_u = -c_2\frac{v}{\sqrt{2}}$, and $m_e = -c_3\frac{v}{\sqrt{2}}$. The parameters c_i are unknown, so the masses of the fermions are arbitrary; however, the couplings are fixed in terms of the fermion masses, as illustrated in Fig. 2.9.

The full theoretical analysis of the SM (within the scope of this thesis) is now complete. The various pieces described in this chapter may be combined into the full SM Lagrangian. Written in terms of Eqs. (2.30), (2.45), (2.81), (2.83), the SM Lagrangian is

$$\mathcal{L}_{\text{SM}} = \mathcal{L}_{\text{QCD}} + \mathcal{L}_{\text{EW}} + \mathcal{L}_{\text{Higgs}} + \mathcal{L}_{\text{Yukawa}}. \tag{2.85}$$

2.2 Open Questions of the Standard Model

The SM is a renormalizable field theory that has been extremely predictive and consistent with experiment. The ATLAS summary plot shows that many SM total production cross measurements have good agreement with the theoretical expectations [9]. Even though the SM has been successful, there are still many open questions that the SM does not provide solutions for.

All known terrestrial matter can be constructed out of the first generation fermions (ν_e, e^-, u, d); however, from experiment, the SM has three generations of quarks and leptons that are heavier copies of the first generation with no obvious role. The SM does not provide an explanation for the existence of these other generations or why they are they heavier than the first generation fermions.

Gravity, the fourth force, is not unified with the other interactions in the SM. General Relativity, which describes gravity, is not a quantum theory, unlike the SM, and there is no way to generate it in the SM as described above. Another difficulty is from the cosmological constant, $\Lambda_{\text{cosm.}}$. The cosmological constant corresponds to the energy of the vacuum; however, the Higgs potential V (see Eq. (2.82)) will contribute to the cosmological constant because it has a non-vanishing expectation value $\langle 0|V(v)|0\rangle = \frac{-\mu^4}{4\lambda}$. The cosmological constant, then, is

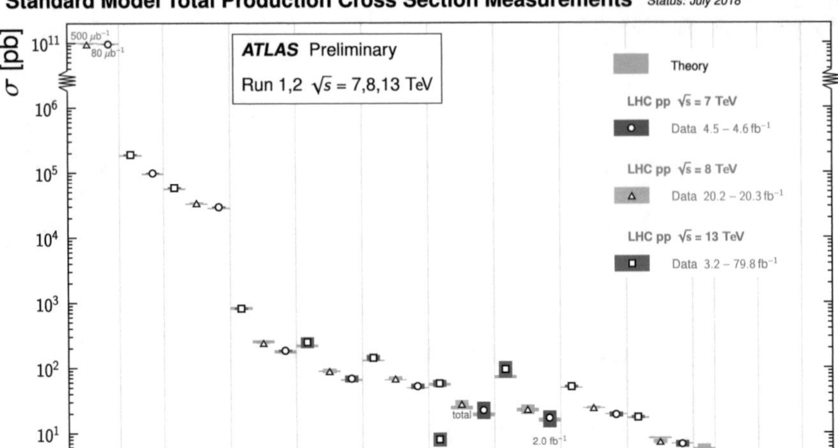

Fig. 2.10 Summary of Standard Model total production cross section measurements, corrected for leptonic branching fractions, compared to the corresponding theoretical expectations [9]

$$\Lambda_{\text{cosm.}} = \Lambda_{\text{bare}} + \Lambda_{\text{SSB}}. \tag{2.86}$$

The term Λ_{SSB} comes from the vacuum expectation at a minimum from the Higgs mechanism,

$$|\Lambda_{\text{SSB}}| = 8\pi G_N |\langle 0|V(v)|0\rangle| \approx 10^{56}\Lambda_{\text{obs.}}. \tag{2.87}$$

Thus, the cosmological constant is 10^{56} times larger than the observed value. To correct this, a constant would be added to Λ_{bare}; however, there is nothing in the theory to suggest that the bare cosmological constant is related to the SSB one and would require fine tuning. Moreover, gravity introduces a new scale, the Planck scale, where $M_{pl} = 2 \times 10^{18}$ GeV, which will become important when discussing the mass of the Higgs boson (Fig. 2.10).

Another open question relates to the Higgs and the hierarchy problem. The SM introduces a Higgs field that gives masses to the W and Z bosons, and the fermions. For the model to be consistent, the mass of the Higgs should be similar to the mass of the W boson, otherwise the Higgs self-interactions would be too strong. The problem is that the bare Higgs mass receives quadratically divergent corrections from loop diagrams, shown in Fig. 2.11. The Higgs mass is calculated as

$$m_H^2 = m_{H,\text{bare}}^2 + \Delta m_H^2, \tag{2.88}$$

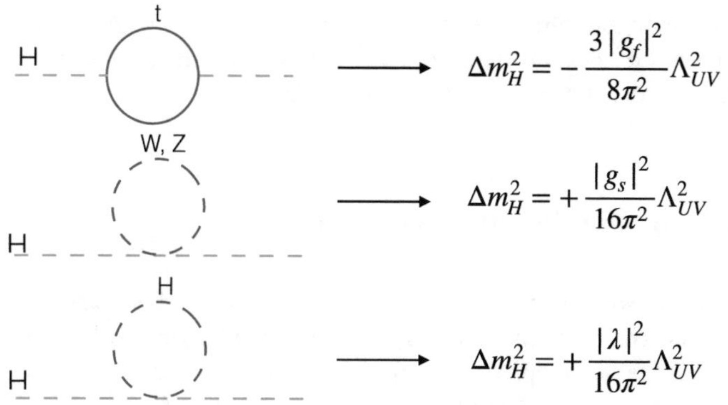

$$\Delta m_H^2 = -\frac{3\,|g_f|^2}{8\pi^2}\Lambda_{UV}^2$$

$$\Delta m_H^2 = +\frac{|g_s|^2}{16\pi^2}\Lambda_{UV}^2$$

$$\Delta m_H^2 = +\frac{|\lambda|^2}{16\pi^2}\Lambda_{UV}^2$$

Fig. 2.11 Loop corrections to the Higgs mass

where the mass correction term, Δm_H^2, is of order Λ_{UV}^2, where Λ_{UV} is a cut-off length scale at which we expect different physics. If Λ_{UV} is the Planck scale, then the Higgs mass would be expected to be of order of the Planck scale, also. This is clearly a problem because the mass is observed to be 125 GeV. Fine-tuning would then be required to get the corrections to precisely cancel and give a small Higgs mass. If minimal corrections to the Higgs mass are required, or $\Delta m_H^2 < m_H^2$, then from the loop corrections in Fig. 2.11,

$$\begin{aligned} \Delta m_H^2 &< m_H^2 \\ \frac{1}{16\pi^2}\Lambda_{UV}^2 &< m_H^2 \end{aligned} \tag{2.89}$$

and replacing m_H with the observed 125 GeVmass, Λ_{UV} would be

$$\Lambda_{UV} < 1\ \text{TeV}. \tag{2.90}$$

In this case of minimal corrections, new physics would appear at around 1 TeV. So, the observed value of the Higgs mass suggests that either there is substantial fine tuning in the SM, or that there should be signatures of new physics at around 1 TeV.

A final open question is that the SM does not provide a candidate for dark matter, which is the matter that makes up most of the universe. The first observation of dark matter comes from galaxy rotation curves which plot the velocity of stars as a function of the distance from the galaxy center. The discrepancy between the experimental and theoretical curves has been explained by the existence of dark matter [10, 11]. Moreover observations of the Bullet Cluster have provided direct evidence of dark matter [12, 13]. The Bullet Cluster consists of two colliding clusters of galaxies which contain stars, visible matter, and dark matter. The stars, observable in visible light, move past one another and are slowed due to gravitational interactions while the baryonic matter, visible in X-rays, interacts electromagnetically, causing them to move slower than the stars. The dark matter is observed using gravitational lensing. The lensing is strongest furthest from the collision which gives evidence that most of the mass of the galaxies are in the dark matter.

2.3 Introduction to Supersymmetry

Given the demonstrated successes of the SM which is based on local gauge symmetries, one could ask if there is another symmetry of the Lagrangian possible that could help resolve some of the open questions described in the previous section. However, relativistic, locally-gauged Lagrangians have limited sets of possibilities. According to the Coleman-Mandula theorem, the only other kind of symmetry that is possible for a Lagrangian with interactions, other than the symmetries of the SM, is called "supersymmetry" (SUSY). This symmetry connects the fermions and the bosons of the Lagrangian. When the symmetry is spontaneously broken, SUSY predicts this existence of a "superpartner" for each particle in the theory: every boson and fermion particle in the Lagrangian has a partner fermion and boson, respectively. So far, no superpartners of the known particles in the Standard Model have been discovered.

Despite the current undecided state of whether the SM is contained within a more supersymmetric theory, SUSY provides many compelling answers to some of the questions raised in the previous section. The prediction of superpartners gives natural candidates for dark matter particles. SUSY also is consistent with the predicted gauge coupling unification at high energies. Also, when SUSY is locally gauged, the gauge mediator is the graviton, the particle that carries the force of gravity. Finally, and most relevantly to the work described in this thesis, supersymmetry at the electroweak scale gives natural loop corrections from the superpartners which cancel out large loop corrections coming from the Planck scale. This naturally resolves the observed discrepancy in energy scales between the electroweak interactions and the Planck scale, without resorting to fine-tuning of theory parameters. For all these reasons and more, SUSY is a compelling theory to be tested at the LHC.

Fermions and bosons are related using the generators, Q, which act on states according to

$$Q|\text{Fermion}\rangle = |\text{Boson}\rangle, \qquad Q|\text{Boson}\rangle = |\text{Fermion}\rangle. \tag{2.91}$$

Therefore, the Q operator changes the spin of a particle and therefore its space-time properties. The Q operator commutes with translations, P, and has the following anti-commutation relationship:

$$\begin{aligned} \{Q_\alpha, \bar{Q}_{\dot{\beta}}\} &= 2\sigma^{\mu}_{\alpha\dot{\beta}} P_\mu \\ \{Q_\alpha, Q_\beta\} &= \{\bar{Q}_{\dot{\alpha}}, \bar{Q}_{\dot{\beta}}\} = 0 \\ [Q_\alpha, P_\mu] &= [\bar{Q}_{\dot{\beta}}, P_\mu] = 0. \end{aligned} \tag{2.92}$$

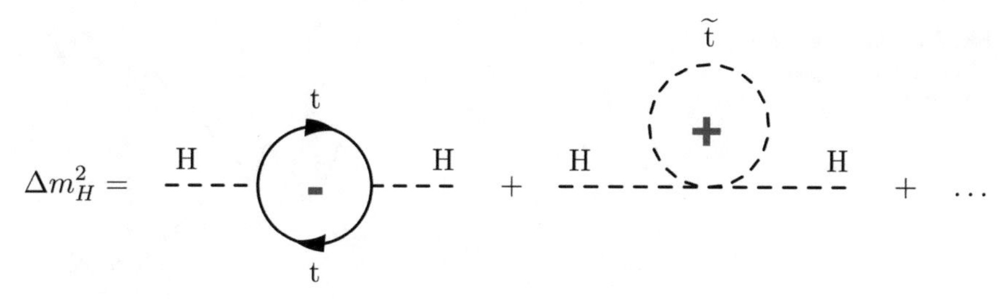

Fig. 2.12 Impact of SUSY particles to the Higgs mass correction

The Q operator, however, does not commute with the Lorentz generators, M,

$$[Q_\alpha, M_{\mu\nu}] = i(\sigma_{\mu\nu})_\alpha^\beta Q_\beta. \tag{2.93}$$

The states of a supersymmetric theory are called "supermultiplets". Each supermultiplet contains both fermion and boson states, which are superpartners of each other. The superpartners of the SM fermions are usually referred to with the letter "s" in front of their name, i.e. the superpartner of a quark is a squark. The superpartners of the SM bosons have an "ino" as their ending, i.e. the superpartners of the Higgs boson are called higgsinos. These superpartners are denoted with a tilde above their corresponding standard model partner particle symbol. For example, the partner of the gluon, g, is the gluino $\tilde{g}$. According to the above algebra, all particles belonging to an irreducible representation of SUSY have the same mass and contains the same number of bosonic and fermionic degrees of freedom, but have spin differing by $\frac{1}{2}$.

The introduction of superpartners to the SM particles cancels the Λ_{UV} corrections to the Higgs mass by adding the contribution from the superpartner as the opposite sign of the SM particle contribution to the Higgs mass correction, as shown in Fig. 2.12. This assumes that SUSY is unbroken with the SM particle and SUSY superpartner having the same mass. Since experiment have not found superpartners with the same mass as the SM particles, SUSY has to be broken and the effective Lagrangian takes the form

$$\mathcal{L} = \mathcal{L}_{\text{SUSY}} + \mathcal{L}_{\text{soft}}. \tag{2.94}$$

As a result, the correction to the Higgs mass has the form

$$\Delta m_H^2 = m_{\text{soft}}^2 \left[\frac{\lambda}{16\pi^2} \ln\left(\Lambda_{\text{UV}}/m_{\text{soft}}\right) + \ldots \right]. \tag{2.95}$$

If $\Lambda_{\text{UV}} = M_p$ and $\lambda = 1$, then m_{soft} and therefore the mass of the lightest SUSY particles should not be greater than a few TeV.

The Higgs boson, which from electroweak symmetry breaking is a neutral scalar in the Standard Model, must also be part of a supermultiplet. However, in the case of the Higgs sector, one supermultiplet is not enough. There must be two Higgs supermultiplets [14] because with only one Higgs supermultiplet, the electroweak

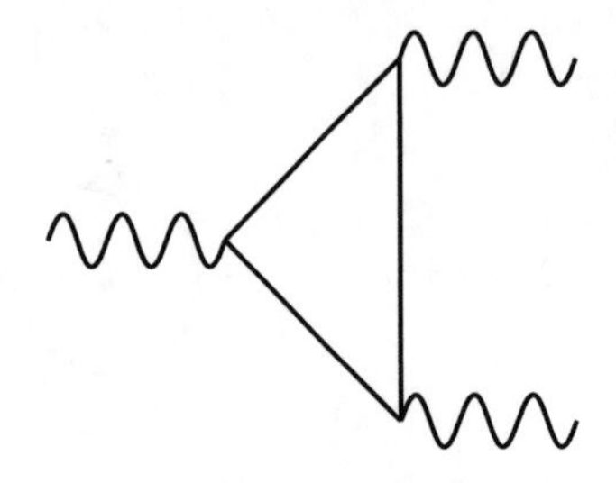

Fig. 2.13 Feynman diagram of an anomalous gauge coupling

gauge symmetry suffers from an anomaly. An anomaly is the breaking of a symmetry due to quantum corrections, which in the context of QFT are loop corrections. For these gauge anomalies, the correction that violates the gauge symmetry is a triangular Feynman diagram with fermions running around the loop and with gauge bosons at the three vertices, as shown in Fig. 2.13.

The triangular diagrams have to cancel in an appropriate way. It can be shown that the condition for the cancellation of gauge anomalies is

$$\mathrm{Tr}[T_3 Y] = 0, \tag{2.96}$$

where $T_3 = \frac{\sigma_3}{2}$ and Y is the hypercharge and the trace is taken over all the left-handed fermions. In the SM, the cancellation of these anomalies is due to the existence of quarks and leptons. In SUSY, the fermionic partner of the Higgs multiplet has isospin $Y = \frac{1}{2}$ or $Y = -\frac{1}{2}$ which leads to a non-zero trace. To solve this issue, there are therefore two Higgs supermultiplets, one with $Y = \frac{1}{2}$ and the other with $Y = -\frac{1}{2}$ so their total contribution satisfies the anomaly cancellation. Another reason to have two Higgs supermultiplets is that the Higgs multiplet with $Y = \frac{1}{2}$ gives masses to the up-type quarks while the Higgs supermultiplet with $Y = -\frac{1}{2}$ gives masses to the down-type quarks.

2.3.1 Minimal Supersymmetric Standard Model

The Minimal Supersymmetric Standard Model (MSSM) is a SUSY model that considers the minimal amount of new particles [14]. Its particle content is summarized in Fig. 2.14. The spin-$\frac{1}{2}$ SM particles have corresponding spin-0 superpartners. The superpartner of the quarks (q) are the squarks ($\tilde{q}$). The superpartners of the leptons (ℓ) are the sleptons ($\tilde{\ell}$), and the partners of the neutrinos (ν) are the sneutrinos ($\tilde{\nu}$). The gauge bosons have fermionic superpartners. The partner of the gluon (g) is the gluino ($\tilde{g}$). The electroweak gauge symmetry, before electroweak symmetry breaking, is associated with the $W^{\pm}$, W^0, and B^0 bosons. The superpartners of the EWK gauge group are the charged winos ($\tilde{W}^{\pm}$), the neutral wino ($\tilde{W}^0$), and the bino ($\tilde{B}^0$). Finally, the partners of the up-type Higgs multiplet (H_u^+, H_u^0) and down-type Higgs multiplet (H_d^-, H_d^0) are the Higgsinos ($\tilde{h}_u^+$, $\tilde{h}_u^0$, $\tilde{h}_d^-$, $\tilde{h}_d^0$). In the SM, after electroweak

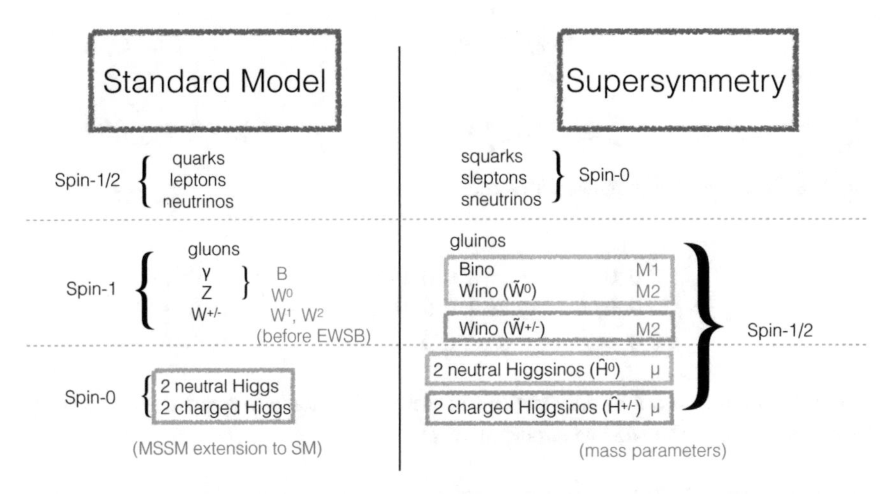

Fig. 2.14 MSSM particle content. The extended Higgs sector with the two chiral multiplets is boxed in green. The superpartners of the neutral gauge bosons and the neutral Higgs bosons are boxed in blue and the superpartners of the charged gauge bosons and the charged Higgs boson are boxed in orange. The mass parameters of the binos, winos, and Higgsinos are also labeled

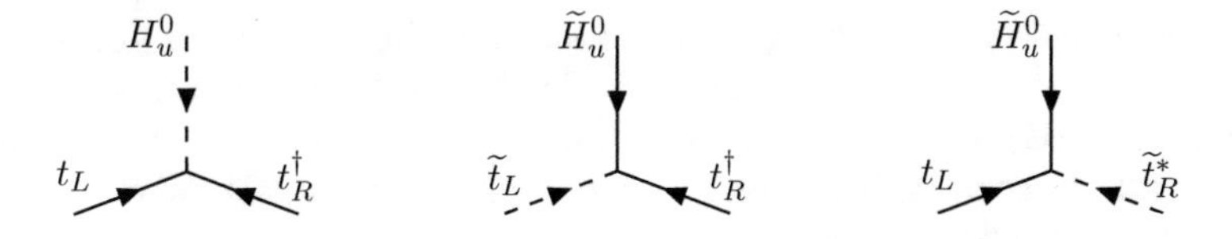

Fig. 2.15 SUSY interactions proportional to y_t

symmetry breaking, the W^0 and B^0 mix to give mass eigenstate, Z^0 and γ. Similarly, in SUSY, the mass eigenstates, winos, binos, and Higgsinos, mix to give the observable states, the charginos ($\tilde{\chi}_1^\pm$, $\tilde{\chi}_2^\pm$, $\tilde{\chi}_3^\pm$, $\tilde{\chi}_4^\pm$) and the neutralinos ($\tilde{\chi}_1^0$, $\tilde{\chi}_2^0$, $\tilde{\chi}_3^0$, $\tilde{\chi}_4^0$). The mixing is described in more detail in Chap. 6.

In the Standard Model, the interaction between particles is governed by the potential. Similarly, in the MSSM, the interactions between particles is governed by a "superpotential", W. The MSSM superpotential is

$$W_{\text{MSSM}} = \overline{u}\mathbf{y_u}QH_u - \overline{d}\mathbf{y_d}QH_d - \overline{e}\mathbf{y_e}LH_d + \mu H_u H_d, \tag{2.97}$$

where H_u, H_d, Q, L, $\overline{u}$, $\overline{d}$, $\overline{e}$ are fields representing the chiral supermultiplets of the particles described above and the Yukawa couplings, $\mathbf{y_u}$, $\mathbf{y_d}$, and $\mathbf{y_e}$ are 3×3 matrices. The Yukawa matrices determine the masses and the CKM mixing angles for the quarks. Since the top quark, bottom quark, and tau leptons are the heaviest SM fermions, the Yukawa matrices can be approximated as

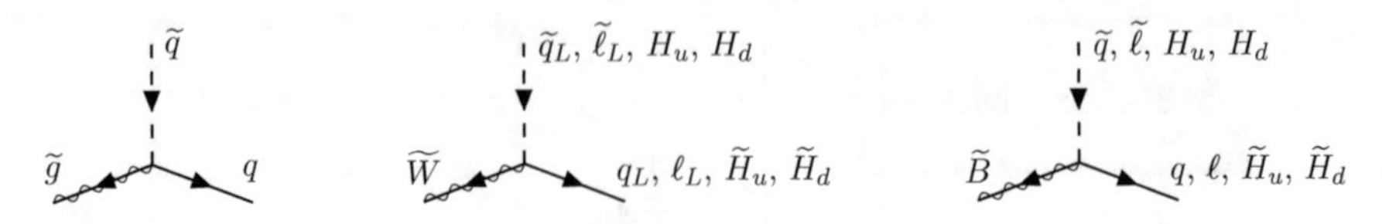

Fig. 2.16 Interactions of gauginos with scalar or fermion pairs

$$\mathbf{y_u} \approx \begin{pmatrix} 0 & 0 & 0 \\ 0 & 0 & 0 \\ 0 & 0 & y_t \end{pmatrix}, \quad \mathbf{y_d} \approx \begin{pmatrix} 0 & 0 & 0 \\ 0 & 0 & 0 \\ 0 & 0 & y_b \end{pmatrix}, \quad \mathbf{y_e} \approx \begin{pmatrix} 0 & 0 & 0 \\ 0 & 0 & 0 \\ 0 & 0 & y_\tau \end{pmatrix}. \tag{2.98}$$

Using this approximation, only the third family of lepton and quarks and the Higgs field contribute to the MSSM superpotential

$$W_{\text{MSSM}} = y_t\left(\bar{t}H_u^0 - \bar{t}bH_u^+\right) - y_b\left(\bar{b}tH_d^- - \bar{b}bH_d^0\right) - y_\tau\left(\bar{\tau}\nu_\tau H_d^- - \bar{\tau}\tau H_d^0\right) + \mu\left(H_u^+H_d^- - H_u^0H_d^0\right). \tag{2.99}$$

These interactions not only represent Higgs-quark-quark or Higgs-lepton-lepton interactions, but also Higgsino-squark-quark and slepton-higgsino-lepton interactions, as shown for the top-quark Yukawa coupling in Fig. 2.15. Examples of couplings of the binos, winos, and gluinos are shown in Fig. 2.16. Winos interact with only left-handed particles and sparticles while binos interact with both left-handed and right-handed particles and sparticles. Gluinos, similar to the gluon, only interacts with quarks and squarks. Each interaction vertex has two sparticle and one particle.

The superpotential also includes terms of the form,

$$\begin{aligned} W_{\Delta\text{L}=1} &= \frac{1}{2}\lambda^{ijk}L_iL_j\bar{e}_k + \lambda'^{ijk}L_iQ_j\bar{d}_k + \mu'^{i}L_iH_u \\ W_{\Delta\text{B}=1} &= \frac{1}{2}\lambda''^{ijk}\overline{u_i}\bar{d}_j\bar{d}_k \end{aligned} \tag{2.100}$$

with terms λ'^{ijk} which violate lepton conservation and terms λ''^{ijk} which violate baryon number conservation. The possibility of these terms would have experimental consequences. In particular, these terms facilitate proton decay, which violates both lepton and baryon number by one unit. If λ' and λ'' couplings are present and unsuppressed, the proton lifetime would be very short; however, since the observation of the proton decay has not been observed, these couplings must be small or do not exist. To solve this, baryon and lepton number conservation can be assumed in the MSSM by adding a new symmetry called "R-parity". This is different from the SM because baryon and lepton conservation in the SM is guaranteed: there are no terms in the Lagrangian that violate either conservation law. This new symmetry, R-parity, is defined as

$$R = (-1)^{3(\text{B-L})+2s} \tag{2.101}$$

where s is the spin of the particle. SM particles have $R = 1$ while SUSY particles have $R = -1$.

There are a few consequences of R-parity conservation. The lightest SUSY particles, called the "LSP", is stable. If the LSP is neutral, it interacts weakly with matter, making it a candidate for dark matter. Also, all other sparticles must eventually decay to a state that contains one LSP. Finally, sparticles are produced in pairs in colliders such as the LHC. The searches discussed in this thesis all assume R-parity conservation.

References

1. Thomson M (2013) Modern particle physics. Cambridge University Press, New York
2. Langacker P (2010) The standard model and beyond. CRC Press, Boca Raton
3. Schwartz MD (2014) Quantum field theory and the standard model. Cambridge University Press, Cambridge
4. Halzen F, Martin AD (1984) Quarks and leptons: an introductory course in modern particle physics. Wiley, New York
5. Georgi H (1984) Weak interactions and modern particle theory. Benjamin/Cummings, Menlo Park
6. Burgard C, Example: standard model of physics. http://www.texample.net/tikz/examples/model-physics/
7. ATLAS Collaboration (2012) Observation of a new particle in the search for the Standard Model Higgs boson with the ATLAS detector at the LHC. Phys Lett B716:1–29. arXiv:1207.7214 [hep-ex]
8. CMS Collaboration (2012) Observation of a new boson at a mass of 125 GeV with the CMS experiment at the LHC. Phys Lett B716:30–61. arXiv:1207.7235 [hep-ex]
9. ATLAS Collaboration (2018) Summary plots from the ATLAS Standard Model physics group. https://atlas.web.cern.ch/Atlas/GROUPS/PHYSICS/CombinedSummaryPlots/SM/
10. Rubin VC, Ford WK Jr (1970) Rotation of the andromeda nebula from a spectroscopic survey of emission regions. Astrophys J 159:379–403
11. Rubin VC, Thonnard N, Ford WK Jr (1980) Rotational properties of 21 SC galaxies with a large range of luminosities and radii, from NGC 4605 /R = 4kpc/ to UGC 2885 /R = 122 kpc/. Astrophys J 238:471
12. Clowe D, Gonzalez A, Markevitch M (2004) Weak lensing mass reconstruction of the interacting cluster 1E0657-558: direct evidence for the existence of dark matter. Astrophys J 604:596–603. arXiv:astro-ph/0312273 [astro-ph]
13. Markevitch M, Gonzalez AH, Clowe D, Vikhlinin A, David L, Forman W, Jones C, Murray S, Tucker W (2004) Direct constraints on the dark matter self-interaction cross-section from the merging galaxy cluster 1E0657-56. Astrophys J 606:819–824. arXiv:astro-ph/0309303 [astro-ph]
14. Martin SP (1998) A Supersymmetry primer. arXiv:hep-ph/9709356 [hep-ph]. [Adv Ser Direct High Energy Phys 18:1 (1998)]

Chapter 3
LHC and the ATLAS Detector

3.1 The Large Hadron Collider

The Large Hadron Collider (LHC) [1] is a two-ring superconducting hadron accelerator and collider located at the European Organization for Nuclear Research (CERN) center at the French-Switzerland border. The LHC is installed in the 27 km tunnel constructed for the Large Electron Positron Collider (LEP) experiment [2] which lies between 45 and 170 m below ground. In 1996, the construction of a 14 TeV LHC was approved [3]. The LEP experiment ran from 1989 to 2000 when it was closed to liberate the tunnel for the LHC. While the LEP experiment collided electrons and positrons, the LHC primarily collides bunches of protons at four interaction points around the ring where four independent experiments are located: ATLAS [4], CMS [5], LHCb [6], and ALICE [7]. ATLAS and CMS are general purpose detectors aiming to run at high luminosities, with peak luminosity of $L = 2 \times 10^{34}$ cm^{-2} s^{-1}. LHCb is designed to study b-hadrons with peak luminosity of $L = 2 \times 10^{32}$ cm^{-2} s^{-1} and ALICE is designed to study heavy ion collisions with peak luminosity of $L = 2 \times 10^{27}$ cm^{-2} s^{-1}. The LHC complex and the four experiments are shown in Fig. 3.1.

The proton beams are created by ionizing hydrogen gas. The protons are accelerated in several stages, increasing their energy by orders of magnitude. The LHC is supplied with protons from the injector chain Linac2 (after LS2 in 2019–2020, LINAC2 will be replaced by LINAC4, designed to deliver double the brightness and intensity of the beam [9]), Proton Synchotron Booster (PSB), Proton Synchotron (PS), and Super Proton Synchotron (SPS) [8]. After ionizing the hydrogen gas, the protons are accelerated up to 100 kV and sent to a Radio Frequency Quadrupole (QRF) which accelerates the beams up to 750 keV. The protons then enter the linear accelerator (linac), LINAC2. The linac is a multi-chamber resonant cavity tuned to a specific frequency. The differences in potential in the cavities accelerate the protons to 50 MeV. After the linac, the protons enter the PSB, a 157 m circular accelerator, where they are accelerated to 1.4 GeV. In the PS, a 628 m circular accelerator, the

© The Editor(s) (if applicable) and The Author(s), under exclusive license
to Springer Nature Switzerland AG 2020

E. Resseguie, *Electroweak Physics at the Large Hadron Collider with the ATLAS Detector*, Springer Theses,
https://doi.org/10.1007/978-3-030-57016-3_3

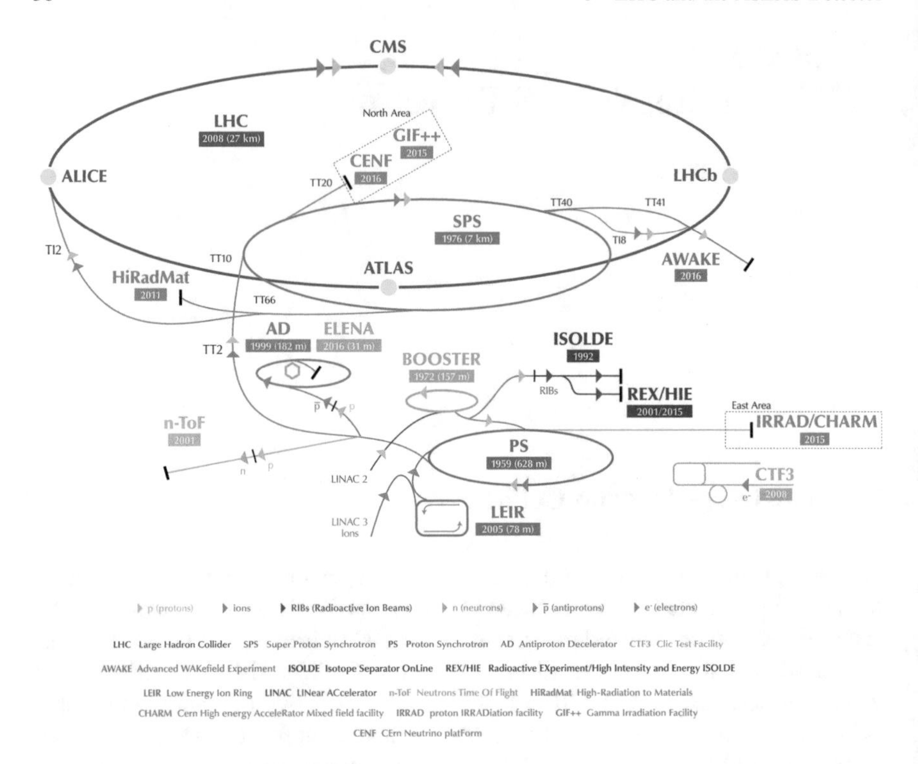

Fig. 3.1 Diagram of the CERN accelerator complex [8]

protons are further accelerated to 25 GeV. The PS is also responsible for providing 81 bunch packets of protons with 25 ns spacing. Triplet bunches from the PS are then injected into the SPB, a 7 km circular accelerator, where they are accelerated to 450 GeV . Finally, the protons are delivered to the LHC where they attain energies up to 7 TeV. The current operation of the LHC accelerates the protons to 6.5 TeV, giving a center-of-mass energy of $\sqrt{s} = 13$ TeV. To accelerate and collide the proton beams, two opposite magnetic dipole fields in both rings is required. The magnets are made out of NbTi cooled to below 1.9 K using superfluid helium and operate at fields above 8 T. Lattice magnets are used to control the particles' trajectory, keeping the beams stable and aligned. There are 1232 dipoles, each 15 m long, and weighing 35 tons. Quadrupole magnets are used to keep the particles in a tight beam so that the particles are bunched together when they reach the detectors, resulting in more collisions. The quadrupoles have four magnetic poles arranged symmetrically to squeeze the beam vertically or horizontally. A LHC dipole magnet is shown in Fig. 3.2. Before the proton beams enter the detector, insertion magnets, made out of three quadrupole magnets called a triplet, are used to squeeze the particles closer together. The triplets tighten the beam, narrowing it from 0.2 mm to 16 μm across. After colliding, the beams are separated by dipole magnets, which help reduce the beam intensity by a factor of 100,000 before it hits a block of concrete and graphite

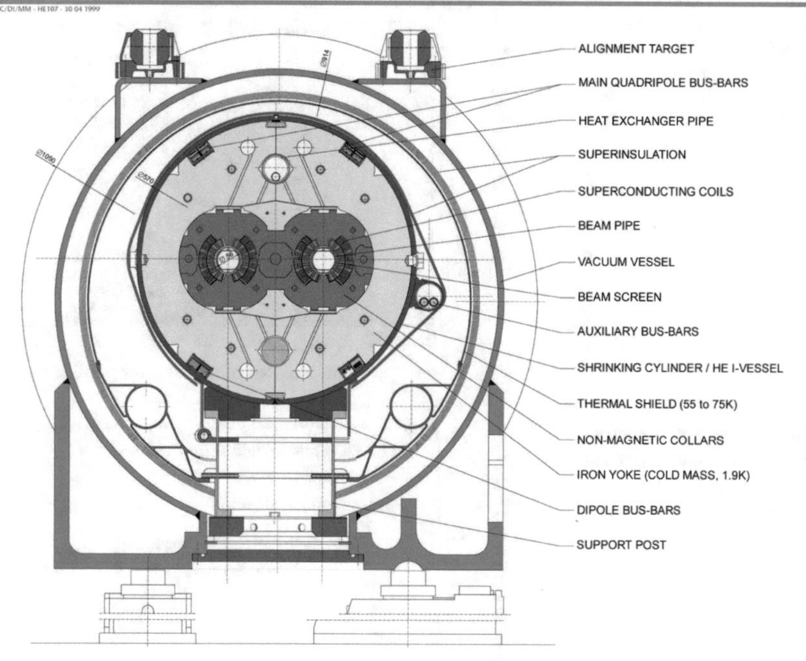

Fig. 3.2 Diagram of an LHC dipole magnet [10]

composite, a process called “beam dump”. Each instance of protons being delivered to the LHC is called a “fill”. These fills last for ten hours until the beam is depleted.

Proton beams in the LHC are composed of 2808 bunches of protons spaced by 25 ns, with each bunch corresponding to 10^{11} protons [8]. As a result, each bunch crossing results in many simultaneous proton-proton (pp) collisions, called an event, at each interaction point. The number of interactions per bunch crossing is called “pileup”, denoted with μ and represents an experimental challenge to determine the interaction vertex of the collision studied. Figure 3.3 shows the number of interactions per bunch crossing for 2015–2018. The average pileup for 2015–2018 is $< \mu >=$ 33.7 and the pileup for 2017–2018 is greater than in 2015–16, which can affect efficiencies of event selections in signal regions. This will be further discussed in Chap. 9. Even though pileup makes reconstruction difficult, it is important to increase the number of collisions to study rare processes such as processes produced by the decay of SUSY particles.

Two other figures of merit are used to specify the number of pp collisions delivered by the LHC. The first measure of interest is the instantaneous luminosity, which is the number of pp collisions per second (divided by the interaction cross-section). The LHC was designed to deliver peak luminosities of $L = 2 \times 10^{34}$ cm^{-2} s^{-1} for the ATLAS detector. As shown in Fig. 3.4, the peak luminosity per fill in 2018 has surpassed this design goal with peak luminosity of $L = 2.1 \times 10^{34}$ cm^{-2} s^{-1}. The

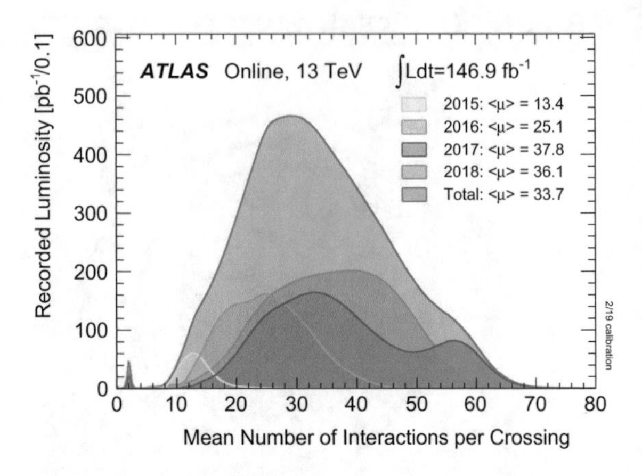

Fig. 3.3 Number of interactions per bunch crossing for 2015–2018 at $\sqrt{s} = 13$ TeV [11]

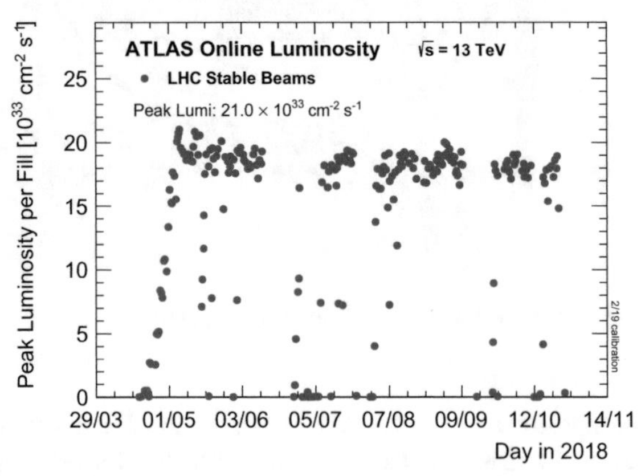

Fig. 3.4 Peak luminosity per fill delivered in 2018 at $\sqrt{s} = 13$ TeV [11]

other figure of merit is the integrated luminosity, or the total number of *pp* collisions delivered by the LHC. Figure 3.5 shows the total integrated luminosity delivered to the LHC in Run 1, in 2011–2012 at $\sqrt{s} = 7, 8$ TeV, and during Run 2, in 2015–2018 at $\sqrt{s} = 13$ TeV. The total integrated luminosity delivered by the LHC to ATLAS in Run 1 is 28 fb^{-1} and the integrated luminosity delivered by the LHC to ATLAS in Run 2 is 156 fb^{-1}. Multiplying the luminosity delivered by the LHC by the cross sections shown in Fig. 2.10, this corresponds to 9 million Higgs bosons produced in 2015–2018. While this appears to be a large number, not all these Higgs bosons are detected or reconstructed.

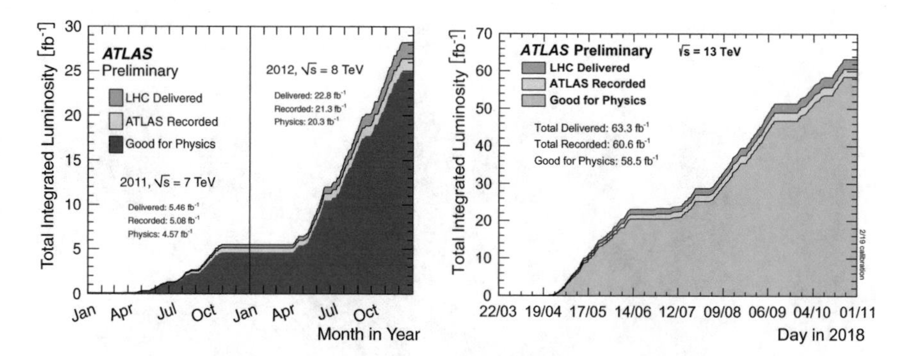

Fig. 3.5 Total integrated luminosity and data quality collected by ATLAS in Run 1 (left) and Run 2 (right) [11, 12]. The luminosity delivered by the LHC to ATLAS is shown in green, the data recorded by ATLAS is shown in yellow, and the good quality data used in physics analyses is shown in blue

3.2 The ATLAS Detector

The ATLAS [4] detector is a general-purpose particle detector located at one of the four interaction points along the LHC ring. The detector has a cylindrical geometry centered around the beam line. ATLAS is 45 m long, 25 m high, and weighs about 7000 tons. It uses a right-handed coordinate system with the interaction point as the origin of the coordinate system. The direction along the beam axis defines the z-axis and the $x-y$ plane is transverse to that beam axis. The positive x-axis is defined as pointing from the interaction point towards the center of the LHC ring, and the positive y-axis is defined as pointing upwards. Usually, cylindrical coordinates are used in the $x-y$ plane with the azimuthal angle ϕ measured around the z-axis. The polar angle θ is the angle away from the beam axis. The pseudorapidity is defined using the polar angle as $\eta = -\ln\tan(\frac{\theta}{2})$. While pseudorapidity is used for massless objects, massive objects use rapidity, defined as $y = \frac{1}{2}\ln(\frac{E+p_z}{E-p_z})$. The angular distance is defined as $\Delta R = \sqrt{\Delta\eta^2 + \Delta\phi^2}$. The transverse momentum, p_T, and transverse energy, E_{T}, are defined in the $x-y$ plane.

The ATLAS detector is composed of many sub-detectors and magnets. The inner detector (ID), used to measure tracks of electrically charged particles, is closest to the beam line and surrounded by a 2 T solenoid field. Pattern recognition, momentum and vertex measurements, and electron identification are achieved in the inner detector. Surrounding the ID are the calorimeters. The electromagnetic calorimeter identifies electrons and photons as well as measures their energies. The hadronic calorimeter measures the energies of the hadrons. Finally, the muon spectrometer measures the energies and tracks of muons. A barrel toroid and two end-cap toroids produce a toroidal magnetic field of 0.5 T and 1 T for the muon detectors. The muon spectrometer defines the dimensions of the ATLAS detector. Figure 3.6 shows a cut-away view of the ATLAS detector and all the sub-detectors.

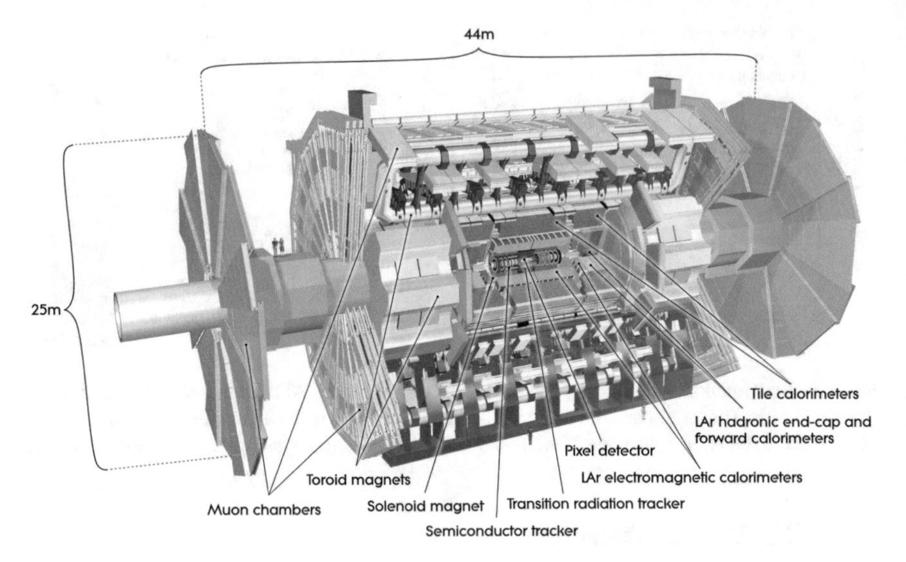

Fig. 3.6 General cut-away view of the ATLAS detector showing all the sub-detectors [4]

3.2.1 The Inner Detector

The Inner Detector (ID) measures the trajectories of charged particles above a given p_T threshold within the pseudorapidity range $|\eta| < 2.5$ and electron identification for $|\eta| < 2.0$. The ID is cylindrical with length of ± 3.512 m and radius of 1.15 m. Along with the trajectory and the 2 T magnetic field provided by the solenoid, the ID can identify the vertex, the momentum of tracks, and the charge of particles. The vertexing capabilities can separate collisions of interest from pileup events. This is also important for decays of *b*-hadrons, which have short lifetimes, leading to displaced vertex of decays.

The inner detector is composed of three sub-detectors, listed in order of increasing radii: the Pixel detector, the Semiconductor Tracker (SCT),and the Transition Radiation Tracker (TRT). Figures 3.7 and 3.8 show a cut-away view of the inner detector and its sub-detectors. The position of a charged particle will be measured 42 times (4 hits in the pixel detector, 8 hits in the SCT, and 30 hits in the TRT).

3.2.1.1 Pixel Detector

The Pixel Detector [15] is the innermost detector of the ID and extends to $|\eta| < 2.5$ and covers radial distances between 50.5 and 150 mm. The Pixel Detector is composed of modules made up of silicon sensors, front-end electronics, and flex-hybrids. Each pixel sensor is $50 \times 250\,\mu$m in the innermost pixel layer and $50 \times 400\,\mu$m in size for the three outermost pixel layers. They provide a resolution of

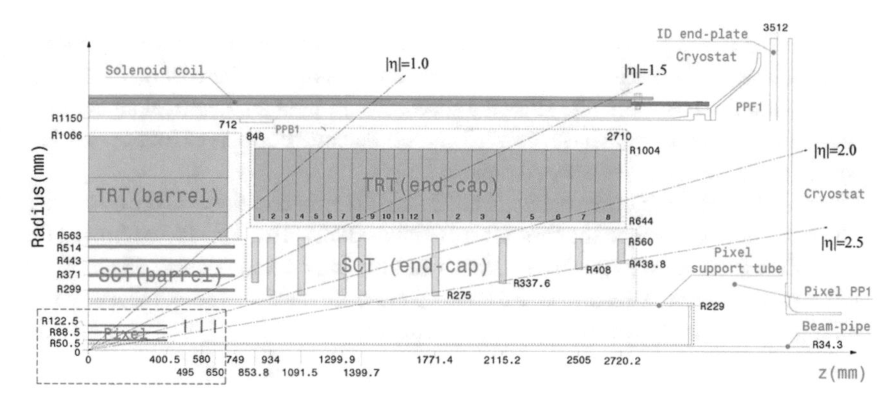

Fig. 3.7 Cut-away view of the ATLAS inner detector [13]

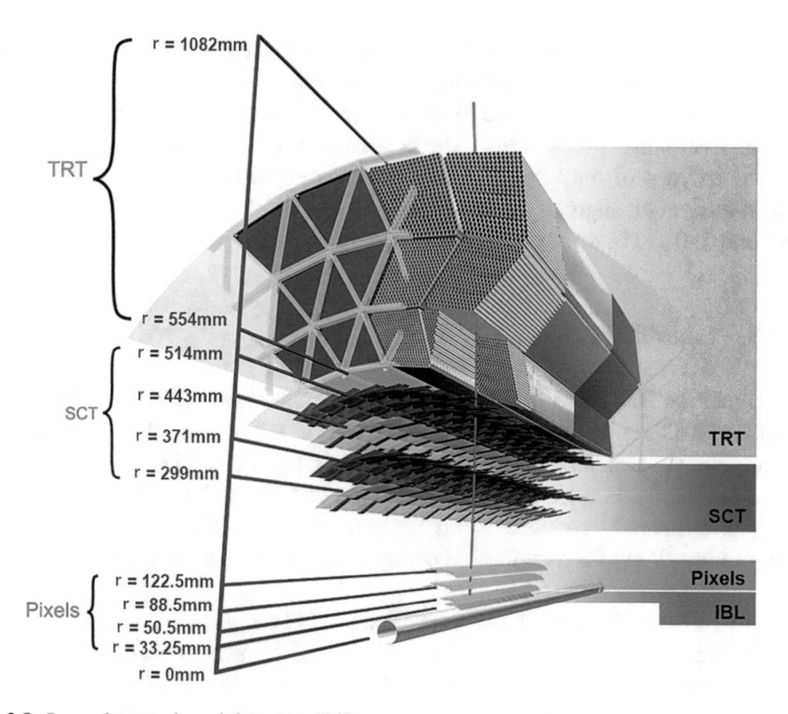

Fig. 3.8 Inner detector barrel detectors [14]

10 μm in the $r - \phi$ plane and 115 μm in the z (r) direction of the barrel (endcap) modules. The pixel detector has 92 million readout channels. When charged particles pass through these sensors, the silicon ionizes, resulting in an electrical signal used to determine which sensors have "hits". This sub-detector consists of four cylindrical barrel layers and three disk-shaped endcap layers.

During the LHC shutdown between 2012 and 2015, an additional layer was added to the Pixel detector, called the Insertable B-Layer (IBL) [16]. This layer allowed the radius of measurement to go from 50 mm from the interaction point to 33 mm from the interaction point. This allows for more precise interaction vertex measurement which improves the b-hadron physics performance.

3.2.1.2 Semiconductor Tracker

The Semiconductor Tracker (SCT) [17] surrounds the Pixel detector and covers radial distances from 299 to 560 mm and covering a range of $|\eta| < 2.5$. The SCT uses the same technology as the Pixel detector; however, the sensor size is larger, 126 mm × 80 μm, called a "strip". The SCT consists of 4088 modules in four barrels and two endcaps of nine disks each. Each layer consists of two sensors arranged back-to-back at an angle of 40 mrad resulting in a resolution of 17 μm in the $r - \phi$ plane and 580 μm in the z plane [13].

3.2.1.3 Transition Radiation Tracker

The Transition Radiation Tracker (TRT) is the outermost subdetector of the ID. The barrel TRT region encompasses the region of $563 < r < 1066$ mm with $|z| < 712$ mm and the endcap region is $644 < r < 1004$ mm and $848 < |z| < 2710$ mm. Unlike the Pixel and the SCT which are made of silicon, the TRT is made of gas-filled proportional drift tubes covering the range of $|\eta| < 2.0$ called straws. The TRT configuration allows for tracking and particle identification using transition radiation (TR).

The barrel is made up of 52,544 straws that are 144 cm in length and oriented parallel to the beam while the endcap is made up of 122,880 straws 37 cm in length, aligned radially to the axis [18–22]. The straws, which are 4 mm in diameter, are made up of a 35 μm thick film of kapton surrounded by an aluminum and a graphite-polyimide layer. The aluminum layer provides good electrical conductivity and the polyimide protects the aluminum from evaporation from gas discharges during operation. The straws are reinforced using carbon fiber bundles. Inside the straw, there is a gold-tungsten anode wire of 30 μm in diameter. A 15,300 V potential is applied between the straw outer casing and the wire, corresponding to a gain of 2.5×10^4 for the chosen gas mixture of xenon (70%), carbon dioxide (27%), and oxygen (3%). Xenon is used for its high efficiency of absorbing TR photon of energy of 6–15 keV. When a charged particle passes through the straw, it ionizes the gas and the electrons drift towards the wire and the ions drift to the straw walls. When the electrons approach

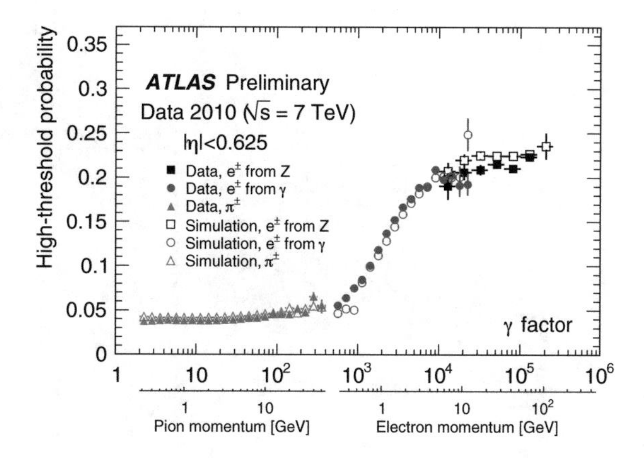

Fig. 3.9 High-threshold turn-on curve, separated into regions according to the reconstructed track η. The value of Lorentz γ factor is calculated using the assumed mass of the candidate, either an electron or pion [22]

the wire, an avalanche of electrons is induced due to the large potential difference, amplifying the signal by a factor of 20,000. This amplified signal is discriminated against two thresholds, a low threshold (LT) at 300 eV and a high threshold (HT) at 6–7 keV. These two thresholds are used for tracking information and to identify the energy absorbed by a TR photon. Timing information is extracted from the straw hits, giving a resolution of 130 μm.

Between each straw there is a polypropylene-polyethylene fiber mat 3 mm thick. Charged particles passing through the mats will have a probability of emitting transition radiation. This radiation is caused by the charge particle moving between media with different index of refraction. It is the energy radiated when a charged particle passes through with plasma frequency, ω_p, [23]

$$I = \alpha z^2 \gamma \hbar \omega_p / 3 \tag{3.1}$$

where z is the atomic charge, and $\hbar\omega_p$ for the TRT is 21 eV. The transition radiation is proportional to the γ factor so electrons will have higher Lorentz factors than pions or muons due to their lighter mass. The HT hit threshold is optimized such that the high threshold probability is higher for electrons than for pions, as shown in Fig. 3.9. This discrimination between pions and electrons is due to the absorption of TR photons by xenon. Before the start of Run 2, gas leaks were discovered in the TRT. Due to cost considerations, the xenon gas was replaced with argon gas since argon provides similar tracking capabilities; however, TR photons cannot be absorbed by argon and, as a result, the argon-filled straws do not contribute to the electron identification.

TRT Electronics and Data Acquisition

The TRT signals are read out by Application Specific Integrated Circuit (ASIC) which perform Amplification, Shaping, Discrimination, and Base-Line Restoration (ASDBLR) [24]. These results are sampled by a second ASIC, Drift Time Measuring Read-Out Chip (DTMROC) which measures timing information and digitizes the output for up to 16 straws. DTMROCs are controlled by Timing, Trigger, and Control (TTC) modules which manage 480 DTMROCs. The readout is performed by the

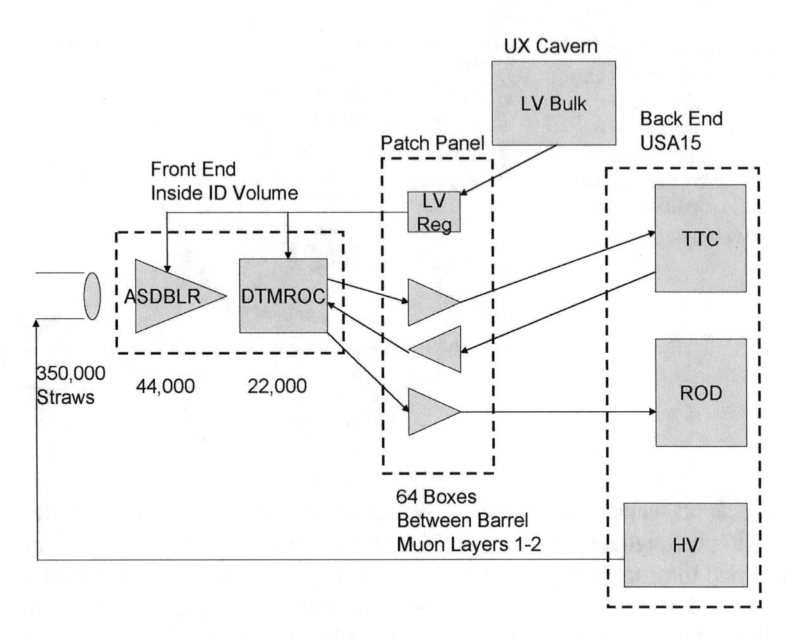

Fig. 3.10 Overview of the TRT electronics [24]

Read Out Drivers (ROD). 10 VME crates contain the TRT's 96 RODs and 48 TTCs. Each VME crate is controlled by a computer running a Linux operating system. A schematic showing the TRT electronics for the readout is shown in Fig. 3.10.

3.2.2 *The Calorimeters*

Two types of calorimeters are used to measure the energy from electromagnetic and hadronic showers. They are the Liquid Argon (LAr) and the hadronic calorimeters which alternate layers of absorber material to induce particle showers and active material to measure the resulting energy. Combined, the two calorimeters cover a region up to $|\eta| < 4.9$. Figure 3.11 shows a cut-away view of the calorimeters.

3.2.2.1 Liquid Argon Calorimeter

The Liquid Argon (LAr) calorimeter is used to measure the energy resulting from electromagnetic showers. High energy electrons predominantly lose energy in matter by Bremsstrahlung, and high-energy photons by e^+e^- pair production. The amount of matter traversed by these interactions is called "radiation length" (X_0). In one X_0, an electron loses all but $1/e$ of its energy by Bremsstrahlung and it represents $7/9$ of the mean free path for the pair production of a high energy photon [23].

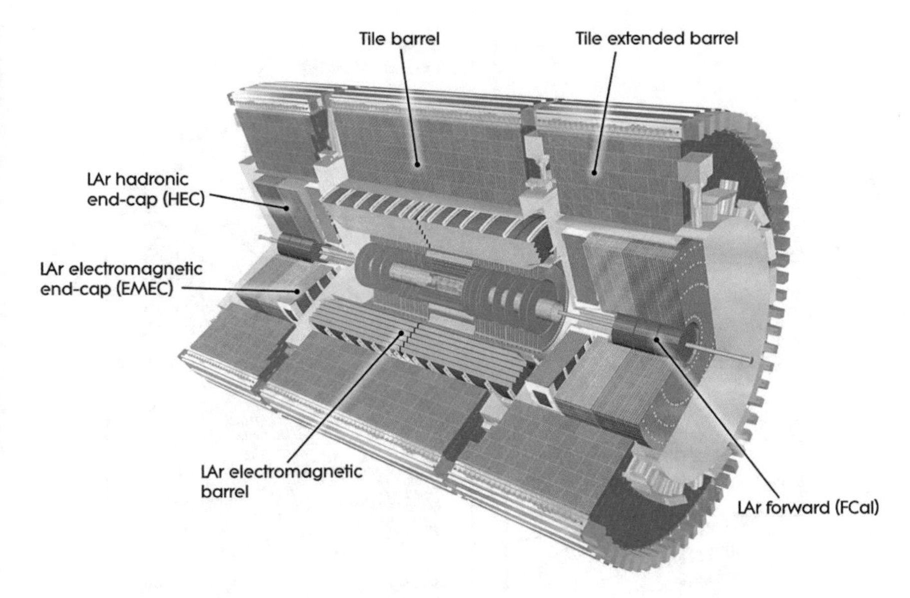

Fig. 3.11 Cut-away view of the ATLAS calorimeters [13]

In order to measure the energy of the particle showers accurately, the shower must be contained with the calorimeter so calorimeter depth is an important design characteristic. Thus, the thickness of the LAr calorimeter is 22 radiation lengths in the barrel and 24 radiation lengths in the endcaps [13]. Lead plates, which have a radiation length of 0.56 cm [23] are used as the absorber. This allows the LAr calorimeter to be about 50 cm long [13].

The active material is liquid argon and the readout of the energy is done using accordion-shaped kapton electrodes [13]. The accordion geometry provides complete ϕ symmetry without azimuthal cracks.

The LAr calorimeter is split into three layers in the barrel, shown in Fig. 3.12, and four layers in the endcap. The barrel layers include a presampler layer of $4.3X_0$ used to correct for the energy lost by the electrons and photons upstream. The second layer of $16X_0$ in length and is where most of the photon and electron energy will be deposited. To measure the energy loss accurately, this region has a granularity of $\Delta\eta \times \Delta\phi = 0.003125 \times 0.1$. The final layer has 2 X_0 in lengths and granularity of $\Delta\eta \times \Delta\phi = 0.05 \times 0.0245$. This layer is used to estimate any energy not sampled by the presampler and the second layer.

3.2.2.2 Hadronic Calorimeters

The hadronic calorimeters are composed of two calorimeters: Tile calorimeter in the range of $|\eta| < 1.7$, and liquid argon calorimeters for $|\eta| > 1.7$ because of the intrinsic radiation hardness of this technology.

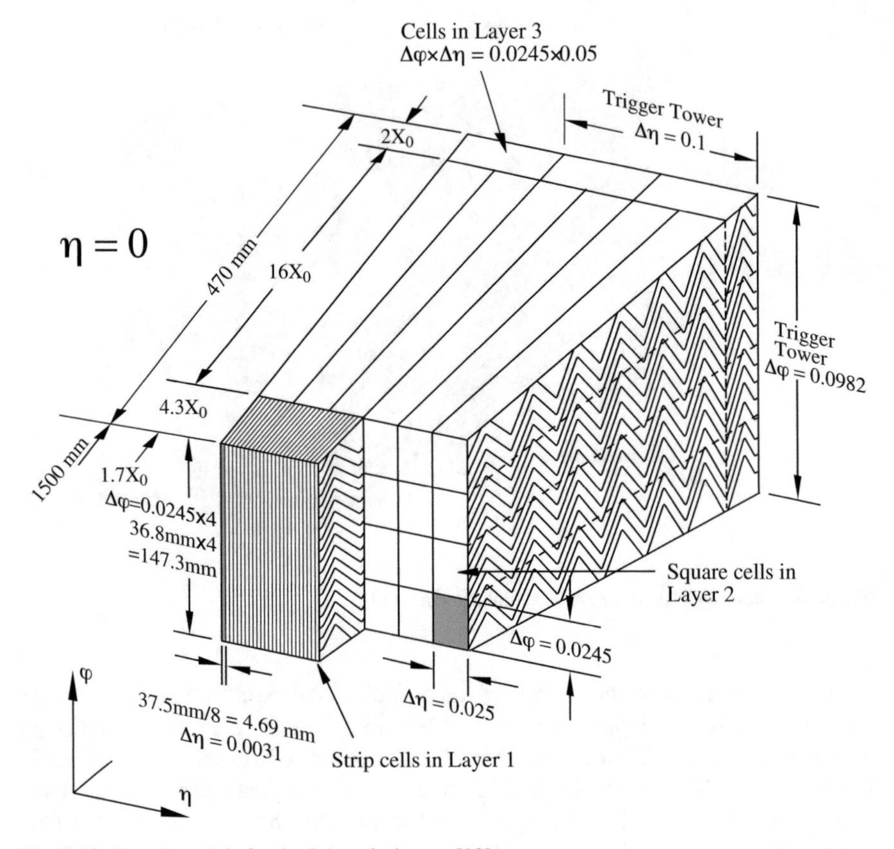

Fig. 3.12 Barrel module for the LAr calorimeter [13]

Similar to the LAr calorimeter, the Tile calorimeter is designed to contain the hadronic showers. The characteristic length for nuclear interaction of a pion, λ_f, in lead or iron is 20 cm [23], which is much larger than for the LAr electromagnetic calorimeter. Thus, steel is used as an absorber for the Tile Calorimeter with scintillating tiles as the active material. The thickness of the Tile calorimeters is 9.7 interaction lengths in the barrel and 10 interaction lengths in the end cap, or extends radially from 2.28 m to 4.25 m [13]. The Tile calorimeter is segmented into three layers of 1.5, 4.1, and 1.8 interaction lengths in the barrel. The granularity of the is $\Delta\eta \times \Delta\phi = 0.1 \times 0.1$ in the first two layers and $\Delta\eta \times \Delta\phi = 0.2 \times 0.1$.

The Hadronic end-cap calorimeter (HEC) is a liquid argon with copper absorbers calorimeter which covers the ranges of $1.5 < |\eta| < 3.2$. The granularity of the HEC is $\Delta\eta \times \Delta\phi = 0.1 \times 0.1$ for $1.5 < |\eta| < 2.5$ and $\Delta\eta \times \Delta\phi = 0.2 \times 0.2$ for $2.5 < |\eta| < 3.2$ [13].

Finally, the Forward Calorimeter (FCal) provides calorimetry for electromagnetic and hadronic showers at ranges of $3.1 < |\eta| < 4.9$. The first layer performs electromagnetic calorimetry with copper as the absorber and the outer two layers perform hadronic calorimetry using tungsten as the absorber. The FCal uses liquid argon as the active material.

3.2.3 The Muon Spectrometer

The muon spectrometer is the outermost sub detector and is designed to detect charged particles that exit the calorimeters. The muon spectrometer measures the particles' momentum up to $|\eta| < 2.7$ and is used to trigger on these particles in the region of $|\eta| < 2.4$. The muons spectrometer is designed to measure muon momenta down to 3 GeV (due to energy loss in the calorimeters). Muon tracking is based on the deflection of muon tracks in the large toroid magnets. Over the range of $|\eta| < 1.4$, magnetic bending is provided by a toroid magnet of 1 T and for $1.6 < |\eta| < 2.7$, the muon tracks are bent using smaller end-cap magnets of 0.5 T. In between $1.4 < |\eta| < 1.6$, a combination of barrel and end-cap magnets are used [13].

The are four types of muon chambers shown in Fig. 3.13: Monitored Drift Tubes (MDT), Cathode Strip Chambers (CSC), Thin Gap Chambers (TGC), and Resistive Plate Chambers (RPC). The MDTs are used to measure the track coordinates up to $|\eta| < 2.7$. The CSCs, which are multiwire proportional chambers with cathodes segmented into strips, are used in the ranges of $2.0 < |\eta| < 2.7$. The trigger system covers ranges up $|\eta| < 2.4$ with TGCs in the region of $|\eta| < 1.05$ and RPCs in the region $1.05 < |\eta| < 2.4$ (up to $|\eta| < 2.7$ for coordinate and energy measurements). These chambers provide bunch crossing identification, well defined p_T thresholds,

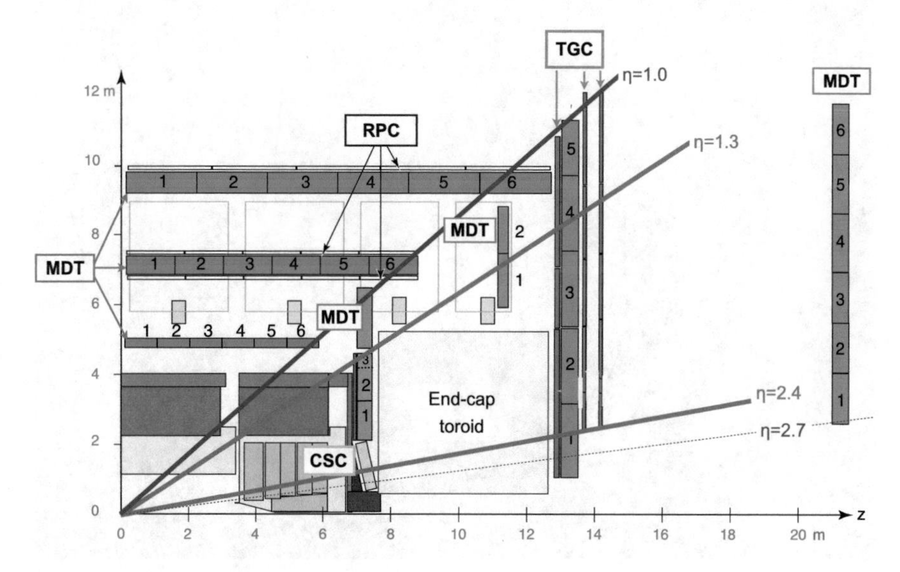

Fig. 3.13 Schematic of the muon system [25]

and measure muon coordinates in the region orthogonal to that determined by the MDTs and CSCs.

3.3 Particle Reconstruction and Identification

Particles are identified in the ATLAS detector by the energy and tracks left by the particles as they traverse the subdetectors, as shown in Fig. 3.14. Particles such as electrons, photons, and jets are identified by the signals left in the detector such as hits or energy deposits. This raw data is converted to physics objects using dedicated algorithms, called "event reconstruction". Particle reconstruction occurs in several stages.

The first stage reconstructs particle trajectories, called tracks, from hits in the Inner Detector and the muon spectrometer (for muons) [13]. The tracking algorithm associates hits from the Inner Detector to particles using pattern recognition and fitting algorithms such as the Kalman Fitter [27] and the ATLAS Global χ^2 Track Fitter [28]. Track reconstruction occurs in three stages [13]. The first stage is a pre-processing stage where the raw data from the Pixel and SCT detectors are converted

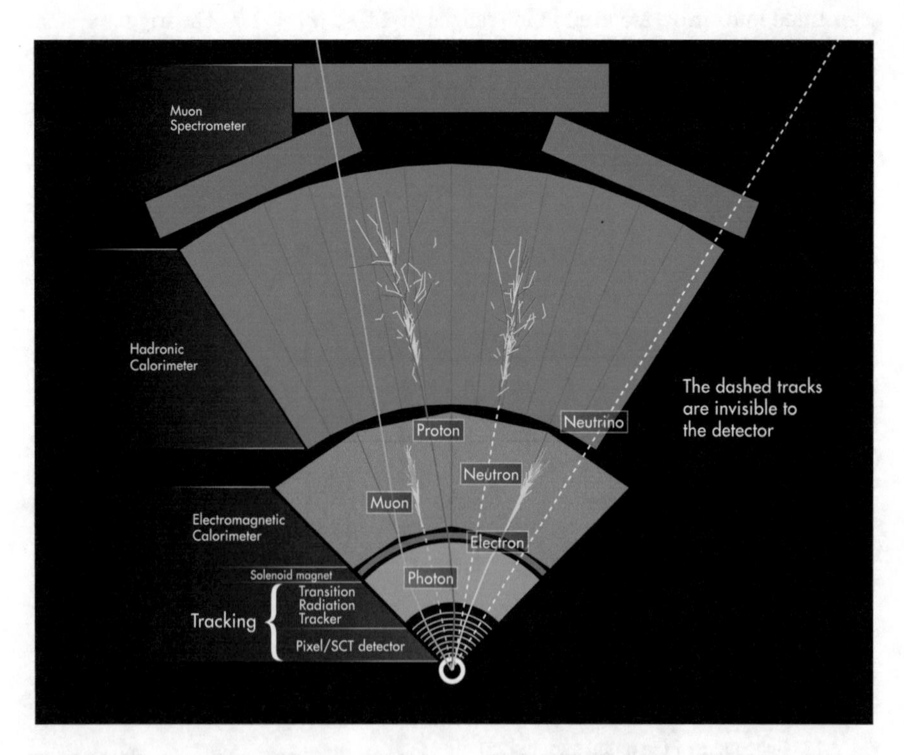

Fig. 3.14 Diagram of how different particle interact in the ATLAS detector [26]

into clusters and TRT timing information is also converted into drift circles. The second stage is a track finding stage where the Kalman Fitter and Global χ^2 Track Fitter are used using pixel and SCT hits. Ambiguities and fake tracks are rejected. The selected tracks are then extended into the TRT using the drift circle information. The tracks are then refitted using the information from all three detectors. A complementary track finding strategy searches for unused track segments in the TRT and then extends those tracks back into the SCT and pixel detectors. This helps the tracking efficiency of secondary tracks from conversions and decays of long-lived particles. Finally, the post-processing stage uses a vertex finder algorithm to reconstruct primary vertices followed by algorithms to reconstruct photon conversions and secondary vertices.

The second stage reconstructs energy clusters within the calorimeters. The electromagnetic calorimeter is divided into a grid of 200×256 elements, called "towers", with granularity of $\Delta\eta \times \Delta\phi = 0.025 \times 0.025$ [29]. Energy clusters are formed using a sliding-window algorithm [30] of size 3×5 towers in $\eta \times \phi$ with energy exceeding 2.5 GeV. A rectangular window is used instead of square window to capture all the energy from Bremsstrahlung. The hadronic calorimeter is used along with the electromagnetic (EM) calorimeter to form clusters called topological clusters [31]. A seed cluster is formed with the signal over noise in the cell greater than 4. Adjacent cells are added to the seed cell if their signal over noise ratio is greater than 2. The process is repeated until the border stops increasing.

Electrons

Electron reconstruction is done by matching tracks to electromagnetic energy clusters within $|\eta| < 0.05$ and $-0.2 < \Delta\phi < 0.05$, and are calibrated using $Z \rightarrow ee$ decays [32].

Prompt electrons, electrons not coming from photon conversions, entering the central region of the detector, $|\eta| < 2.47$, are selected using a likelihood (LH) based identification [29]. The inputs to the likelihood are the hits in the Inner Detector and the energy clusters. The identification criteria in the likelihood are optimized in bins of η and E_T.

Photons

Photons are reconstructed using the sliding-window algorithm [13] using clusters in the electromagnetic calorimeter with either no associated Inner Detector tracks for unconverted photon candidates or with SCT tracks but not Pixel hits for converted photons.

Muons

Muons are reconstructed at first independently in the Inner Detector (ID) and the Muon Spectrometer (MS) [33]. The combined ID and MS is performed using different algorithms depending on the information provided by the ID, the MS, and the calorimeters.

- Combined (CB) muon: a MS track is matched to an ID track, and the measurements of the momenta are combined.

- Segment-tagged (ST) muon: the ID track is matched to at least one MS track. The momentum measurement is taken from the ID.
- Calorimeter-tagged (CT) muon: the ID track is matched to an energy deposit in the calorimeter. The criteria for CT muons are optimized for $|\eta| < 0.1$ and $15 < p_T < 100$ GeV.
- Extrapolated (ME) muon: the muon trajectory is reconstructed from the MS track and a loose requirement on originating at the interaction point. ME muons are used to extend the acceptance for muon reconstruction in the region $2.5 < |\eta| < 2.7$, not covered by the ID.

Muons are calibrated using $Z \rightarrow \mu\mu$ and $J/\psi \rightarrow \mu\mu$ decays.

Jets

Quarks and gluons hadronize as they travel in the detector and collectively are referred to as jets. Jets are reconstructed using the anti-k_t algorithm [34, 35] with the radius parameter $R = 0.4$ using topological clusters as inputs. The equations describing this algorithm are

$$d_{ij} = \min\left(\frac{1}{p_{\mathrm{T},i}^2}, \frac{1}{p_{\mathrm{T},j}^2}\right) \times \frac{R_{ij}^2}{R} \tag{3.2}$$

where d_{ij} is the calculated distance, $p_{\mathrm{T},i}$ is the p_T of the ith jet, R_{ij} is the ΔR between the ith and jth jet, and $R = 0.4$. This algorithm is dominated by high-p_T jets and the algorithm will cluster larger p_T jets first. The d_{ij} between similarly separated soft particles will be larger so soft particles will cluster with larger p_T jets than with themselves.

The jet energy scale (JES) and resolution (JER) are first calibrated to particle level using MC simulation and then through Z+jet, γ+jet, and multijet measurements

***b*-Jets**

Bottom quarks hadronize before they decay [36]. The lifetime of the resulting b-hadrons is around picoseconds resulting in a decay length, length the b-hadron will travel before decaying, of about 0.5 mm. If the b-hadron is produced relativistically, the decay length will be even longer. The b-hadrons will decay to multiple final states. Thus, the signature of a b-jet is many jet tracks, a secondary (or displaced) vertex, and jets forming the invariant mass of a b-hadron.

A multivariate algorithm, MV2c10, using a boosted decision tree (BDT) is used for the b-jet reconstruction [37]. The inputs to the BDT explain the features of the long b-hadron lifetime: jet p_T and η, presence of a secondary vertex, information from track impact parameters, the presence of displaced secondary vertices, and the reconstructed flight paths of b- and c-hadrons inside the jet.

Missing Energy

Some particles, such as neutrinos, do not interact in the detector and escape undetected. Their presence is inferred as a momentum imbalance, called missing transverse momentum $\mathbf{p}_{\mathrm{T}}^{\mathrm{miss}}$, with magnitude calling missing energy, $E_{\mathrm{T}}^{\mathrm{miss}}$. The missing transverse momentum is calculated as the negative of the sum of the momenta of all particles [38]. The missing momentum comes from two contributions. The first one is from reconstructed and calibrated particles and jets described above, called "hard" objects. The second contribution from "soft-event" signals consisting of reconstructed charged-particle tracks associated with the scatter vertex but not with the hard objects.

3.4 Trigger System

Collisions of protons take place every 25 ns, corresponding to 10^9 collisions per second. Due to limited storage space, less than 2,000 events per second can be saved, thus selective triggering is required. A two-level triggering system is used to select events [39–41]. The first-level, Level-1 or L1, is implemented in hardware and reduces the input event rate from 40 MHz to 100 kHz. These trigger decisions are made using calorimeter signals and muon trigger chamber (RPC and TGC) signals. Thus, the L1 trigger decision is based on the energy deposition in the calorimeters and muon track segments. The L1 trigger is followed by a software-based trigger, high-level trigger or HLT, which further reduces the rate to 1 kHz. The L1 output defines a rectangular Region of Interest (RoI). Reconstruction algorithms are run on the RoI to select objects are close to physics analysis software. The output rate decreases exponentially as a function of time due to the decrease in luminosity during a fill. Figure 3.15 shows the trigger rate for different types of trigger during one fill in 2018.

The trigger menu contains different types of triggers. Primary triggers are used for physics measurements and are run unprescaled. These triggers use topological information and p_T thresholds of different particles (electrons, muons, $E_{\mathrm{T}}^{\mathrm{miss}}$, jets, etc.) to select events. Most of the trigger bandwidth is used for the physics triggers. 15% of the HLT bandwidth is dedicated to support triggers. They are used for efficiency and performance measurements and run at a small rate of 0.5 Hz each. Backup triggers, with tighter selection and lower rates, are used in case the rate of the

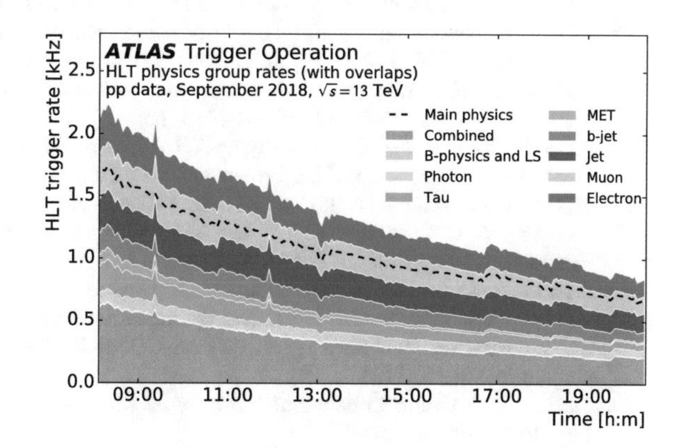

Fig. 3.15 HLT trigger rates during a fill in 2018 with peak luminosity of $L = 2.0 \times 10^34\ \mathrm{cm}^{-2}\mathrm{s}^{-1}$ and $< \mu >= 56$ [42]

primary triggers becomes too high. There are also calibration triggers that are used for detector calibrations. They run a high rate but store very small events, keeping only the information relevant to perform the calibration.

References

1. Evans LR, Bryant P (2008) LHC machine. JINST 3:S08001. https://cds.cern.ch/record/1129806. This report is an abridged version of the LHC Design Report (CERN-2004-003)
2. LEP design report. CERN, Geneva, 1984. https://cds.cern.ch/record/102083. Copies shelved as reports in LEP, PS and SPS libraries
3. LHC Study Group Collaboration, Pettersson TS, Lefvre P (1995) The large hadron collider: conceptual design. Technical report CERN-AC-95-05-LHC. https://cds.cern.ch/record/291782
4. ATLAS Collaboration (2008) The ATLAS experiment at the CERN large hadron collider. JINST 3:S08003
5. Collaboration CMS (2008) The CMS experiment at the CERN LHC. JINST 3:S08004
6. LHCb Collaboration, Alves AA, Jr. et al (2008) The LHCb detector at the LHC. JINST 3:S08005
7. ALICE Collaboration, Aamodt K, et al (2008) The ALICE experiment at the CERN LHC. JINST 3:S08002
8. Mobs E, The CERN accelerator complex. Complexe des accelerateurs du CERN. https://cds.cern.ch/record/2197559. General photo
9. Arnaudon L, et al (2006) Linac4 technical design report. Technical report CERN-AB-2006-084. CARE-Note-2006-022-HIPPI, CERN, Geneva. http://cds.cern.ch/record/1004186. Revised version submitted on 2006-12-14 09:00:40
10. AC Team (1999) Diagram of an LHC dipole magnet. Schema d'un aimant dipole du LHC. https://cds.cern.ch/record/40524
11. Luminosity Public Results Run 2. https://twiki.cern.ch/twiki/bin/view/AtlasPublic/LuminosityPublicResultsRun2
12. Luminosity Public Results. https://twiki.cern.ch/twiki/bin/view/AtlasPublic/LuminosityPublicResults
13. ATLAS Collaboration. Expected performance of the ATLAS experiment - detector, trigger and physics. arXiv:0901.0512 [hep-ex]
14. ATLAS Collaboration (2017) Performance of the ATLAS track reconstruction algorithms in dense environments in LHC run 2. Eur Phys J C77, no 10:673. arXiv:1704.07983 [hep-ex]
15. ATLAS Collaboration (2008) ATLAS pixel detector electronics and sensors. JINST 3:P07007
16. ATLAS Collaboration (2010) ATLAS insertable B-layer technical design report. Technical report CERN-LHCC-2010-013. ATLAS-TDR-19. https://cds.cern.ch/record/1291633
17. ATLAS Collaboration (2014) Operation and performance of the ATLAS semiconductor tracker. JINST 9:P08009. arXiv:1404.7473 [hep-ex]
18. ATLAS TRT Collaboration (2008) The ATLAS Transition Radiation Tracker (TRT) proportional drift tube: design and performance. JINST 3:P02013
19. ATLAS Collaboration (2008) The ATLAS TRT end-cap detectors. JINST 3:P10003
20. ATLAS TRT Collaboration (2008) The ATLAS TRT barrel detector. JINST 3:P02014
21. Fratina S, Klinkby E (2010) The geometry of the ATLAS transition radiation tracker. Technical report ATL-COM-INDET-2010-002, CERN, Geneva. https://cds.cern.ch/record/1232064
22. ATLAS Collaboration (2011) Particle identification performance of the ATLAS transition radiation tracker. Technical report ATLAS-CONF-2011-128, CERN, Geneva. https://cds.cern.ch/record/1383793
23. Particle Data Group Collaboration, Tanabashi M, et al (2008) Review of particle physics. Phys Rev D 98, no. 3:030001

24. ATLAS Collaboration (2008) The ATLAS TRT electronics. JINST 3:P06007
25. ATLAS Collaboration (2015) Performance of the ATLAS muon trigger in pp collisions at $\sqrt{s} = 8$ TeV. Eur Phys J C75:120. arXiv:1408.3179 [hep-ex]
26. Pequenao J, Schaffner P (2013) How ATLAS detects particles: diagram of particle paths in the detector. https://cds.cern.ch/record/1505342
27. Fruhwirth R (1987) Application of Kalman filtering to track and vertex fitting. Nucl Instrum Meth A 262:444–450
28. Cornelissen TG, Elsing M, Gavrilenko I, Laporte JF, Liebig W, Limper M, Nikolopoulos K, Poppleton A, Salzburger A (2008) The global chi**2 track fitter in ATLAS. J Phys Conf Ser 119:032013
29. ATLAS Collaboration (2019) Electron reconstruction and identification in the ATLAS experiment using the 2015 and 2016 LHC proton-proton collision data at $\sqrt{s} = 13$ TeV. Submitted to: Eur Phys J. arXiv:1902.04655 [physics.ins-det]
30. Lampl W, Laplace S, Lelas D, Loch P, Ma H, Menke S, Rajagopalan S, Rousseau D, Snyder S, Unal G (2008)Calorimeter clustering algorithms: description and performance. Technical report ATL-LARG-PUB-2008-002. ATL-COM-LARG-2008-003, CERN, Geneva. https://cds.cern.ch/record/1099735
31. ATLAS Collaboration (2017) Topological cell clustering in the ATLAS calorimeters and its performance in LHC run 1. Eur Phys J C77:490. arXiv:1603.02934 [hep-ex]
32. ATLAS Collaboration (2019) Electron and photon energy calibration with the ATLAS detector using 2015–2016 LHC proton–proton collision data. JINST 14:P03017. arXiv:1812.03848 [hep-ex]
33. ATLAS Collaboration (2016) Muon reconstruction performance of the ATLAS detector in proton proton collision data at $\sqrt{s}$ =13 TeV. Eur Phys J C 76, no. 5:292. arXiv:1603.05598 [hep-ex]
34. Cacciari M, Salam GP, Soyez G (2008) The anti-k_t jet clustering algorithm. JHEP 04:063 arXiv:0802.1189 [hep-ph]
35. Cacciari M, Salam GP, Soyez G (2012) FastJet user manual. Eur Phys J C 72:1896 arXiv:1111.6097 [hep-ph]
36. Schwartz MD (2018) TASI lectures on collider physics. In: Proceedings, theoretical advanced study institute in elementary particle physics : anticipating the next discoveries in particle physics (TASI 2016): Boulder, CO, USA, June 6-July 1, 2016, pp 65–100. arXiv:1709.04533 [hep-ph]
37. ATLAS Collaboration (2018) *Measurements of b-jet tagging efficiency with the ATLAS detector using* $t\bar{t}$ events at $\sqrt{s} = 13$ TeV. JHEP 08:089. arXiv:1805.01845 [hep-ex]
38. ATLAS Collaboration (2018) Performance of missing transverse momentum reconstruction with the ATLAS detector using proton-proton collisions at $\sqrt{s} = 13$ TeV. Eur Phys J C78, no 11:903. arXiv:1802.08168 [hep-ex]
39. ATLAS TDAQ Collaboration (2016) The ATLAS data acquisition and high level trigger system. JINST 11, no 06:P06008
40. ATLAS Collaboration (2017) Performance of the ATLAS trigger system in 2015. Eur Phys J C77, no 5:317. arXiv:1611.09661 [hep-ex]
41. ATLAS Collaboration (2018) Trigger menu in 2017. Technical report ATL-DAQ-PUB-2018-002, CERN, Geneva. https://cds.cern.ch/record/2625986
42. Trigger Operation Public Results. https://twiki.cern.ch/twiki/pub/AtlasPublic/TriggerOperationPublicResults/

Chapter 4
HL-LHC Inner Detector Upgrade

4.1 Motivation for the Detector Upgrade

ATLAS has collected data on and off since 2011 till present time. Run 1 started in 2011 when data was collected at $\sqrt{s} = 7$ TeV and in 2012 at $\sqrt{s} = 8$ TeV. The amount of data delivered in those years corresponds to almost 30 fb^{-1}. During those years, there was on average 10–40 collisions per bunch crossing. Run 2, collected data at $\sqrt{s} = 8$ TeV, started in 2015 and ended in 2018. During Run 2, the amount of collisions per bunch crossing increased to 40–60. The total amount of data collected by ATLAS up to 2019 is about 200 fb^{-1} with another 300 fb^{-1} expected to be collected in Run 3, scheduled to run from 2021 to 2023. The ATLAS experiment was designed to operate for 10 years at a constant instantaneous luminosity of 1.0×10^{34} $\text{cm}^{-2}\text{s}^{-1}$ and to remain functional for an integrated luminosity of 700 fb^{-1} [1].

The High Luminosity LHC (HL-LHC) will operate at a peak luminosity of 7.5×10^{34} $\text{cm}^{-2}\text{s}^{-1}$, which corresponds to approximately 200 collisions per beam crossing. HL-LHC will operate for ten years and ATLAS is expected to collect 3,000 fb^{-1} of data [2]. The current ATLAS Inner Detector was not designed to operate in these conditions. The Pixel detector was designed to withstand radiation damage up to 400 fb^{-1}, a factor of 10 lower than the expected HL-LHC luminosity, while the SCT is constructed to operate up to 850 fb^{-1}. With the increase in the number of proton-proton collisions to 200, the increase in hits in the TRT would cause the occupancy to approach 100% and nearby particles would not be resolved in the SCT. Thus, without replacement of the Inner Detector, detector performance in HL-LHC would be degraded resulting in the compromise of the physics reach. The current timeline for the LHC running until HL-LHC is shown in Fig. 4.1.

© The Editor(s) (if applicable) and The Author(s), under exclusive license
to Springer Nature Switzerland AG 2020

E. Resseguie, *Electroweak Physics at the Large Hadron Collider with the ATLAS Detector*, Springer Theses,
https://doi.org/10.1007/978-3-030-57016-3_4

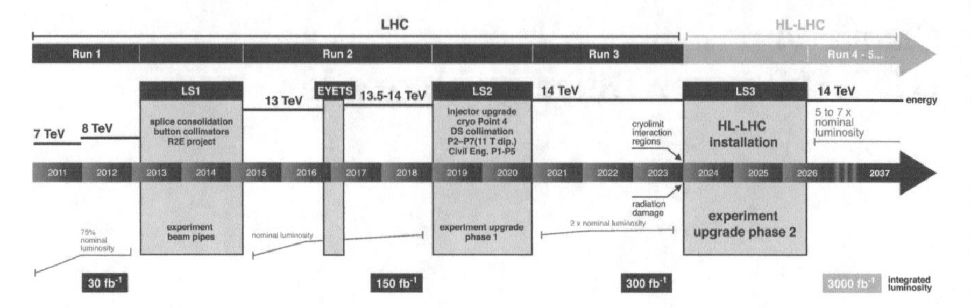

Fig. 4.1 Timeline for the LHC program including HL-LHC [2]

4.2 Inner Tracker: ITk

Since the current Inner Detector will not be able to sustain the running conditions of HL-LHC, it will be replaced with a new detector, the Inner Tracker (ITk), around 2024, after the end of Run 3. Unlike the current Inner Detector, the ITk will use silicon sensors for all its subdetectors. Moreover, the current Inner Detector has three subdetectors, the Pixel detector, the SCT, and the TRT, while the ITk will only have two subdetectors: the Pixel and Strips detectors. The Pixel detector, just like in the current Inner Detector, is closest to the beamline and the Strips detector starts at 40 cm from the beamline. The layout for ITk can be found in Fig. 4.2. The current Inner Detector has coverage up to $|\eta| \leq 2.7$ while the ITk extends the coverage to $|\eta| \leq 4.0$.

The Pixel detector will consist of approximately 5 billion sensors with 2,500 μm^2 area per sensor [3]. The design of the hybrid pixel module is similar to the one used in the present Pixel detector. The hybrid Pixel module is composed of two parts: a silicon sensor with a front-end chip fabricated in CMOS (Complementary metal-oxide-semiconductor) technology (bare module), and a flexible printed circuit board (PCB) called a module flex.

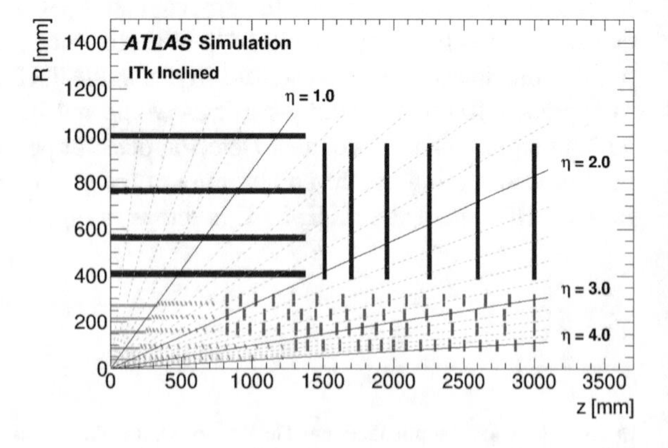

Fig. 4.2 Schematic of the layout of the ITk detector. The Pixel detector is shown in red while the strips detector is in blue [2]

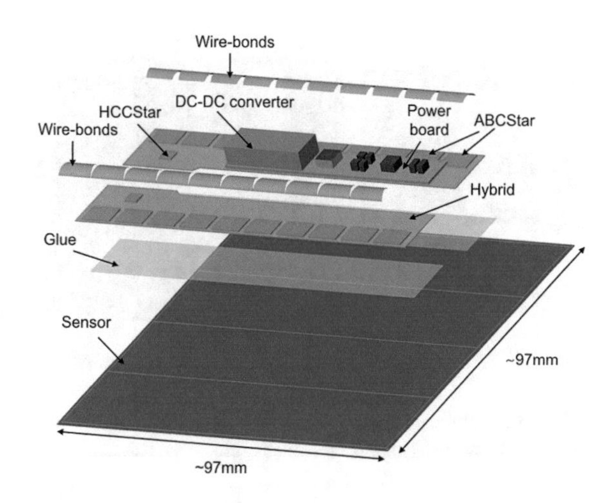

Fig. 4.3 Schematic of an ITk strip module

The Strips detector will consist of approximately 18,000 silicon strips sensors, corresponding to 60 million readout channels [2]. The basic unit in the Strips detector is also called a hybrid. It consists of one sensor and one PCB, which hold the readout Application-specific Integrated Circuits (ASICs). The modules are constructed by gluing kapton flex hybrids to silicon sensors, as shown in Fig. 4.3. These modules will be installed on the ITk in supports called staves and petals.

4.3 Strips Read-Out System

Custom-made ASICs are developed for the ITk Strips detector:

- ATLAS Binary Chip (ABC): converts incoming charge signal into hit information. The prototype is called ABC130 and the production chip is called the ABCStar.
- Hybrid Controller Chip (HCC): interfaces between the ABC130 and the bus-tape. The prototype is called HCC130 and the production chip is called the HCCStar.

The HCC130 is the interface between the stave service bus (stave side) and the front end ASICs on the hybrid. Each hybrid in the inner barrel will service one half of a sensor, 2,560 strips with 10, 256 channel ABC130s. The HCC130 gathers the serialized data from the front end chips (ABC130) and serves as the communication point between the End of Stave Controller (EOSC) and the front-end chips. Figure 4.4 shows a diagram of the readout from the strips detector.

Five ABC130 are daisy chained and can send the data across two data-loops, resulting in four data lines to the HCC130. Since the design of the prototype chips, HCC130 and ABC130, the ATLAS upgrade trigger requirement has doubled from L0/L1 rates of 500 kHz/ 200 kHz to 1 MHz/ 400 kHz. However, latency studies have shown that there can be a chain of at most three ABC130 in high occupancy regions of the Strip detector. As a result, the "star" architecture was developed: with point-

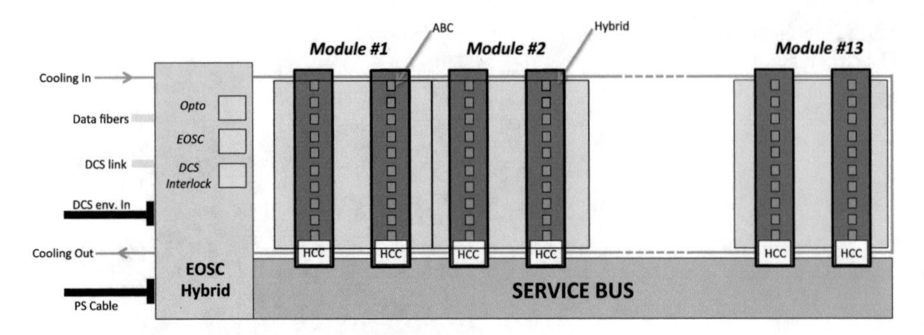

Fig. 4.4 Readout system for the Strip detector with the ABC130 and HCC130 [4]

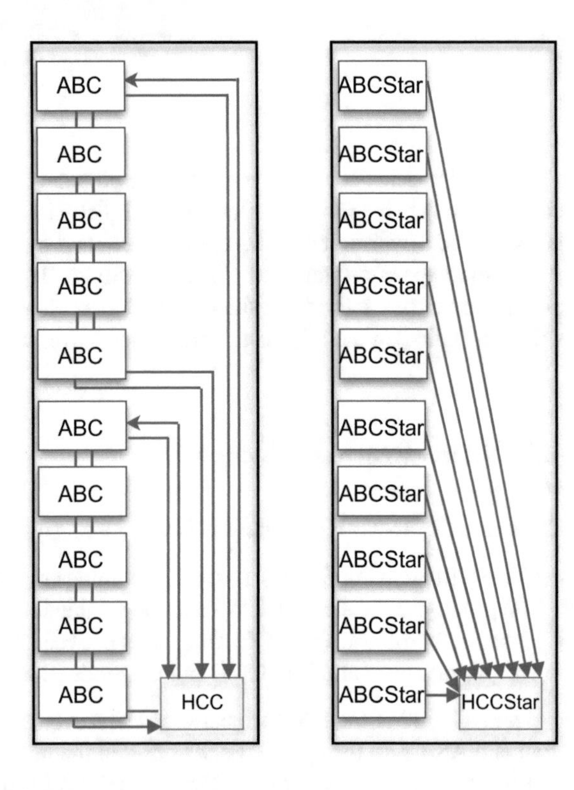

Fig. 4.5 Diagrams of the original daisy chain signal routing (left) and the star design (right) [2]

to-point connections between each ABCstar and the HCCstar on the hybrid. The two designs are shown in Fig. 4.5.

Another ASIC was developed, the Autonomous Monitor and Control Chip (AMAC), which provides monitoring and interrupt functionality. This chip's design is based on the HCC130's Analog Monitor (AM), which will be described further in a later section. The final readout system with the HCCstar, ABCstar, and AMAC is shown in Fig. 4.6.

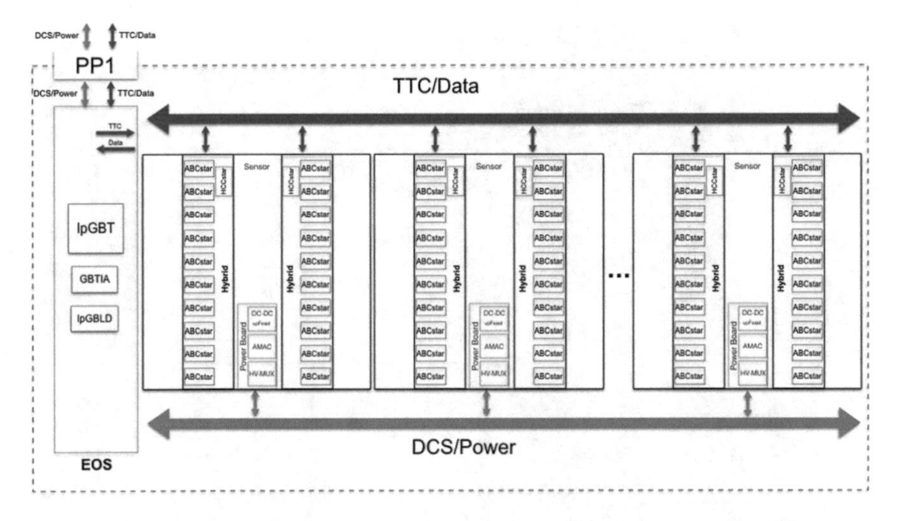

Fig. 4.6 Readout system for the Strip detector with the HCCstar, ABCstar, and AMAC [2]

There are different size modules depending on their location on the Strips detector. In the barrel, two strips lengths are used: long strips used for lower occupancy and shorter strips at smaller radii where the occupancy is expected to be larger. The short strips contain two hybrids, each with ten ABCStars and long strips contain one hybrid and ten ABCStars.

4.4 HCC130 Testing

The goals of the program are to test and characterize HCC130 chips to ensure correct implementation, electrical robustness, and radiation hardness, for the HCCStar and AMAC chip designs.

4.4.1 HCC130 Signals

To communicate with the HCC130, signals entering and leaving the HCC130 need to be understood. All control, clock, and data lines use Low Voltage Differential Signaling (LVDS). Signals can also be separated into two categories: hybrid side and stave side. All signals entering and leaving the HCC130 are shown in Fig. 4.7.

The stave-side signals are generated at the End-of-Stave (EoS) and bussed to the HCC130. The stave-side clock, called BC stave, is a 40 MHz clock sent by the EoS and serves as a reference clock for the internal HCC130 clocks, ePLL. The ePLL generates 40 MHz, 160 MHz, 320 MHz, and 640 MHz clocks to receive data from

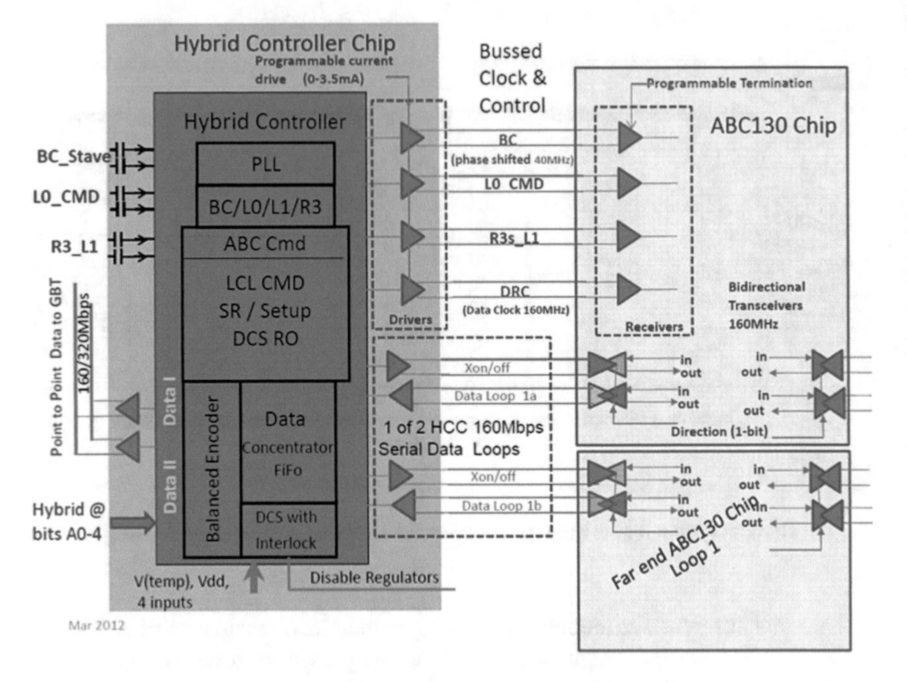

Fig. 4.7 Diagram of the signals sent and received between the HCC130 and the ABC130 chips [4]

the ABC. Stave side control signals are L0COM and R3SL1, which are multiplexed. The L0COM line is composed of two 40 Mbps multiplexed signals: level-0 trigger signal, and the command input. When both are idle, a 40 MHz clock signal can be seen on this line. The L0 signal is active high meaning that it is sent with the falling edge of the BC stave clock while the command line is active low, or sent with the rising edge of the BC clock. Similarly, the R3SL1 line has time-multiplexed signals: the Regional Readout Request signal (R3) and the level-1 trigger (L1). R3 is active high while L1 is active low. When both signals are idle, a 40 MHz clock is seen on the line.

The hybrid side signals are sent from the HCC130 to the ABC chips. The hybrid BC clock is generated from the incoming BC stave clock or from the ePLL. This clock's phase is programmable. Additional clocks, generated from the ePLL, are the Data Readout Clock (DRC), and the FastClusterFinder clock (FCLK). The DRC, set to 80 or 160 MHz, is used to clock the data transmission from the ABC to the HCC130. This data may be incoming from any of the four data input ports are sorted into one output stream. The FCLK can be set to 80, 160, 320, or 640 MHz. Additional signals are replicated stave-side signals: L0CMD and R3SL1 and are also time-multiplexed.

4.4.2 Testing Setup

To test the HCC130, two boards were designed: an passive board which holds the HCC130 and protection diodes, and an active board. The active board holds level shifting transceivers to control the common mode, CERN transceiver to create LVDS signals, and the MicroZed, a Xilinx system-on-chip (SoC) microcontroller with a FPGA. Both of these boards are pictured in Fig. 4.8.

The passive board is connected to the active board via four connectors. The left side and the right side of the passive board are not mirrors of each other. The left side has the interface signals: BC stave, the L0/cmd, R3/L1 signals, as well as data line one while data line two is on the right side of the tester. The interface signals are used to control the HCC130 and to read out the data returning from the HCC130. On the active board, the LVDS signals leaving the HCC130 are treated with a comparator before going to the MicroZed to be read out which tests the robustness of the LVDS signal. The common mode of these comparators can be controlled using a Digital-to-Analog converter (DAC). Figure 4.9 shows the schematic of how the signal is processed with a comparator. The LVDS signals that are sent from the HCC130 from the MicroZed (stave signals and the data) first pass through the CERN transceivers, which is an 8 channel LVDS transceiver chip [5]. MicroZed signals are not LVDS so this ensures that LVDS signals are generated with the appropriate common mode voltage to be sent to the HCC130.

The voltages on the active board are 5 V, 3.3 V, and −3.3 V. The MicroZed requires 2.5 V, 3.3 V, and 5 V. The general purpose input-output (GPIO) signals (such as the signals setting the pad IDs) are set at 3.3 V and the LVDS signals are set at 2.5 V. The HCC130 requires up to 1.5 V. The ± 3.3 V are used to operate the op-amps used

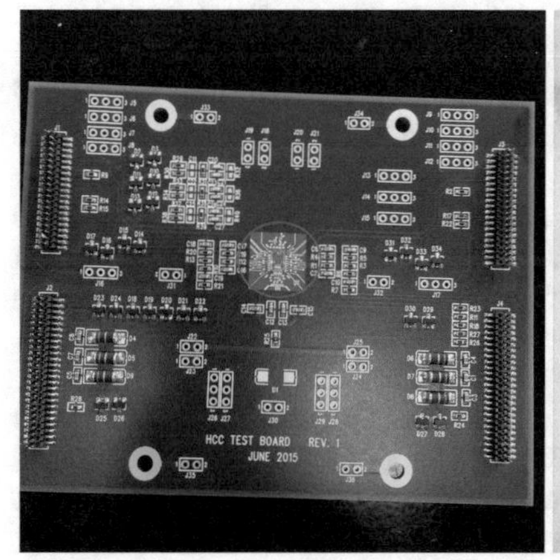

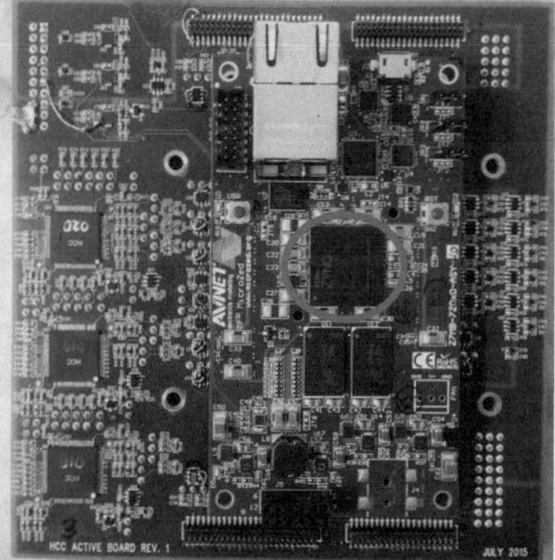

Fig. 4.8 Photographs of the passive board (left) and active board (right) used in the testing of the HCC130 chip. The HCC130 chip is circled in red in the picture on the left, the MicroZed Zynq FPGA chip is circled in red in the picture on the right

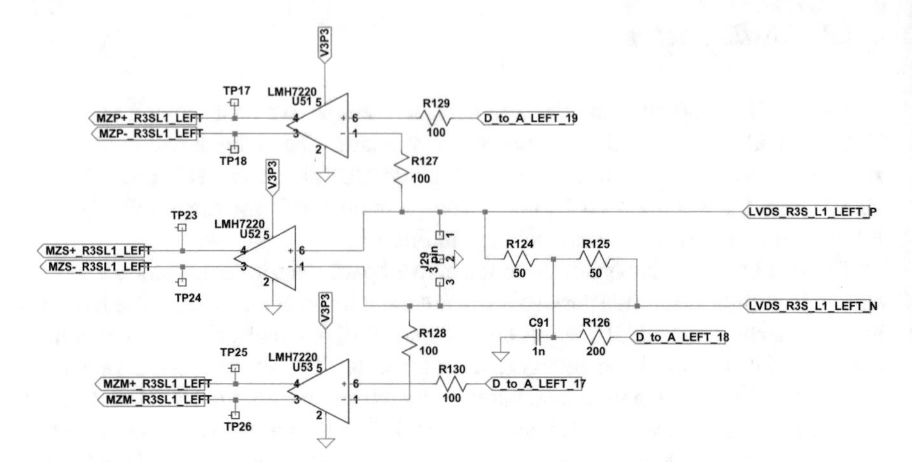

Fig. 4.9 Spice schematic for the comparator used to read signals from the HCC130. The common mode value can be varied using a DAC

in the comparators. Finally the DAC uses 3.3 V while the ADCs use 5 V, 3.3 V, and 2.5 V.

4.4.3 Description of the FPGA Code

To test the HCC130, all signals shown in Fig. 4.7 must be emulated in the MicroZed. The MicroZed contains a Zynq Processing System (PS) and a Zynq Programmable Logic (PL) component. The PS includes Ethernet capabilities, USB 2.0, and a UART. The PL separates each logic block into blocks called "IP blocks" and is programmed in VHDL. Three IP blocks are defined to test the HCC130: HCC130 interface, 2 ABC130 chain emulator IP blocks, and ADC and DAC controls. Summary of all the blocks are shown in Fig. 4.10.

The HCC130 Interface (IF) IP block emulates the stave side for the HCC130 chip. The outgoing signals from the FPGA are BC_STAVE, the BC clock signal at 40 MHz, L0COM, the command and L0 trigger, R3SL1, the R3 an L1 triggers. The incoming signals are Data1 and Data2, the two data return lines from the HCC130. All these signals are differential and are summarized in Table 4.1.

Four clock signals are also generated in this block. One 40 MHz clock signal is used to generate the outgoing BC_STAVE signal. Another 40 MHz clock signal is used as an internal BC signal to generate the command and triggers. An 80 MHz clock is used to multiplex the two signals on the L0COM and R3SL1 lines. Finally, a 160 MHz clock signal is used to read the two DATA input lines.

The L0COM and R3SL1 signals are multiplexed using a multiplexer and a flip-flop. The multiplexer selects during the first signal when the BC clock is high and

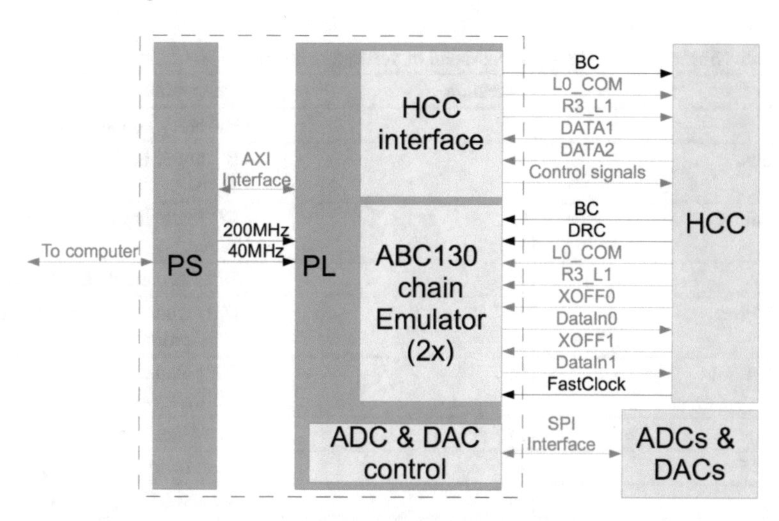

Fig. 4.10 Cartoon illustrating the logic blocks programmed in the MicroZed for the testing of the HCC130. The blue boxes represent physical chips on the boards such as the HCC130, ADCs, and DACs. The box in dashed line represents the logic in the MicroZed. The Processing System (PS) controls the communication between the external user and the blocks in the Programmable Logic (PL). The PL sends and receives data from the HCC130 and to the DACs and ADCs

Table 4.1 Summary of the signals emulated or verified in the HCC130 Interface IP block.

Line	Direction	Description
BC_STAVE	out	BC clock signal at 40 MHz
L0COM	out	command and L0 trigger
R3SL1	out	R3 and L1 trigger line
DATA1	in	data return line
DATA2	in	data return line

the second signal when the BC signal is low. The signal is sampled by a flip flop with the 80 MHz clock.

The HCC130 Chain Emulator (CE) IP block emulates the hybrid side of the HCC130 chip. It represents one chain of ABC130 chips. There are two CE blocks to test the left and right side of the HCC130 since the chip reads data from the ABC130s on both sides. The CE does not implement the full functionality of the ABC130 but only the required signals for testing the HCC130. The signals generated or read by the CE block are summarized in Table 4.2.

Similar to the signals in IF block, the L0COM and R3SL1 are multiplexed signals. The data signals, DATA1 and DATA2, are implemented using a serializer to generate sample data as an ABC130 chain would send using a 60 bit parallel in/ serial out (PISO) register. The data control register has two bits, which trigger the transmission of the data. And the Data status register shows with two flags, Xonoff1 and Xonoff2,

Table 4.2 Summary of the signals emulated or verified in the HCC130 Chain Emulator IP block.

Line	Direction	Description
BC	in	BC clock signal at 40 MHz
BC_plus	in	BC upper limit verification signal
BC_minus	in	BC lower limit verification signal
DRC	in	Data read clock at 160 MHz
DRC_plus	in	Data read clock upper limit verification signal
DRC_minus	in	Data read clock lower limit verification signal
FCLK	in	Fast clock signal at 640 MHz
L0COM	in	Command and L0 trigger
R3SL1	in	R3 and L1 trigger line
R3SL1_plus	in	R3 and L1 upper limit verification signal
R3SL1_minus	in	R3 and L1 lower limit verification signal
Xonoff1	in	Data transmission control line
Xonoff2	in	Data transmission control line
DATA1	out	Data output line
DATA2	out	Data output line

if the transmission of each line is completed. The clock used to serialize the data is a copy of the DRC input clock.

The Data-read clock (DRC) and fast clock (FCLK) are set in the HCC130 registers. The DRC source can be chosen to be 80 MHz or 160 MHz and the speed of the fast clock ranges from 80 MHz to 640 MHz. These clocks are compared using a comparator, using the same setup shown in Fig. 4.9, and read in the CE block.

Finally, ADCs were programmed to read in voltages from the testing board and HCC130, and DACs were used to set voltages for the common mode, DVDD (voltage power the HCC130 chip), on-board voltages used in voltage regulators or to power other chips, and voltages used for testing in the Analog Monitor, described in the next section. Both the ADCs and DACs were programmed using Finite-State-Machine (FSM) to replicate the clocking and signals in their specification sheets.

The ADC code drives the LTC2308 ADCs [6] on the active board. The HCC130 testing board uses 3 ADCs which can be selected by the user. There are 8 channels in the ADC and 16 possible channel configuration commands that can be sent to the ADC. Each analog input is used as a single-ended input, as opposed to differential and it is also run in the unipolar mode instead of the bipolar mode. The user can set the $O/\bar{S}$ (odd/sign) bit, the address selection bits (S1 and S0) to select which ADC channel to read out. After the user selects the channel and the ADC, a start signal is sent to the ADC IP block to start the conversion. A command (called SDI) is sent to the ADC containing the $S/\bar{D}$ bit (selecting single-ended mode), $O/\bar{S}$ bit, the 3-bit channel, the unipolar bit, and the sleep mode bit (set to 0). The output, SDO, from the previous conversion is returned from the AD while SDI is sent to prepare the ADC for another conversion. A done bit will be returned from the ADC after the conversion is complete, signaling that the output (SDO) can be sent back to the user.

The DAC code drives the LTC2656 DACs [7] on the active board. The HCC130 testing board uses 5 DACs which can be selected by the user. The input word to the DAC is 24 bits long and is comprised of a bit 4 command word, a 4 bit address word, and a 16 bit data word. The address word also has 9 options to address any of the 8 channels, as well as an option to address all of them. The command word is the analog voltage the user wishes to send to the DA. A digital-to-analog transfer function is used to calculate it:

$$V_{\text{out (ideal)}} = \frac{k}{2^N} \times 2 \times (V_{\text{REF}} - V_{\text{REFLO}}) + V_{\text{REFLO}} \tag{4.1}$$

In this case, V_{REFLO} is set to ground (0 V) and V_{REF} is the common mode, 0.6 V. N represents the resolution of the DA, which in this case is 12 bits. K is the decimal equivalent of the binary DAC input and that will be the command word to be sent to the DAC.

Similar to the ADC, after the user selects a DAC to address (1–5), inputs an address and a word, a $\bar{C}S/LD$ (chip select, load input) signal will become low triggering the beginning of the conversion. The DAC register will then be set to the voltage specified by the transfer function. After this is completed, a done signal is returned. There is also the possibility to clear the DAC. A logic low at this input clears all registers and causes the DAC voltage outputs to drop to 0V. The asynchronous DAC update pin ($L\bar{D}AC$) is set to high as all the commands are sent at the rising edge of the clock.

4.4.4 Functional Tests of the HCC130

Functional and parametric tests were performed on the HCC130 to determine reliability and range of operation. The HCC130 was found to draw 100 mA at startup and up to 160 mA when setting the DRC at 160 MHz.

The tests ran to determine the functionality of the HCC130 are:

- Checking the startup pattern produced by the HCC130, shown in Fig. 4.11
- Setting the clocks frequencies and reading them back, shown in Fig. 4.12
  - DRC at 80 MHz and at 160 MHz
  - BC stave and HCC130 internal clock (ePLL) at 40 MHz
  - FCLK at 80 MHz, 160 MHz, and 320 MHz
- Register read: verify contents of the registers at 80 MHz and at 160 MHz

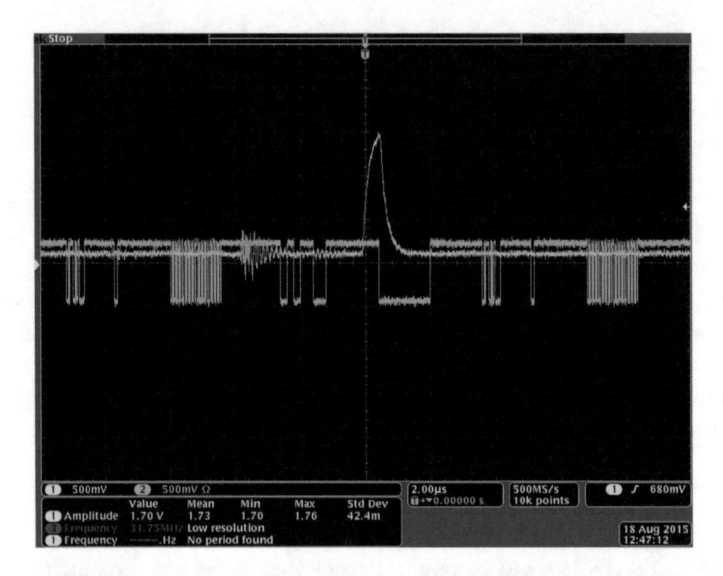

Fig. 4.11 Oscilloscope trace of the pattern sent from the HCC130 at startup (blue) with a trigger (yellow)

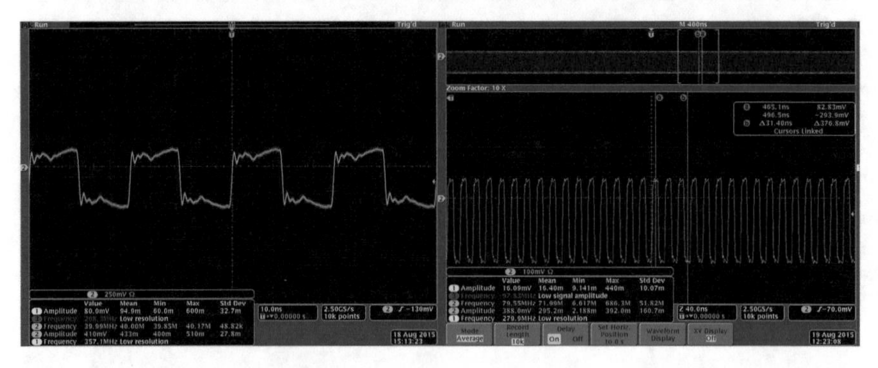

Fig. 4.12 Oscilloscope trace of BC (left) and DRC (right) clocks from the HCC130. The BC clock is at 40 MHz and DRC is set to 80 MHz

- Sending ABC data on all 4 data lines and reading the data from the HCC130 at 80 MHz and at 160 MHz
- Sending triggers and commands to the HCC130, shown in Fig. 4.13

This testing protocol was used to test over 150 HCC130 at Lawrence Berkeley National Laboratory.

Parametric tests were also conducted and they include:

- Checking levels of signals by varying the common mode
- Sweeping over clock phases
- Checking register settings

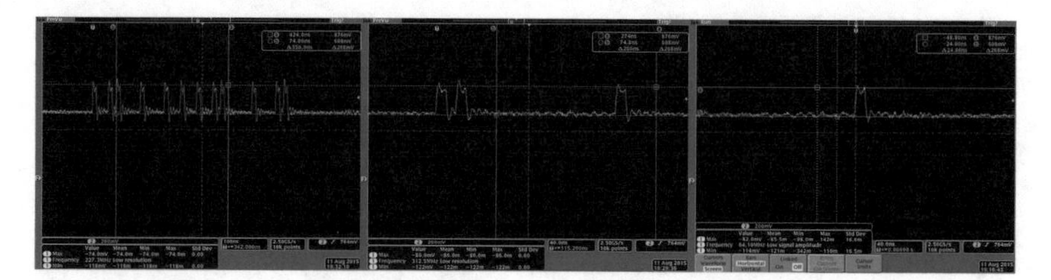

Fig. 4.13 Oscilloscope traces of L0 (left), L1 (middle), and R3 (right) triggers from the HCC130

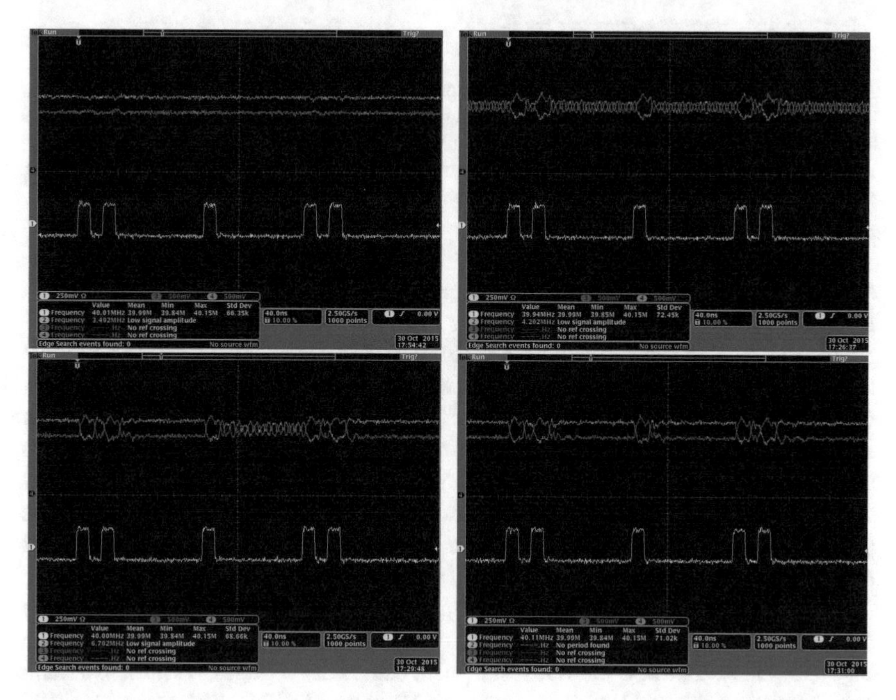

Fig. 4.14 Oscilloscope traces of L1 trigger for different common mode voltages. The yellow signal is the differential signal coming from the HCC130. The green and pink signals are outputs of the comparators and are only half of a differential signal read with a single ended probe

- Testing the Analog Monitor and Interlock

The levels of three signals (BC clock, DRC, and R3SL1) coming from the HCC130 can be checked using a comparator, shown in Fig. 4.9. A DAC sets the common mode and the optimal common mode value is found to be 0.6 V. To test the robustness of the system, the common mode value can be varied to check for what ranges of voltages can the trigger data or clock values be decoded. This process is shown in Fig. 4.14. At too high or too low DAC values, no signal or an unstable oscillating signal is seen. As the voltage becomes closer to the common mode voltage, the signal can be

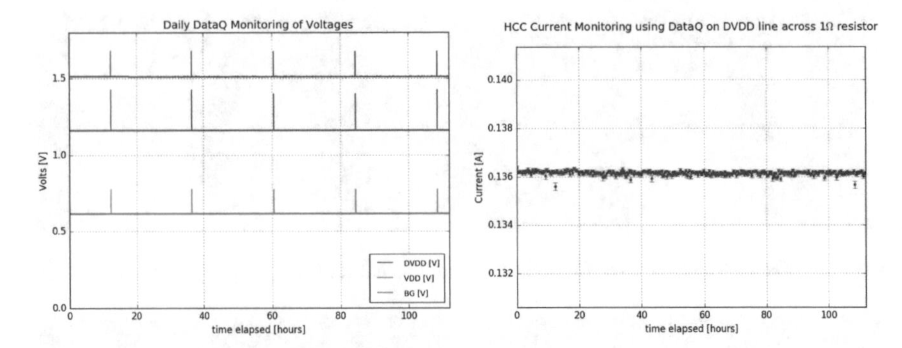

Fig. 4.15 Shown on left, during a data taking period of three days, DVDD,VDD, and Bandgap (BG) were monitored. DVDD remained constant throughout the entire time. VDD and BG were constant except for daily board resets set by the testing protocol. There were no additional voltage resets on the board. Shown on right, the current, measured on the DVDD line across a 1 Ω resistor, remained at 136 mA during the three-day data taking period

resolved and once the optimal voltage is reached, the signal is clearly decoded with no oscillations.

The ranges of the common mode used for the R3SL1 trigger is found to be 511 mV to 669 mV, the range of the common mode used for the BC and DRC clocks is 512–667 mV. The current coming from the HCC130, measured across a 100Ω resistor used in the comparator, is found to be 1.5 mA. The range of phases of BC stave with respect to the L0/Command where commands sent to the HCC130 were correctly decoded by the HCC130 and sent out was found to be -1 ns to 12 ns (where 0 ns is the default rising edge of the clock) for a distance of 4 ft between our active board and the passive board where the HCC130 is located.

The HCC130 was ran without interruption for over one hundred hours. During that time, bandgap, VDD, DVDD, and the current were found to be stable as shown in Fig. 4.15. Daily, the HCC130 was reset which causes the voltages to return to default values before being set by DAC values for VDD and DVDD, and in the HCC130 register for the bandgap voltage.

4.4.5 Testing the Analog Monitor (AM)

The HCC130 contains a voltage based analog monitor (AM) with a 10 bit sensitivity for four internal and three external values. The AM uses a clock-driven integrating ramp generator to compare with the seven monitored quantities, and a counter is used to determine the point where the reference equals the sensed value. The value of the counter is compared to the high and low limits and is stored in monitor registers in the HCC130. A schematic of the quantities measured by the AM is shown in Fig. 4.16.

The quantities monitored by the AM are:

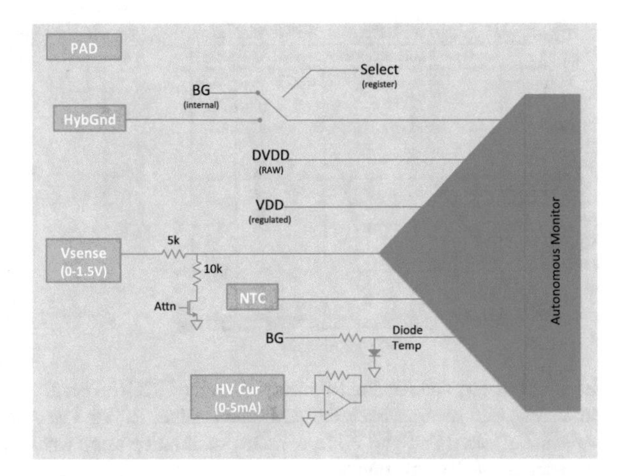

Fig. 4.16 Schematic of the quantities monitored by the AM

- External ground: measures the ground level of the hybrid. Either the hybrid ground or the internal bandgap can be selected.
- DVDD (raw VDD): measures the voltage used to power the HCC130.
- VDD (regulated VDD): measures VDD and has a 2/3 gain.
- Voltage sensor (Vsense): external voltage with selectable 0–1 V (with no attenuator applied) or 0–1.5 V range (with the attenuator applied).
- External temperature (NTC): can monitor the temperature using an external Negative Temperature Coefficient sensor (NTC).
- Internal temperature: the temperature monitor is based on reading voltages across a diode referenced to 1 V through a resistor. Temperature will vary with the bandgap setting.
- Current sensor (HVCurrmon): measures a current across an op-amp with range of 5 μA to 5 mA.

The AM needs to be calibrated in order to know which counter value corresponds to which voltage at the input. The input voltage pin, Vsense, is used for this purpose as it is the only channel that directly inputs a voltage into the HCC130 and can have a large range of input voltages from 0 V to 1.5 V. The result of the calibration is shown in Fig. 4.17. After calibration, the input and output voltages match perfectly showing that the calibration is valid. The maximum value the AM can reach is 1023 counts (about 820 mV) and once a voltage exceeds that value, the AM was designed to keep reporting 1023; however, during testing, it was discovered that after the maximum count of 1023, the counter rolls over, appearing to start counting from 0 again. This was corrected for the Autonomous Monitor and Control Chip AMAC and the HCCstar. The AM needs to be calibrated as well as for each value of band-gap, DVDD (VDD raw), and VDD (VDD regulated) (Fig. 4.18). The Analog Monitor appears operational for a range of DVDD from 1.25 V to 1.6 V, and for a range of VDD between 1.125 V to 1.35 V.

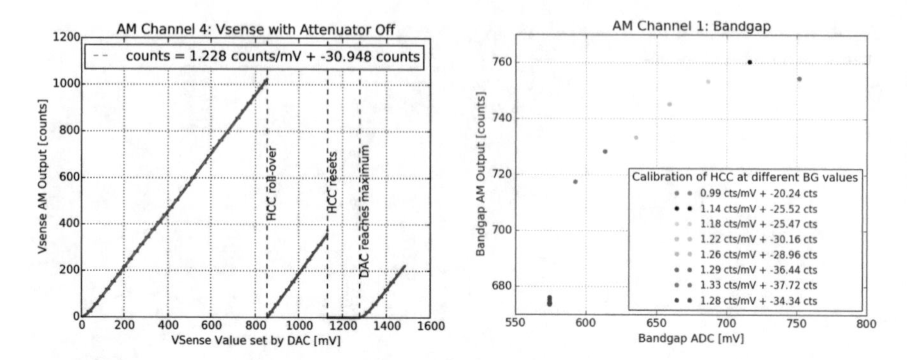

Fig. 4.17 Left: The Analog Monitor within the HCC130 is calibrated by varying the input voltage to Vsense and reading back the AM count value. Right: The AM is read for different band-gap values. For each band-gap value, a Vsense calibration ramp was taken and the slopes and intercepts are seen on each of the points

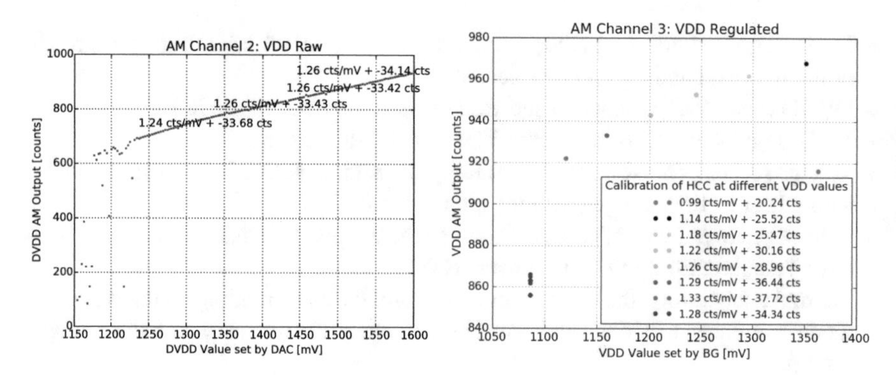

Fig. 4.18 Left: This plot shows the DVDD Analog Monitor output as a function of input voltage. For various values of DVDD, the AM was calibrated and the resulting slopes and intercepts are seen on this figure. Right: This plot shows the VDD AM output as a function of input voltage. For various values of VDD, the AM was calibrated and the resulting slopes and intercepts are seen on this figure

To test the NTC, a thermistor from Amphenol [8] is used which is a low cost, highly sensitive to changes in temperature, sensor. Another reason it is used for testing is that it suitable for PCB and probe mounting so it could be easily used with the HCC130 testing setup. The thermistor circuit used is shown in Fig. 4.19. The thermistor, based on its specification, can be represented as a 10 kΩ resistor. An additional resistor is used to decrease the input voltage to the NTC pin in the HCC130 to about 500 mV, with the input voltage being around 600 mV. An external voltage monitor, DATAQ [9], is used to monitor the voltage into the AM. The measurement is done by submerging the thermistor into ice water and take both analog voltage measurements from the DATAQ and digital measurements from the AM as the water

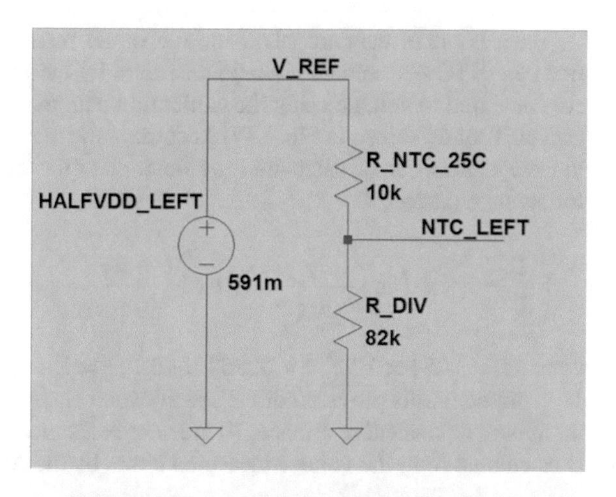

Fig. 4.19 Schematic of the circuit used for temperature measurement testing in the AM for the HCC130. Not figured are the capacitors in parallel with R_{div} to minimize noise. The thermistor is represented by $R_{\text{NTC 25C}}$. NTC is the signal in the HCC AM and this signal is also probed by an external voltage monitor, DATAQ

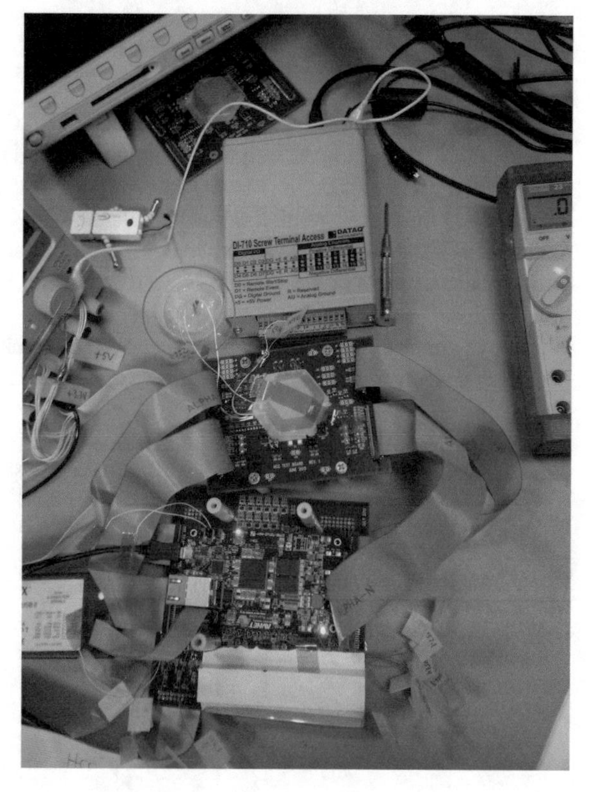

Fig. 4.20 Picture of the NTC test setup. The DATAQ is the grey box at the top, Amphenol thermistor in a volume of ice water circled in red, the passive board with the HCC130 under hexagonal protection dome at the center, and the active board with the MicroZed at the bottom

heats up to room temperature. The setup is pictured in Fig. 4.20, where the red circle shows the thermistor submerged in water.

Three hours of data are taken and the values measured are the HCC130 counts from the NTC AM and the voltage measured by the DATAQ. The HCC130 counts are converted to voltage using the calibration data measured by varying the voltage sent on Vsense, shown in Fig. 4.17. According the thermistor specification sheet [8], the temperature can be calculated as a function of the thermistor resistance using the following equation.

$$\frac{1}{T} = a + b\left(\ln \frac{R_{\mathrm{T}}}{R_{\mathrm{NTC\,25C}}}\right) + c\left(\ln \frac{R_{\mathrm{T}}}{R_{\mathrm{NTC\,25C}}}\right)^2 + d\left(\ln \frac{R_{\mathrm{T}}}{R_{\mathrm{NTC\,25C}}}\right)^3 \tag{4.2}$$

where $a = 3.354 \times 10^{-3}, b = 2.562 \times 10^{-4}, c = 2.139 \times 10^{-6}$, and $d = -7.253 \times 10^{-7}$ are constants provided in the specification sheet. The resistor, $R_{\mathrm{NTC\,25C}}$, is the thermistor equivalent resistance, 10 kΩ. The resistance, R_{T} is the only unknown and is calculated from the voltage measured either by the AM or the DATAQ, as:

$$R_{\mathrm{T}} = \left(\frac{V_{\mathrm{ref}}}{V_{\mathrm{AM}}} - 1\right) R_{\mathrm{div}} \tag{4.3}$$

where V_{ref} is the common mode voltage, measured at 0.6 V, R_{div} is 82 kΩ, and V_{AM} is the voltage using the AM output counts and the AM calibration equation.

The measured voltage versus the calculated temperature is shown in Fig. 4.21. The DATAQ line agrees with the line calculated using AM values, showing that the calibration of the AM works as expected. A fit function is determined which would allow measurements of the AM to be translated into external temperatures on the board.

The Analog Monitor can be used to set up an interlock system. High and low limits can be programmed for any input channel of the AM and an interlock can be set on

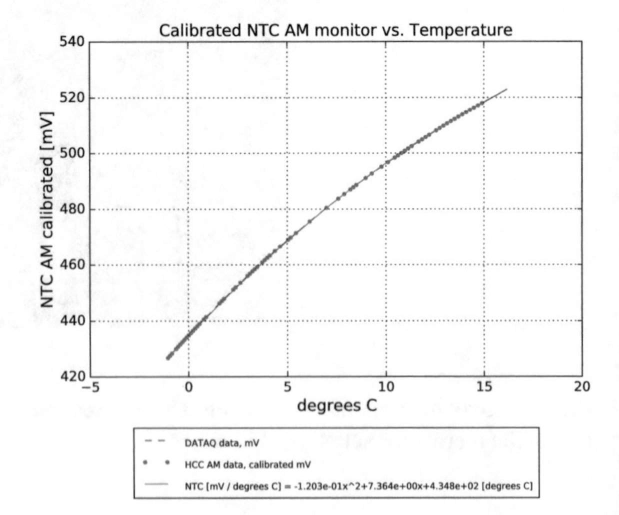

Fig. 4.21 Calibrating the NTC AM using a thermistor. The blue line represents the voltage measured using the DATAQ and the calculated temperature. The green dots represented the voltage measured from the AM and the calculated temperature. The red line is the best fit function

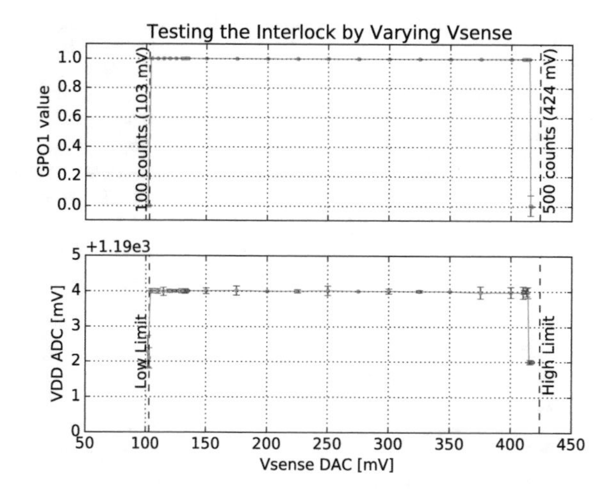

Fig. 4.22 Testing the interlock system on a general purpose output pin, GPO1, by setting limits on Vsense. GPO1 and VDD ADC values for different values input values of Vsense. Only at either the high or low limits set on Vsense is the interlock triggered and the GPO1 signal goes to 0 and VDD drops in voltage. For the high limit, the interlock is triggered 20 mV below the limit; however, when the test is conducted without reading the AM, the interlock is triggered only when the high limit is reached

the General Purpose Outputs (GPO), Regulator Enable, or clock lines. The limits were set to 100 counts and 500 counts for the Vsense input line. The signal being monitored is GPO1 which changes from on to off once the interlock is triggered. For the low limit, the interlock was triggered when the low limit was reached. For the high limit, the interlock was triggered about 20 mV below the interlock limit. This is due to the fact that during the test, for each value of Vsense, the AM output was read out, perturbing the measurement. When the AM is not being read out during the test, the interlock is triggered only when Vsense reaches the limit. These results are shown in Fig. 4.22. For the AMAC, additional protection was added so that the interlock would only be triggered at the set voltage regardless of whether the AM was read out or not.

4.5 HCC130 Irradiation

4.5.1 Motivation: Current Increase at Low Total Ionizing Dose

X-ray irradiations of the ABC130 have shown a digital current increase at low total ionizing doses (TID), called a "TID, or current, bump", around 1–2 MRad, as shown in Fig. 4.23. ABC130 chips are irradiated at different dose rates from 0.6 kRad/hr to

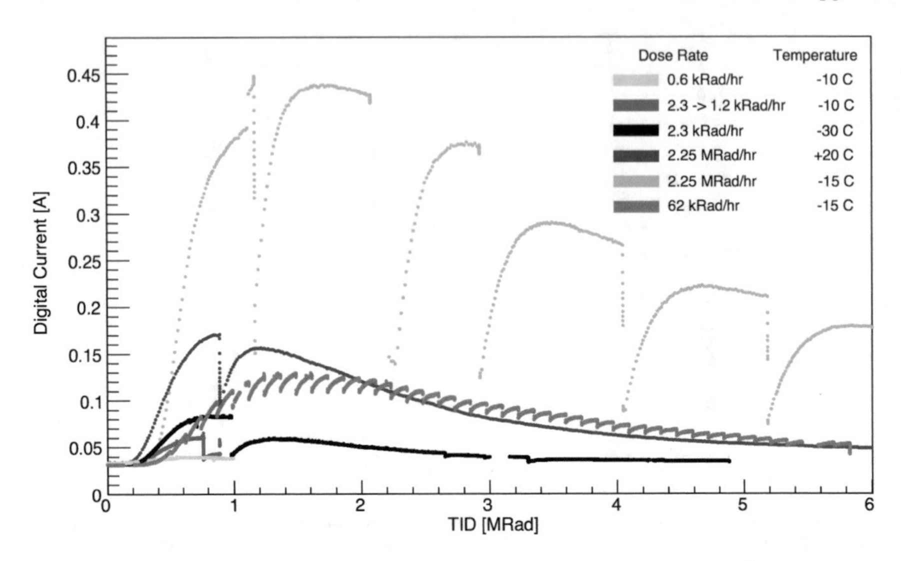

Fig. 4.23 Digital current versus TID for the ABC130 during X-ray irradiation at different dose rate and temperatures [2]

62 kRad/hr and at different temperature varying from −30°C to +20°C. The increase in current is dependent on dose rate and temperature and the largest increase in current appears at low dose rates and for low temperatures. This behavior is also observed during an irradiation campaign with ^{60}Co source where there is a 2.5 times current increase at −25°C and at 2 kRad/hr, which is the highest HL-LHC dose rate for the Strips detector.

This increase in current at low total ionizing dose is a due to damage to the 130 nm CMOS (Complementary metal-oxide-semiconductor) during irradiation [10–12]. To reduce leakage current between adjacent devices, CMOS technology includes the use of shallow trench isolation (STI). The STI is created by depositing a dielectric material such as silicon dioxide into the silicon before the transistors are added. Since the STI still has some thickness, about 100 nm, the CMOS is still sensitive to radiation [11]. During irradiation, positively-charged holes accumulate in the STI, which cause a shift in the threshold voltage in the chip while negatively charged electrons drift faster outside the oxide. In order for charge to be conducted from the source to the drain in CMOS technology, the voltage applied between the gate and the source must be larger than the threshold voltage. The threshold voltage is proportional to the accumulated charge so the increase in charge in the STI causes the charge to be able to move between the source and the drain even if the transistor is turned off, a process called "leakage current". This leakage current in many transistors in the ABC130 results in an overall increase in current. Over time, there is an increase in interface states, which results in a decrease in current, known as the rebound effect. The leakage current and the overall ASIC current decreases back to almost pre-irradiation levels. This process is illustrated in Fig. 4.24.

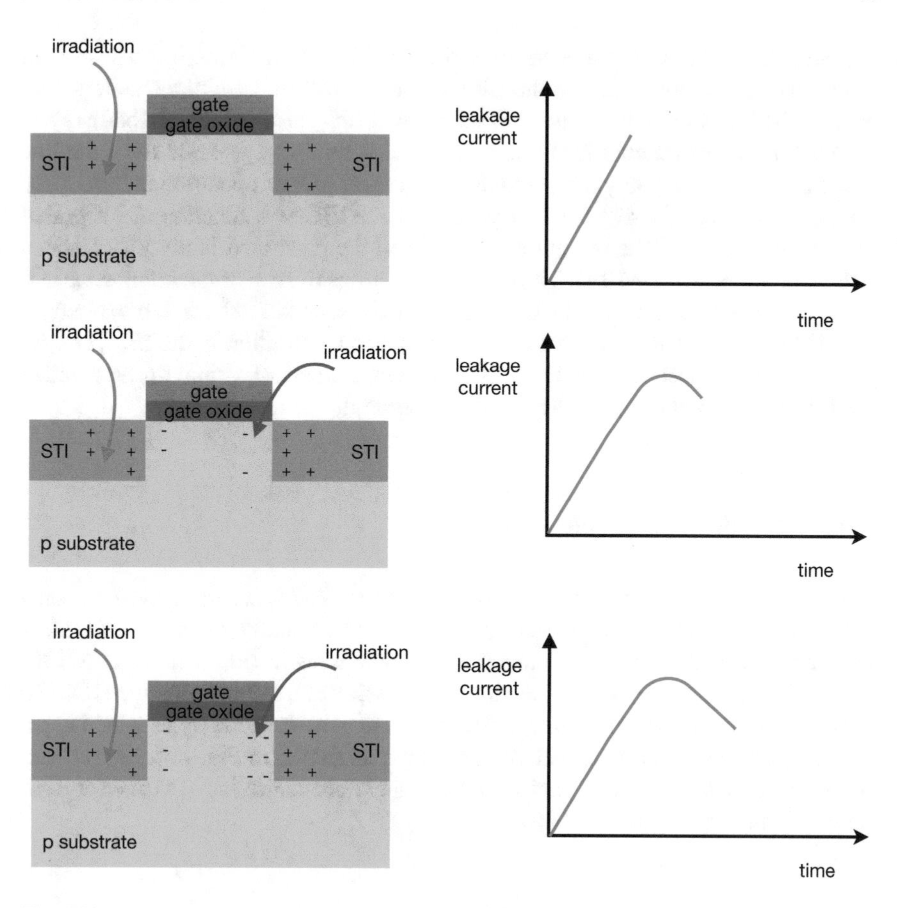

Fig. 4.24 Illustration of the effect of radiation on NMOS transistor

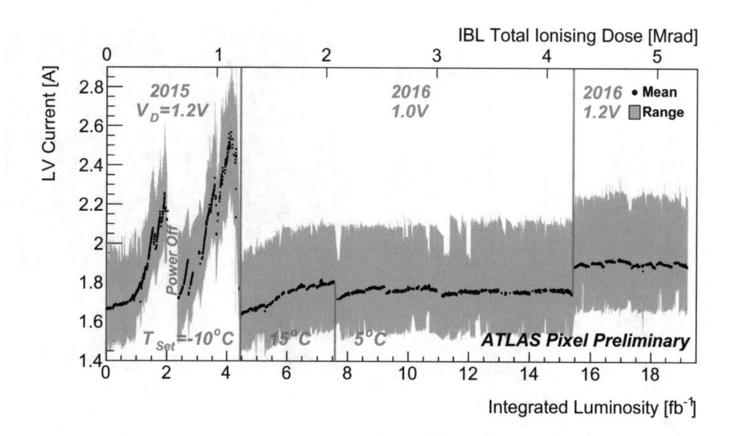

Fig. 4.25 Average LV current in the IBL's FE-I4 chips during stable beam as a function of integrated luminosity and TID [13]. The set temperature (T_{Set}) corresponds to the actual module temperature

The TID current bump was not only observed in the ABC130, it was also seen in the IBL. The readout chip in the IBL is made of 130 nm CMOS technology and called FE-I4 (front-end integrated circuit, version four) [14]. Although the IBL was designed to be operational for up to an integrated luminosity of 300 fb^{-1}, the low voltage (LV) unexpectedly increased during the first year of operation in 2015 [15]. Figure 4.25 shows the average low voltage of the IBL as a function of integrated luminosity and TID. The current increased until the integrated luminosity reached 4 fb^{-1}, or 1.2 Mrad at -10°C. The current then drops to 10% of its initial value and remains stable during 2016. The current increase resulted in a rise in temperature in the FE-I4 which resulted in a temperature-dependent distortion in the IBL [16]. As a result of the impact of the TID on IBL operation, extensive testing to characterize the TID current bump for the ITk ASICs is required.

4.5.2 *Facility and Setup*

The HCC130 is irradiated with ^{60}Co for 4 months at -10°C at Brookhaven National Laboratory (BNL) to investigate the current bump at low total ionizing dose (TID) as seen in the ABC130 and to test the reliability of communication with the HCC130.

In order to reach the low temperature, the passive board containing the HCC130 is placed in a cold box, composed of Styrofoam in between a Plexiglas window and a metal plate with a commercial Peltier cooler, as shown in Fig. 4.26. The cooler, composed of a fan, Peltier, regulator, and a temperature controller was tested at BNL for total ionizing doses of 7 MRad.

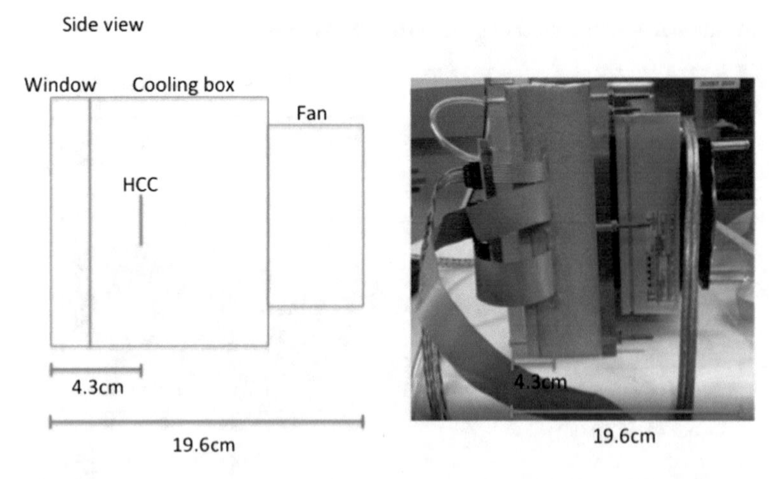

Fig. 4.26 Schematic and picture of the side view of the cold box containing the passive board with the HCC130. The cold box is made up a clear Plexiglas window, a Styrofoam box housing the HCC130, and the cooler with a fan

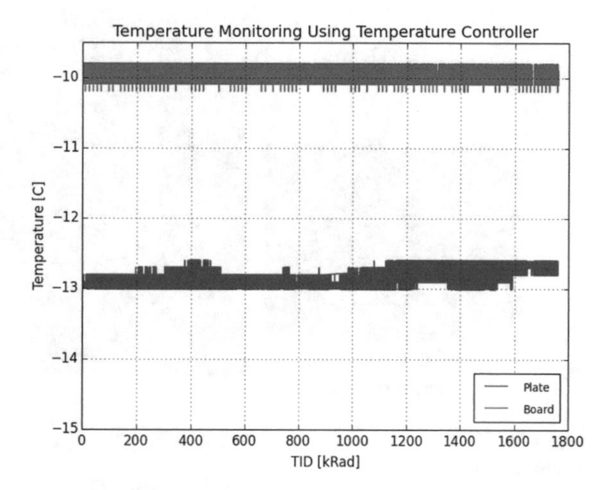

Fig. 4.27 Monitoring the temperature on the passive board and on the cooler plate during HCC130 irradiation

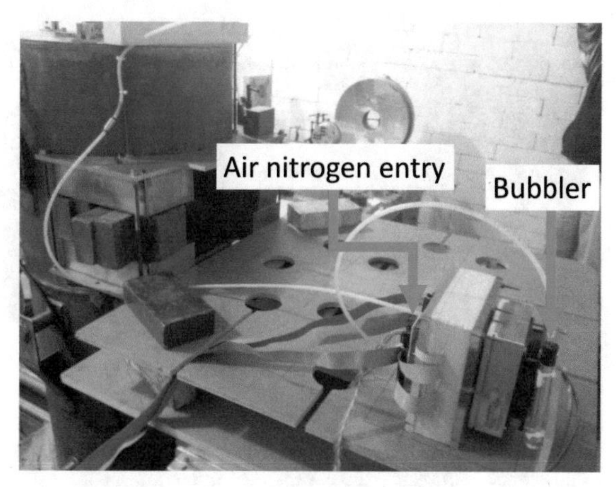

Fig. 4.28 Picture of the nitrogen flow setup and monitoring before HCC130 radiation

In the cold box, there are also thermistors and a humidity sensor. The thermistors measure the temperature on the passive board and on the cooler plate, kept colder than the passive board, to ensure that the HCC130 is kept at a constant -10°C during irradiation as seen in Fig. 4.27.

Since the HCC130 is being cooled, nitrogen is continuously flowed to ensure that there is no condensation on the chip. To visually check that the nitrogen is being flowed, a bubbler is used. This setup is pictured in Fig. 4.28. The humidity is measured using a humidity sensor. Since this sensor would not survive irradiation, the humidity is monitored only for a few hours after the HCC130 is cooled to -10°C. It is measured during this time to be constant at 25% with a flow rate of 0.2 ft^3/hr.

The cold box containing the HCC130 is 42" in radius away from the source. At that distance, the dose rate, as shown in Fig. 4.29 is 0.6 kRad/hr; therefore, the total ionizing dose collected after about three months of running is 2.3 MRad.

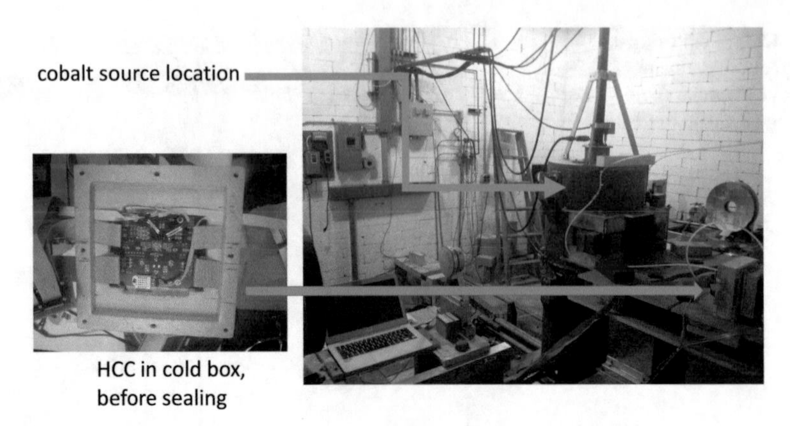

Fig. 4.29 Picture of the location of the HCC130 with respect to the ^{60}Co source inside the BNL source room

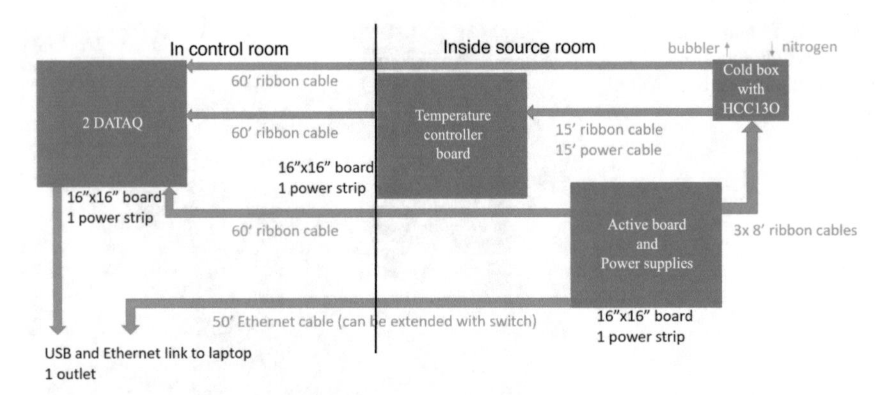

Fig. 4.30 HCC130 irradiation setup detailing the distance from the cold box housing the HCC130 of all other equipment. This is separated into equipment located inside the source room and in the control room, in the experimental hall

The remainder of the setup can be split between equipment in the source room and in the control room as shown in Fig. 4.30. The equipment inside the source room, at least 8 ft from the HCC130 location and under lead shielding, contains the active board and the power supplies used to supply the voltages needed to power the active board and the HCC130. The board with the temperature controller, about 12 ft from the HCC130, is also placed within the lead shielding, pictured in Fig. 4.31.

The control room in the experimental hall, outside the source room, is about 50 ft away and requires 60 ft of cabling to read out the voltages monitored on the HCC130 as well as a 50 foot Ethernet cable which allows communication and data transfer with the active board.

Fig. 4.31 Picture of the lead shield used to protect controller boards during the HCC130 irradiation, the active board and the temperature controller board

4.5.3 Registers and Voltages Monitored

Since the decay ^{60}Co produces γ rays, SEUs are not expected, but to confirm this, all 17 control registers in the HCC130 are read and the first 17 bits in register 36 are read, which correspond to whether or not an SEU occurred in the first 17 control registers. These registers are read out every 5 min. Register 36 is a clear-on-read register and therefore, since this register is read every minute, the limit on SEU would have a granularity of a minute.

Four channels of the AM are monitored: BG (bandgap), VDD (regulated voltage), DVDD (voltage used to power the HCC130), and Vsense (pin where a known voltage was injected into the HCC130). The HCC130 AM can be calibrated by changing the voltage on the Vsense pin and reading out the counts produced by the HCC130. By ramping the voltage throughout irradiation, it would then be possible to see the impact of the TID on the calibration values. Each AM channel is read ten times.

The current consumed by the HCC130 is also monitored to see if the current bump was also seen at low TID by the HCC130 as it had been seen by the ABC130. The current is measured by adding a 1 Ω resistor in series with the DVDD line and measuring the voltage drop across this resistor.

The AM channels are read out once daily while the registers were read out many times daily. The current and voltages were read out every 10 s during the irradiation with an external analog to digital data logger (DATAQ, shown in Fig. 4.30). In the case of an incorrect register readback, or if a register cannot read after a few attempts, the

HCC130 would be reset and reinitialized. The HCC130 is also reset daily to simulate detector operator.

4.5.4 Pre-irradiation Calibrations

Before irradiation commences, the HCC130 is run to test the stability of the system and to understand the expected variations in the AM values. The voltages are monitored by both the AM and the DATAQ and are shown in Fig. 4.32. Each AM value is the average of the ten read out values and the DATAQ values are averaged over a time period of 5 min. The voltages in the DATAQ are stable to within 1 mV and in the AM, the variation is 1 count.

The voltage, set by a DAC, is increased and sent to the Vsense creating a ramp in voltage. The AM output are plotted against the DATAQ measured voltage in order to determine the calibration of the AM. The slope and intercept of the best fit calibration line as well as the ramp are shown in Fig. 4.33. The ramps and fits are also very stable during the pre-irradiation run.

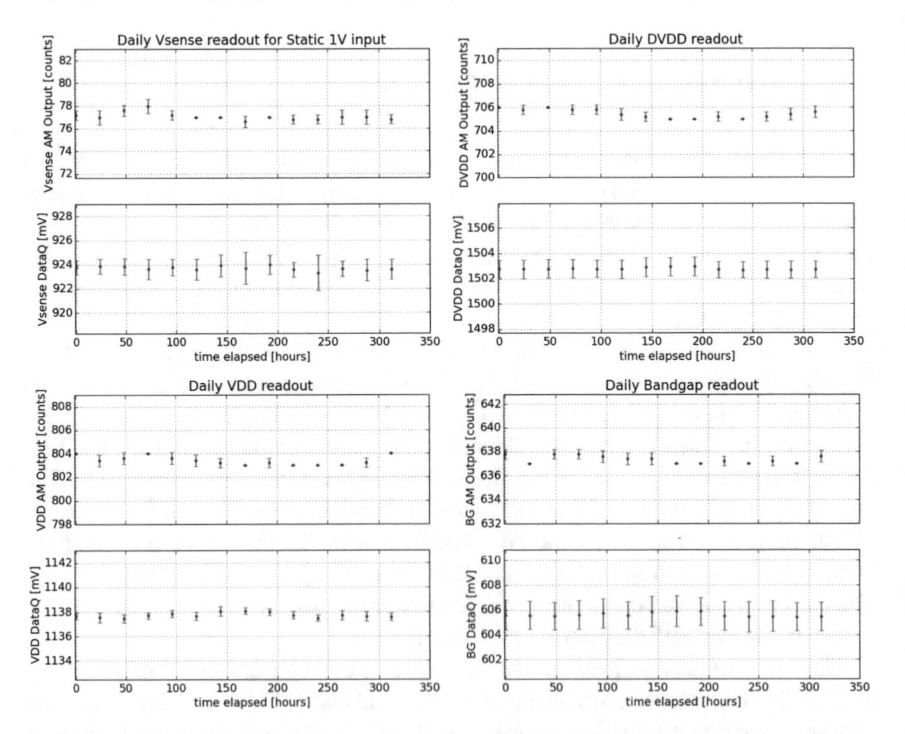

Fig. 4.32 Monitoring of the voltages Vsense, with an input voltage of 1V, (top left), DVDD (top right), VDD (bottom left), and bandgap (bottom right) with both the AM and the DATAQ before the HCC130 irradiation

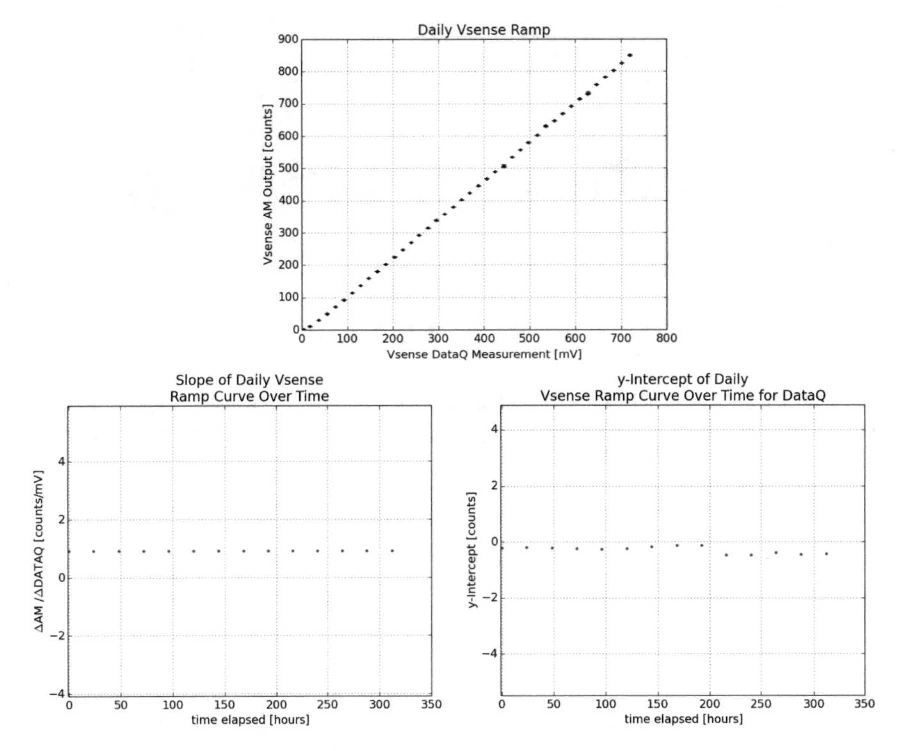

Fig. 4.33 Calibrating the AM before irradiation by varying the input voltage and reading out the AM output. Resulting ramps are on the bottom, with the resulting slopes (top left), and y-intercepts (top right)

4.5.5 Results

The current is plotted as a function of elapsed hours since the beginning of data taking and of TID in Fig. 4.34. Before irradiation, the current is stable at 0.132 A on average. After irradiation, the current rises and continues to rise for about 2 mA. It reaches peak current and begins to drop back to 0.133 A after 1.5 MRad. The HCC130 therefore does not experience a sharp increase in current at low TID. The periodic dips in current are due to short timescale artifacts from the data acquisition sequence, primarily current fluctuations from the daily reading of the analog monitor block inside HCC130, and a daily reset of the host controller and of the HCC130.

The bandgap AM readout was also monitored during irradiation. As shown in Fig. 4.35, the AM readout before irradiation was stable at 638 counts and rapidly rose during the first 500 kRad by 25 counts. During irradiation, the BG had an average value of 664 counts. The designers of the bandgap expect a variation of 30–40 mV due to TID, which is consistent with the measurements when the calibration of 1.10 mV/count is taken into account. A 30 mV variation out of nominal 600 mV will cause a 0.5% overall measurement error.

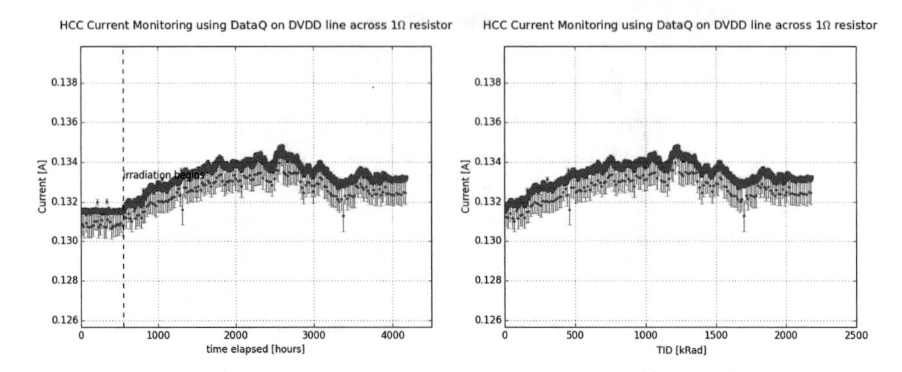

Fig. 4.34 HCC130 current as a function of elapsed hours since the beginning of data taking (left), and of TID (right). The plots represent average current with the fluctuations due to the reading of the AM and daily reset of the HCC130

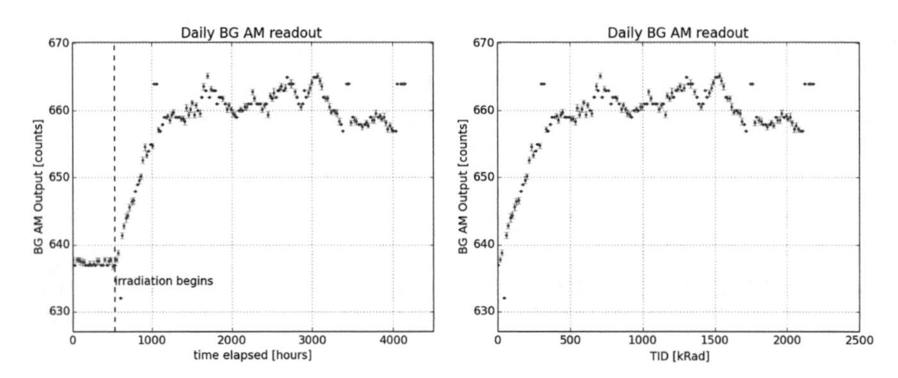

Fig. 4.35 HCC130 Bandgap AM counts as a function of elapsed hours since the beginning of data taking (left), and of TID (right). Before irradiation, the bandgap AM count is stable. Irradiation causes the AM counts to jump 25 counts and oscillates around that count value.

Calibration of the AM was monitored during irradiation by ramping an external voltage from 0–1 V on the Vsense pin and reading out the AM count value. The resulting ramps are shown in Fig. 4.36. The slopes remained constant but the intercepts decreased by 5 counts during irradiation; however, after 500 kRad, the intercepts stabilized at around -19 counts. Based on the BG readout and the ramps, it appears that after the initial increase in AM counts during the first 500 kRad, the AM readout values remained constant for the remainder of the irradiation.

Finally, register read errors and SEUs were monitored during irradiation. As shown by Fig. 4.37, there were no SEUs or errors in register reads. There were also no register read errors for the remaining registers 4–17 shown in Appendix A.1.

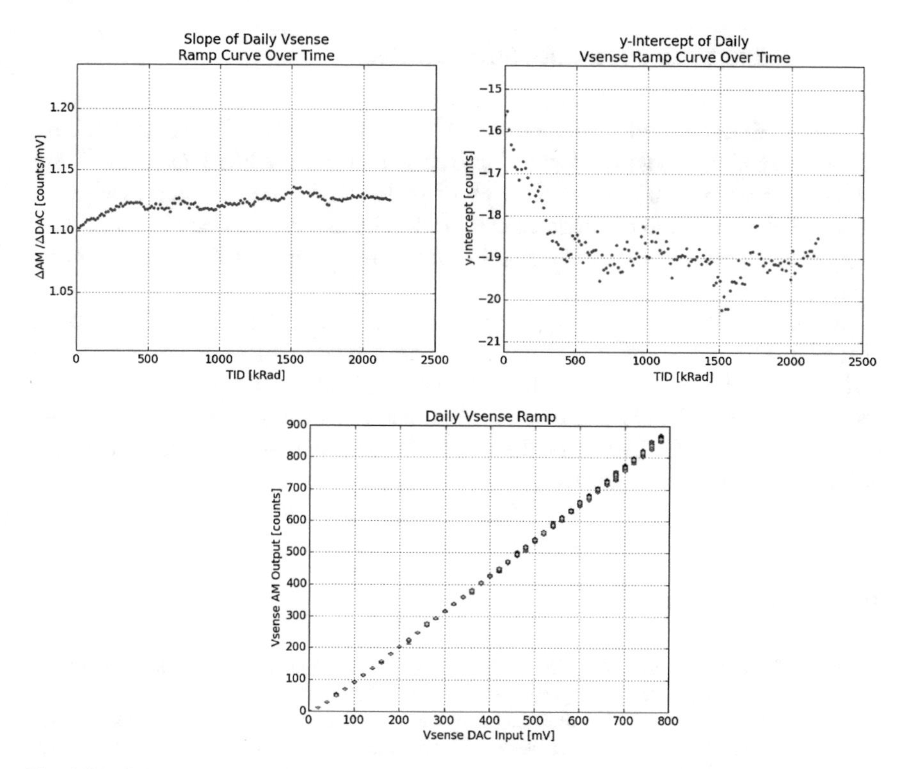

Fig. 4.36 Calibrating the AM during irradiation by varying the input voltage and reading out the AM output. Resulting ramps are on the bottom, with the resulting slopes (top left), and y-intercepts (top right)

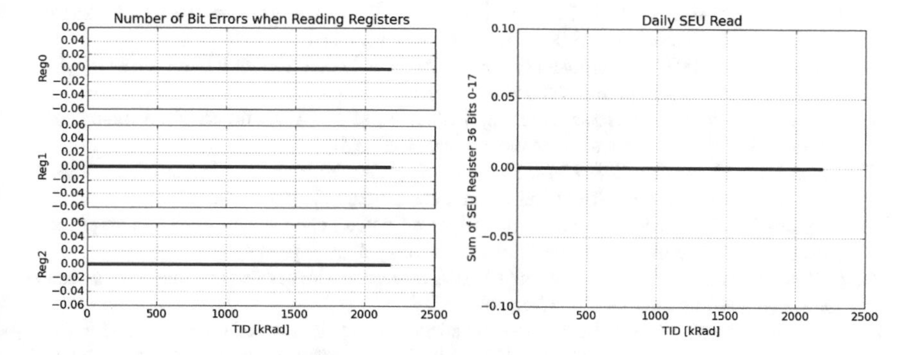

Fig. 4.37 Number of error reads for each of the control registers 0–3 (left) and the sum of SEUs that occur in registers 0–17 (right) during the HCC130 irradiation

4.6 Future Improvements and Studies

The HCC130 testing and irradiation has been able to determine that HCC130 chip is very stable and does not experience a large current increase at low TID. A few design errors were uncovered such as the lack of protection in the interlock mechanism causing the interlock to be triggered early if the channel the interlock is placed on is read back and the max count of the AM is not properly kept at 1023 but instead rolls over.

While the HCC130 is being tested, the HCCstar and the AMAC are being designed. The HCCstar's AM is much simpler and has a more complex data read and receive architecture than the HCC130. The HCC130's AM capabilities are transferred to the AMAC. Future studies would include conducting the same kinds of tests performed on the HCC130 but on these new chips since those chips are the ones that will be in the ITk Strip detector.

References

1. ATLAS Collaboration (2008) The ATLAS experiment at the CERN large hadron collider. JINST 3:S08003. https://doi.org/10.1088/1748-0221/3/08/S08003
2. ATLAS Collaboration (2017) Technical design report for the ATLAS inner tracker strip detector. Tech Rep CERN-LHCC-2017-005. ATLAS-TDR-025, CERN, Geneva. https://cds.cern.ch/record/2257755
3. ATLAS Collaboration (2017) Technical design report for the ATLAS inner tracker pixel detector. Tech Rep CERN-LHCC-2017-021. ATLAS-TDR-030, CERN, Geneva. https://cds.cern.ch/record/2285585
4. Anghinolfi F, et al (2018) HCC DRAFT Specification V1.5.11. https://twiki.cern.ch/twiki/pub/Atlas/ItkAsicDocumentation/HCC_Specification_V1.5.11.pdf
5. Karagounis M (2008) Development of the ATLAS FE-I4 pixel readout IC for b-layer Upgrade and Super-LHC. https://cds.cern.ch/record/1158505
6. Linear Technology, Low Noise, 500ksps, 8-Channel, 12-Bit ADC. https://www.analog.com/media/en/technical-documentation/data-sheets/2308fc.pdf
7. Linear Technology, *Octal 16-/12-Bit Rail-to-Rail DACs with 10ppm/C Max Reference*, https://www.analog.com/media/en/technical-documentation/data-sheets/2656fa.pdf
8. Amphenol, NTC Type MS- Thermometrics Epoxy-Coated Thermistor. https://www.amphenol-sensors.com/en/thermometrics/ntc-thermistors/epoxy/3330-type-ms
9. DATAQ instruments, Low-cost, portable, USB/Ethernet data logger system. https://www.dataq.com/resources/pdfs/datasheets/710ds.pdf
10. Faccio F, Cervelli G (2005) Radiation-induced edge effects in deep submicron CMOS transistors. IEEE Trans Nucl Sci 52:2413–2420. https://cds.cern.ch/record/1024467,https://doi.org/10.1109/TNS.2005.860698
11. Faccio F, Barnaby HJ, Chen XJ, Fleetwood DM, Gonella L, McLain M, Schrimpf RD (2008) Total ionizing dose effects in shallow trench isolation oxides. 2007 Reliability of Compound Semiconductors (ROCS) Workshop. Microelectron Reliab 48(7):1000–1007. https://doi.org/10.1016/j.microrel.2008.04.004, http://www.sciencedirect.com/science/article/pii/S0026271408000826

12. Gonella L, Faccio F, Silvestri M, Gerardin S, Pantano D, Re V, Manghisoni M, Ratti L, Ranieri A (2007) Total ionizing dose effects in 130-nm commercial CMOS technologies for HEP experiments. Nucl Instrum Meth A 582:750–754. https://doi.org/10.1016/j.nima.2007.07.068
13. ATLAS Collaboration, IBL LV currents and sensor leakage currents. https://atlas.web.cern.ch/Atlas/GROUPS/PHYSICS/PLOTS/PIX-2016-006/
14. ATLAS Collaboration (2010) ATLAS insertable B-layer technical design report. Tech Rep CERN-LHCC-2010-013. ATLAS-TDR-19. https://cds.cern.ch/record/1291633
15. La Rosa A, Oide H, Dette K, Bindi M, Rozanov A, Dann NS, Ferrere D, Giordani M, Garcia-Sciveres M (2016) Radiation induced effects in the ATLAS Insertable B-Layer readout chip, working draft for the IBL LV paper. Tech Rep ATL-COM-INDET-2016-082, CERN, Geneva. https://cds.cern.ch/record/2227740
16. ATLAS Collaboration, Time dependent alignment corrections to IBL distortions. https://atlas.web.cern.ch/Atlas/GROUPS/PHYSICS/PLOTS/IDTR-2015-011/

Chapter 5
$W^{\pm}Z$ Cross-Section Measurement at $\sqrt{s} = 13$ TeV

Diboson cross section measurements provide important tests of the electroweak (EWK) sector of the Standard Model (SM) by measuring precisely the triple and quartic gauge couplings. Any deviation in the measurement of the strength of these couplings would provide evidence for physics beyond the SM.

Diboson refers to WZ, WW, and ZZ processes. The production of WZ is a good probe for diboson processes. Figure 5.1 shows the leading order diagrams for the WZ production in proton-proton collisions. The s-channel diagram has a triple electroweak gauge boson interaction gauge, which is a feature of the non-abelian structure of the group describing EWK interactions, but is also sensitive to new physics. Deviations in this coupling would lead to an enhancement in the production cross section. Limits on anomalous triple gauge couplings, as well as on anomalous quartic gauge couplings, which can be probed via vector boson scattering production, have been placed in Run 1 by ATLAS [1, 2] and CMS [3–5].

The WW process also receives contributions from leading order triple gauge coupling; however, the WZ process is an experimentally cleaner signature with fewer backgrounds. The WW process has large background contribution from $t\bar{t}$ produced with associated jets. In order to reduce this background, a jet veto needs to be imposed, leading to larger jet systematic uncertainties. Moreover, the final leptonic decay of WW is $\ell\nu\ell\nu$ which contains two invisible particles, while the WZ process contains only one invisible particle if the W boson decays leptonically, making the WW process kinematically more challenging to reconstruct than WZ. Finally, the cross section times branching ratio is larger for WZ than for ZZ production process.

As discussed in Sect. 2.1.4, final states with either the W or Z bosons decaying to hadrons have the largest branching ratio; however, they also have large backgrounds from many processes such as multijet, $t\bar{t}$, Z+jet, and W+jet production. The final state where both the W and Z bosons decay leptonically, also known as the three lepton channel, is therefore used to measure the WZ cross section. Z decays to neutrinos are not considered because they are not detected in the ATLAS detector. The three lepton channel includes decays to electrons and muons. Decays to taus are

© The Editor(s) (if applicable) and The Author(s), under exclusive license to Springer Nature Switzerland AG 2020

E. Resseguie, *Electroweak Physics at the Large Hadron Collider with the ATLAS Detector*, Springer Theses,
https://doi.org/10.1007/978-3-030-57016-3_5

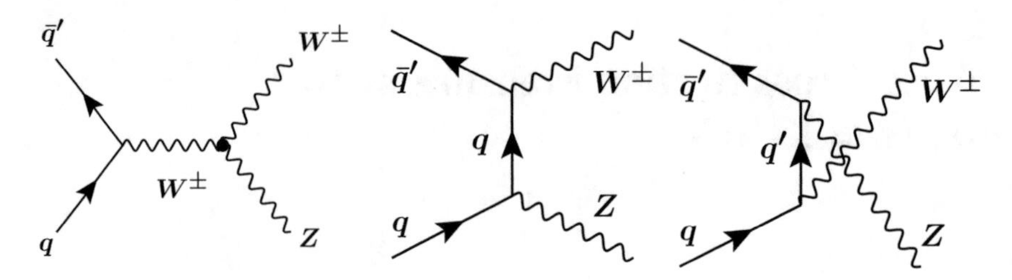

Fig. 5.1 Leading-order diagrams for $W^{\pm}Z$ production in pp collisions

not considered directly. Taus have a 65% branching ratio to hadrons. The remaining 35% branching ratio is to leptons (electrons and muons). Leptonic τ decays have final states of the form $\mu\bar{\nu}_{\mu}\nu_{\tau}$ or $e\bar{\nu}_{e}\nu_{\tau}$. These states are indistinguishable from promptly produced electrons or muons, except for the additional missing energy from the tau neutrino. Thus, events with leptonically decaying taus contribute to the signal regions of the four channels used in the measurement region. This contribution is estimated using simulation and accounted for using correction factors during the calculation of the cross section.

The cross section of the WZ production decaying to three leptons was measured in proton-antiprotons collisions and published by the CDF and D0 collaborations [1, 2], as well as in proton-proton collisions at the LHC by ATLAS at $\sqrt{s} = 7, 8$ TeV [6, 7], and by CMS at $\sqrt{s} = 7, 8, 13$ TeV [3–5].

The analysis presented in this chapter measures the fiducial cross section of the WZ production decaying to three leptons in the eee, $ee\mu$, $\mu\mu e$, $\mu\mu\mu$ decay channels using 3.2 fb^{-1} of ATLAS data collected at a center-of-mass energy of 13 TeV. The four decay channels are combined using a χ^2 minimization technique. The measurement is also extrapolated to the total phase space to determine the inclusive cross section. The paper on which these results are based also present the cross section as a function of jet multiplicity [8].

5.1 Cross Section Methodology

The cross section is the probability of an event occurring. The number of events produced depend on the cross section and the luminosity. Thus, the total number of WZ events produced is determined using the equation

$$N_{WZ} = \mathcal{L} \cdot \sigma \tag{5.1}$$

N_{WZ} is determined by the total number of data events observed, N, minus the number of background events, B. Detector effects such as lepton identification, trigger, and others are accounted for with a correction factor, C, and the finite acceptance of events is accounted with for with an acceptance factor, A. Combining all this, the total cross section becomes,

$$\sigma = \frac{N - B}{\mathcal{L} \cdot \mathrm{C} \cdot \mathrm{A}} \tag{5.2}$$

5.1.1 Fiducial Cross Section

A fiducial cross section measurement (differential or inclusive) is calculated such that the measurement can be directly compared with theory, with little dependence on the underlying model ("model independent" cross section measurement). The phase space selection for this measurement can be either geometric or kinematic. The geometric selections ensures that events will be within the detector while a kinematic selection reduces background in the measurement region that are difficult to model. The fiducial phase space selection can be further subdivided as a function of one or more kinematic properties. This cross section measurement is known as a differential cross section.

After events are selected within the fiducial phase space, background events are subtracted. The remaining events must be corrected to account for detector-level effects such as lepton identification, trigger efficiency, or resolution on kinematic variables, a process known as unfolding. Unfolding is done with as few assumptions on the model as possible. For an inclusive cross section, unfolding is a single number, called a correction factor, C_{WZ}. The procedure is more complicated for a differential cross section because of event migration between bins.

The fiducial cross section is calculated as

$$\sigma^{\text{fid.}}_{W^{\pm}Z\rightarrow\ell'\nu\ell\ell} = \frac{N_{\text{data}} - N_{\text{bkg}}}{\mathcal{L} \cdot \mathrm{C}_{WZ}}. \tag{5.3}$$

The correction factor is the ratio of the number of events in the fiducial phase space using reconstructed events over the number of events derived from the Monte Carlo simulation in the fiducial phase space, so that

$$\mathrm{C}_{WZ} = \frac{N^{\text{signal}}_{\text{reco}}}{N^{\text{signal}}_{\text{fiducial}}}. \tag{5.4}$$

This will be discussed further in Sect. 5.7. To compare with theory, events generated using a Monte Carlo generator have the fiducial selection applied. The resulting cross section can be compared with the experimentally measured cross section.

5.1.2 Total Cross Section

The fiducial cross section can be extrapolated into a cross section in the total phase space. This extrapolation takes into account the branching ratios of W and Z bosons to three leptons as well as an acceptance factor, A_{WZ}. This acceptance factor is a truth-level factor that extrapolates from the fiducial phase to the total phase space. The factor is given by

$$\mathrm{A}_{WZ} = \frac{N^{\text{signal}}_{\text{fiducial}}}{N^{\text{signal}}_{\text{total}}}. \tag{5.5}$$

Determining this factor will be discussed further in Sect. 5.7. The total cross section is calculated as equation (5.6):

$$\sigma^{\text{tot}}_{WZ} = \frac{\sigma^{\text{fid.}}_{W^{\pm}Z\to\ell'\nu\ell\ell}}{\mathrm{BR}_{W\to\ell\nu}\cdot\mathrm{BR}_{Z\to\ell\ell}\cdot\mathrm{A}_{WZ}} \tag{5.6}$$

5.2 Phase Space Definition

The fiducial phase space is chosen to be closely define to the selection described in Sect. 5.4 and is identical to the one used in the WZ cross section measurement collected at 8 TeV by the ATLAS detector [7]. The fiducial selection is applied to both events in data and to the Monte Carlo simulation predictions.

In the simulation, "dressed" leptons are used to calculate event kinematics. Leptons are dressed by including contributions from final-state-radiation (FSR) photons within a cone of
$\Delta R \equiv \sqrt{(\Delta\eta)^2 + (\Delta\phi)^2} < 0.1$. Dressed leptons and final state neutrinos that do not originate from hadrons or τ decays first need to be assigned to the parent bosons, W and Z, before applying the fiducial selection which depends on the assignment. The procedure to assign the leptons is called a "resonant shape" algorithm which maximizes an estimator for all choices of $m_{\ell\ell}$ and $m_{\ell\nu}$,

$$P_k = \left(\frac{1}{M^2_{(\ell\ell)k} - M^2_Z + i\Gamma_Z M_Z}\right)^2 \times \left(\frac{1}{M^2_{(\ell\nu)k} - M^2_W + i\Gamma_W M_W}\right)^2, \tag{5.7}$$

where $k = 1, 2$, Γ_Z and Γ_W are the total width of the W and Z, as reported by the Particle Data Group. This boson assignment is used when applying the fiducial selections on events from any generator in order to ensure uniform treatment across generators.[1]

[1]Some generators, such as POWHEG, have a built-in algorithm to assignment daughter particles to parent bosons; however, this procedure differs across generators. The SHERPA generator does not explicitly assign the leptons to either the W or Z bosons due to the quantum-mechanical ambiguity of events with three leptons and the associate interference effect. A uniform assignment procedure

Table 5.1 Summary of fiducial and phase space selection. Kinematics are calculated with dressed leptons

Fiducial phase space $W^{\pm}Z \rightarrow \ell'\nu\ell\ell$ production	Total phase space WZ production
$81.2 < m_{\ell\ell} < 101.2$ GeV	$66 < m_{\ell\ell} < 116$ GeV
Z leptons: $p_{\mathrm{T}} > 15$ GeV, $\lvert\eta\rvert < 2.5$	
W lepton: $p_{\mathrm{T}} > 20$ GeV, $\lvert\eta\rvert < 2.5$	
$\Delta R(\ell_Z, \ell_W) > 0.3$	
$\Delta R(\ell_Z^{\mathrm{lead}}, \ell_Z^{\mathrm{sublead}}) > 0.2$	
$m_{\mathrm{T}}^{W} > 30$ GeV	

Table 5.1 summarizes the fiducial and total phase space definitions. The total phase space, after bosonic assignment, is defined with a requirement on the mass of the Z boson, $66 < m_{\ell\ell} < 116$ GeV, to distinguish WZ production from $W\gamma^*$. The fiducial phase space has a tighter requirement on the Z mass, lepton η and p_{T} requirements, and a cut on m_{T}^{W}, calculated between the lepton and missing transverse energy vector in the data, and with the neutrino in the simulation.

5.3 Data Set and MC Samples

The proton proton collision data corresponds to an integrated luminosity 3.2 fb^{-1} collected at a center-of-mass energy of 13 TeV. WZ and background processes are simulated using Monte Carlo generators. The propagation of events through the ATLAS detector is simulated using GEANT4 [9], digitized, and reconstructed [10].

Table 5.2 summarizes the Monte Carlo (MC) used specifying the generator used to simulate the events. The WZ production is generated with both POWHEG [11–14]+PYTHIA [15–17] and SHERPA [18, 19] generators. The POWHEG samples are used to determine the fiducial and total cross sections. The SHERPA samples are used as a comparison for the differential cross section measurement. For events where the data-driven method, "Fake Factor", is used, the MC is used for cross checks. Top-like backgrounds, which include $t\bar{t}$, singletop, WW+jets, are estimaetd using MC normalized to data.

---

ensures that differences in fiducial cross section predictions are not due to nonphysical assignment of leptons.

Table 5.2 Summary of the background processes and methods used to estimate them. For the data-driven method, the MC is used as a cross check

Process	Estimation	Generator	sub-process
WZ	MC	POWHEG +PYTHIA8	$WZ \to \ell\nu\ell\ell$
	MC	SHERPA	$W^{-}Z \to \ell\nu\ell\ell$ SF
	MC	SHERPA	$W^{-}Z \to \ell\nu\ell\ell$ OF
	MC	SHERPA	$W^{+}Z \to \ell\nu\ell\ell$ SF
	MC	SHERPA	$W^{+}Z \to \ell\nu\ell\ell$ OF
ZZ	MC	POWHEG +PYTHIA8	$q\bar{q} \to ZZ \to \ell\ell\ell\ell$
	MC	SHERPA	$gg \to \ell\ell\ell\ell$
$t\bar{t} + V$	MC	MADGRAPH +PYTHIA8	$t\bar{t} + W$
	MC	MADGRAPH +PYTHIA8	$t\bar{t} + Z, Z \to \ell\ell$
tZ	MC	MADGRAPH +PYTHIA6	$tZ \to \ell\ell\ell$
VVV	MC	SHERPA	$VVV \to$ $3\ell3\nu, 4\ell2\nu, 5\ell1\nu,$
			$3\ell3\nu, 6\ell0\nu, 4\ell2\nu$
Z+jets	Fake Factor	POWHEG +PYTHIA8	$Z \to \ell\ell$
$Z + \gamma$	Fake Factor	SHERPA	$Z + \gamma \to \ell\ell\gamma$
$t\bar{t}$, Wt, top	NF	POWHEG +PYTHIA6	$t\bar{t} \geq 1\ell$
			$Wt \geq 2\ell$
			top (s-channel)

5.4 Object and Event Selection

5.4.1 Object Selection

Electrons and muons are identified using identification, isolation, and tracking criteria described in Sect. 3.3. Three levels of object selection are used for electrons and muons, described in Table 5.3. Each level "baseline", "Z", and "W" leptons applies the selection of the previous levels along with additional criteria. The baseline leptons use the looser identification criteria and lower lepton p_{T} in order to provide a higher efficiency of identifying and removing processes decaying to four prompt leptons (four-lepton veto requirement). Leptons associated with the W and Z boson must satisfy stricter criteria. The selection on the lepton associated with the W boson has the most stringent criteria to suppress the reducible backgrounds, Z+jets and Z+γ, which have a fake lepton associated with the W.

Baseline electrons must have $p_{\mathrm{T}} > 7$ GeV and fall within the inner detector, $|\eta| < 2.5$, as well as within the electromagnetic calorimeter, $|\eta^{\mathrm{cluster}}| < 2.47$. The electrons must also satisfy the `LooseAndBLayerLLH` quality criteria, which has an efficiency of 84–96% for electrons with $10 < p_{\mathrm{T}} < 80$ GeV . Electrons need to pass the impact parameter cuts of $|z_0 \sin\theta| < 0.5$ and $|d_0/\sigma_{d_0}| < 5$, designed to

Table 5.3 Summary of the three levels for electron and muon criteria. Each new level contains the selection of the previous level

Cut	Value/description	
	Baseline Electron	Baseline Muon
Acceptance	$p_T > 7$ GeV, $\|\eta^{\text{cluster}}\| < 2.47$, $\|\eta\| < 2.5$	$p_T > 7$ GeV, $\|\eta\| < 2.5$
Identification	`LooseAndBLayerLLH`	`Loose`
Isolation	`LooseTrackOnly`	`LooseTrackOnly`
Impact parameter	$\|z_0 \sin\theta\| < 0.5$ mm,	$\|z_0 \sin\theta\| < 0.5$ mm,
	$\|d_0/\sigma_{d_0}\| < 5$	$\|d_0/\sigma_{d_0}\| < 3$
	Z Electron	Z Muon
Acceptance	$p_T > 15$ GeV, exclude $1.37 < \|\eta^{\text{cluster}}\| < 2.47$	$p_T > 15$ GeV
Identification	`MediumLH`	`Medium`
Isolation	`GradientLoose`	`GradientLoose`
	W Electron	W Muon
Acceptance	$p_T > 20$ GeV	$p_T > 20$ GeV
Identification	`TightLH`	
Isolation	`Gradient`	

suppress fake electrons from pileup jets. Finally, the electrons must satisfy a track isolation requirement, `LooseTrackOnly`.

Electrons satisfying the Z electron criteria must fulfill the additional criteria of a tighter p_T threshold, $p_T > 10$ GeV, a tighter identification criteria, `MediumLH`, which has an efficiency of 72–93% for electrons with $10 < p_T < 80$ GeV, and tighter isolation, `GradientLoose`. A calorimeter crack veto is also applied, where $1.37 < |\eta^{\text{cluster}}| < 2.47$ is excluded, which allows for more efficient ZZ rejection.

Electrons passing the W electron criteria must pass the p_T threshold, $p_T > 20$, and even tighter ID and isolation criteria, `TightLH` and `Gradient` respectively, which has 68–88% efficiency for electrons with $10 < p_T < 80$ GeV.

Baseline muons must have $p_T > 7$ GeV and fall within the inner detector, $|\eta| < 2.5$. They must also satisfy the `Loose` quality criteria and pass the impact parameter cuts of $|z_0 \sin\theta| < 0.5$ and $|d_0/\sigma_{d_0}| < 3$. Finally, the muons must satisfy a track isolation requirement, `LooseTrackOnly`, which has a 99% uniform efficiency.

Muons satisfying the Z electron criteria must fulfill the additional criteria of a tighter p_T threshold, $p_T > 10$ GeV, a tighter identification criteria, `Medium` and tighter isolation, `GradientLoose`, which has an efficiency of at least 95% for muons with $p_T > 25$ GeV. Muons passing the W electron criteria must pass the p_T threshold, $p_T > 20$ GeV.

Jets are reconstructed from topological clusters using the anti-k_t algorithm with distance parameter $\Delta R = 0.4$. Jets are required to have $p_T > 25$ GeV and fulfill the pseudorapidity requirement of $|\eta| < 4.5$. To suppress jets originating from pileup, jets are further required to pass a JVT cut ($JVT > 0.59$) if the jet p_T is within $20 < p_T < 50$ GeV and it resides within $|\eta| < 2.4$ [20].

Separate algorithms are run in parallel to reconstruct electrons, muons, and jets. A particle can be reconstructed as one or more objects. To resolve these ambiguities, a procedure called "overlap removal" is applied. For electrons, this overlap removal is applied in two steps. At the baseline selection, an electron that shares a track with a muon, and the sub-leading p_{T}electron from two overlapping electrons are removed. The second step removes W or Z electrons if they are within $0.2 < \Delta R < 0.4$ of a jet. For muons, overlap removal is applied to W or Z muons to separate prompt muons from those originating from the decay of hadrons in a jet. A W or Z muon is removed if it is within $\Delta R < 0.4$ of a jet that at least 3 tracks.

The missing transverse momentum, with magnitude $E_{\mathrm{T}}^{\mathrm{miss}}$, is calculated as the negative vector sum of the transverse momenta of the calibrated selected leptons and jets, and the sum of transverse momenta of additional soft objects in the event, which are reconstructed from tracks in the inner detector or calorimeter cell clusters.

5.4.2 Event Selection

Table 5.4 summarizes the event selection. After passing the event cleaning cuts and the primary vertex requirement, having a reconstructed vertex with at least two tracks, events are required to pass the lowest unprescaled single lepton triggers. To minimize the loss of efficiency due to the turn-on of the triggers, the leading lepton p_{T} is required to be greater than 25 GeV.

Event kinematics depend on the assignment of the leptons to the parent boson. Leptons associated with the Z boson have to satisfy the Z lepton criteria. They must be the same flavor and have opposite charges (SFOS) with an invariant mass within 10 GeV of the mass of the Z boson to suppress non-resonant backgrounds such as $t\bar{t}$. If more than one pair can be formed, the pair whose invariant mass is closest to the mass of the Z boson is chosen as the Z lepton pair. The leptons associated with the Z boson are referred to as $\ell_{\mathrm{Z}}^{\mathrm{lead}}$ and $\ell_{\mathrm{Z}}^{\mathrm{sublead}}$.

The remaining third lepton is associated to the W boson and must satisfy the W lepton criteria described in the section above. The W lepton is referred to as ℓ_W. A requirement on the transverse mass of the W boson is applied to select W bosons. The transverse mass of the W boson is calculated with the W lepton and the missing transverse energy vector, and defined as:

$$m_{\mathrm{T}}^{W} = \sqrt{2 p_{\mathrm{T}}^{W} E_{\mathrm{T}}^{\mathrm{miss}} (1 - cos(\Delta\phi))}, \tag{5.8}$$

where $\Delta\phi$ is the angle between the W lepton and the missing transverse energy vector.

The transverse mass is required to be above 30 GeV. This suppresses backgrounds that have low missing energy such as Z+jets, Z+γ, and ZZ. No explicit missing transverse energy is required to have a selection identical to the 8 TeV WZ cross section measurement [7].

Table 5.4 Signal event selection. Leptons are associated to Z boson if they form a SFOS pair that minimizes their invariant mass with respect to the mass of the Z boson

Event selection	
Event cleaning	Reject LAr, Tile, and SCT corrupted events and incomplete events
Primary vertex	Reconstructed vertex with at least two tracks
Trigger	eee events must pass any of the electron triggers
	$\mu\mu\mu$ events must pass any of the muon triggers
	$\mu\mu e$, $ee\mu$ events must pass any of electron or triggers
Single electron triggers (MC)	HLT_e24_lhmedium_L1EM18VH || HLT_e60_lhmedium || HLT_e120_lhloose
Single electron triggers (data)	HLT_e24_lhmedium_L1EM20VH || HLT_e60_lhmedium || HLT_e120_lhloose
Single muon triggers (MC and data)	HLT_mu20_iloose_L1MU15 || HLT_mu50
Lepton multiplicity	Exactly 3 baseline leptons, at least 1 SFOS pair
Z lepton criteria	leptons associated with Z boson pass Z lepton criteria
W lepton criteria	lepton associated with W boson passes W lepton criteria
Lead lepton p_{T}	$p_{\mathrm{T}}^{\ell 1} > 25$ GeV
$m_{\ell\ell}$ requirement	$\lvert m_{\ell\ell} - m_Z \rvert < 10$
m_{T} requirement	$m_{\mathrm{T}}^{W} > 30$ GeV

5.5 Overview of Backgrounds

The backgrounds in this analysis can be classified into two groups: reducible backgrounds, containing at least one "fake" lepton, and irreducible backgrounds with at least three prompt leptons in the final state.

The reducible backgrounds originate from Z+jets, $Z+\gamma$, $t\bar{t}$, Wt, and WW production processes. The reducible backgrounds can be split further into Z+jets/$Z+\gamma$ processes and "top-like" processes, and their treatment is different. Z+jets/$Z+\gamma$ processes, shown in Fig. 5.2, form a Z boson and the fake lepton results from the jet or photon being misidentified as an electron. The fake electron in Z+jet can also come from the final state lepton due the decay of a b- or c-quark. Fake leptons coming from $Z+\gamma$ come from a photon converting to electrons. This photon comes from initial-state-radiation (ISR) instead of final-state-radiation (FSR). The analysis requires a lepton pair with an invariant mass close to the mass of the Z boson and the leptons resulting from the FSR scenario typically have an invariant mass below that requirement. Fake leptons are often poorly modeled in the simulation because the rate at which each source (heavy flavor jet, light flavor jets, and photon conversions) fakes a lepton is difficult to model. This background is usually estimated using a data-driven technique called the "Fake Factor" method [21, 22].

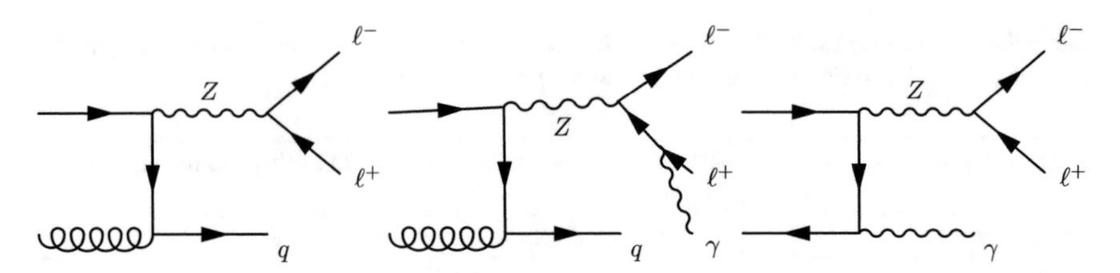

Fig. 5.2 Z+jet and $Z + \gamma$ production diagrams at leading order. The Z+jet diagrams also include FSR. In each of these diagrams, a jet or a photon fake a lepton

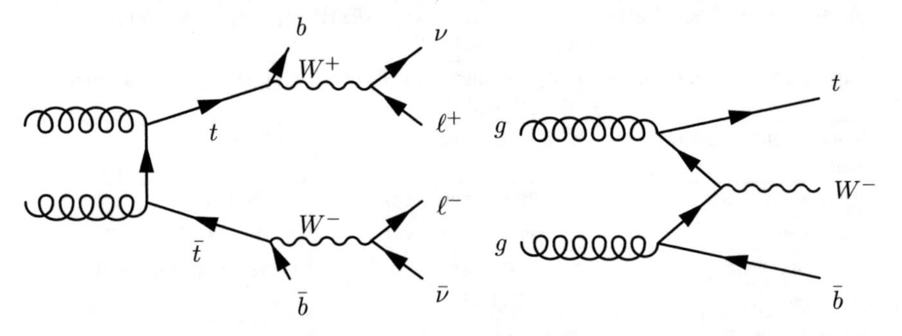

Fig. 5.3 "Top-like" production processes where the decay from the b-quark results in a fake lepton. WW is not picture here but has a diagram similar to WZ production diagrams in Fig. 5.1

The other fake background comes from $t\bar{t}$, Wt, and WW, or "top-like" processes, shown in Fig. 5.3. These fakes result from the decay of heavy flavor (b- or c-quark) decays. The final state leptons do not usually form an invariant mass consistent with the Z. Moreover, these events can have both same flavor events as well as opposite flavor events.

Both Z+jets/$Z + \gamma$ and top-like backgrounds are called reducible because their contribution in the signal region can be minimized by make more stringent requirements on the lepton identification and isolation criteria.

The other class of backgrounds is called "irreducible" and includes contribution from ZZ, VVV, $t\bar{t}V$, tZ and double parton scattering (DPS) events. These backgrounds have three prompt leptons in the fiducial volume and are estimated using simulation (MC). The dominant irreducible background comes from the ZZ process. Figure 5.4 shows the leading order diagrams for the ZZ production in proton-proton collisions. Diagrams with gluon-gluon in the initial state are not shown because the cross section of $gg \rightarrow ZZ$ is smaller than for $pp \rightarrow ZZ$, as seen in Table 5.2. This background is reduced by applying a veto on events with at least four baseline leptons. ZZ can still enter the fiducial volume by two means: either the fourth lepton falls outside the detector acceptance (too low lepton p_{T} or large $|\eta|$), which is irreducible, or the lepton falls within acceptance and is not identified. For the cases where the lepton is not identified, a correction for the MC modeling of prompt leptons that fail identification is applied.

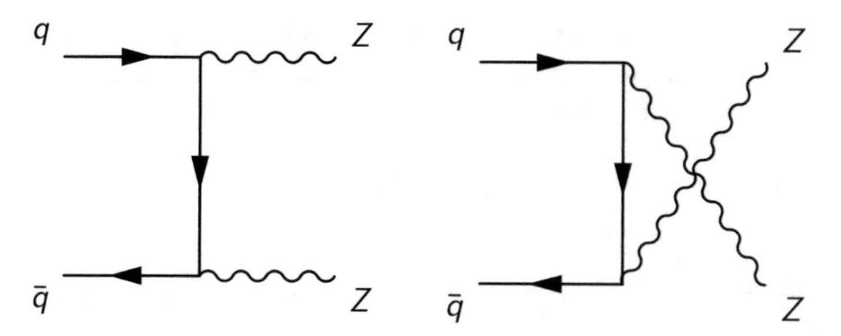

Fig. 5.4 Leading-order diagrams for ZZ production in pp collisions. Only diagrams with pp in the initial state are shown because their cross section than for the gluon-gluon initial state

5.6 Background Estimation

This section will explain the techniques used to estimate the data-driven fake background estimates and the corrections applied to the ZZ simulation to account for lepton mismodeling effects. Other processes, VVV, $t\bar{t}V$, tZ, and DPS are estimated using simulation.

The Z+jets/$Z+\gamma$ fake background has in the the final two prompt leptons with an invariant mass consistent with the mass of the Z boson and one "fake" or non-prompt lepton from light flavor or heavy flavor (lepton from the b- or c-quark decays) jets or photon conversion from $Z+\gamma$ processes. This background is estimated using the "Fake Factor" method. An extrapolation factor, or the Fake Factor (FF), between the leptons passing the signal selection criteria and leptons passing an inverted selection is measured in a region enriched with Z+jets, and $Z+\gamma$ events. This factor is applied to a region identical to the signal region except for the quality of the leptons: two leptons pass the Z lepton criteria, and one passing the inverted criteria.

Top-like backgrounds, where $t\bar{t}$ is the dominant contribution, are measured in a region with different flavor opposite charge (DFOS) events which minimize contribution from WZ events. This region is used to derive a normalization factor that is then applied to the top-like background simulation in the signal region.

5.6.1 *Z+jet/Z+γ Background*

5.6.1.1 Fake Factor Methodology

The Fake Factor method uses two sets of lepton identification criteria. The first is the same criteria used to identify signal leptons in the analysis and will be referred to as "signal", "tight", or "ID" leptons. The second is criteria is an orthogonal criteria where one identification or isolation criteria is inverted to enrich in fake leptons and is referred to as "loose" or "anti-ID" lepton criteria.

Using these ID and anti-ID lepton selections, a Fake Factor, F, is calculated in a region enriched in the process that processes fake leptons with a similar composition to the fakes in the signal region. The Fake Factor is then defined as:

$$F = \frac{N_{\text{ID}}}{N_{\text{anti-ID}}} \tag{5.9}$$

In this analysis, the Fake Factor is calculated as a function of lepton p_{T} and therefore becomes:

$$F(i) = \frac{N_{\text{ID}}(i)}{N_{\text{anti-ID}}(i)}, \tag{5.10}$$

where i is the ith p_{T} bin. This is an idealized case where the measurement has no contamination from other sources. Realistically, there will be other processes present in the Fake Factor measurement region, which contain three prompt leptons, and need to be subtracted out from the observed data. The Fake Factor takes the form:

$$F(i) = \frac{N_{\text{ID,data}}(i) - N_{\text{ID,prompt MC}}(i)}{N_{\text{anti-ID,data}}(i) - N_{\text{anti-ID, prompt MC}}(i)}, \tag{5.11}$$

The Fake Factor is then applied to a control region identical to the signal region except that instead of having three signal leptons, the lepton selection is replaced by the anti-ID selection for the fake lepton (in this case, the W lepton). The number of reducible events due to Z+jets and $Z + \gamma$ in the signal region is

$$N_{\text{SR}}^{\text{reducible}} = \sum_{i} N_{\text{CR}}^{i} \cdot F(i) \tag{5.12}$$

The Fake Factor derivation will now be discussed for the one, two, and three lepton cases.

The number of events with tight or loose leptons are of the form N_{T} and N_{L}, where the subscript indicates whether the lepton passes the ID criteria ("T") or the anti-ID criteria ("L"). In the three lepton case, these will take the form of N_{TTT}, N_{TLT}, N_{LLT}, N_{LLL}, etc., where the labeling of leptons is ordered by decreasing p_{T}.

Similarly, the number of events with real or fake leptons are of the form N^{R} and N^{F}, where the subscript indicates whether the lepton is real ("R") or fake ("F"). In the three lepton case, these will take the form of N^{RRR}, N^{RRF}, N^{FFR}, N^{FFF}, etc., where the labeling of leptons is also ordered by decreasing p_{T}.

The efficiency for the real lepton to pass the"tight" criteria is denoted as ε^{R} and the efficiency for a fake lepton to satisfy the tight criteria is denoted ε^{F}. Also, the efficiency for the real lepton to pass the"loose" criteria is denoted as $\bar{\varepsilon}^{\text{R}}$ and the efficiency for a fake lepton to satisfy the tight criteria is denoted $\bar{\varepsilon}^{\text{F}}$.

The derivation for the one lepton case will now be discussed in more detail. The number of tight and loose leptons can be related to the number of real and fake leptons using the efficiencies by the following equation:

$$\begin{pmatrix} N_{\mathrm{T}} \\ N_{\mathrm{L}} \end{pmatrix} = \begin{pmatrix} \varepsilon^{\mathrm{R}} & \varepsilon^{\mathrm{F}} \\ \bar{\varepsilon}^{\mathrm{R}} & \bar{\varepsilon}^{\mathrm{F}} \end{pmatrix} \begin{pmatrix} N^{\mathrm{R}} \\ N^{\mathrm{F}} \end{pmatrix} \tag{5.13}$$

To make a single analytic equation, both sides of the equation are multiplied by the row vector:

$$\left(1 - \tfrac{\varepsilon^{\mathrm{F}}}{\bar{\varepsilon}^{\mathrm{F}}}\right), \tag{5.14}$$

which results in:

$$N_{\mathrm{T}} - \frac{\varepsilon^{\mathrm{F}}}{\bar{\varepsilon}^{\mathrm{F}}} N_{\mathrm{L}} = \varepsilon^{\mathrm{R}} N^{\mathrm{R}} + \varepsilon^{\mathrm{F}} N^{\mathrm{F}} - \frac{\varepsilon^{\mathrm{F}}}{\bar{\varepsilon}^{\mathrm{F}}} \bar{\varepsilon}^{\mathrm{R}} N^{\mathrm{R}} - \frac{\varepsilon^{\mathrm{F}}}{\bar{\varepsilon}^{\mathrm{F}}} \bar{\varepsilon}^{\mathrm{F}} N^{\mathrm{F}} \tag{5.15}$$

The fraction $\frac{\varepsilon^{\mathrm{F}}}{\bar{\varepsilon}^{\mathrm{F}}}$ can be replaced by the symbol F, representing the Fake Factor, and canceling terms, this gives:

$$N_{\mathrm{T}} - F N_{\mathrm{L}} = \varepsilon^{\mathrm{R}} N^{\mathrm{R}} - F \bar{\varepsilon}^{\mathrm{R}} N^{\mathrm{R}}. \tag{5.16}$$

Moreover, additional substitutions can be made ($\bar{\varepsilon}^{\mathrm{R}} N^{\mathrm{R}} = N_{\mathrm{L}}^{\mathrm{R}}$ and $\varepsilon^{\mathrm{R}} N^{\mathrm{R}} = N_{\mathrm{T}}^{\mathrm{R}}$), which gives the final result

$$N_{\mathrm{T}}^{\mathrm{F}} = N_{\mathrm{T}} - N_{\mathrm{T}}^{\mathrm{R}} = F(N_{\mathrm{L}} - N_{\mathrm{L}}^{\mathrm{R}}). \tag{5.17}$$

Solving for the Fake Factor, F,

$$F = \frac{N_{\mathrm{T}}^{\mathrm{F}}}{N_{\mathrm{L}} - N_{\mathrm{L}}^{\mathrm{R}}} = \frac{N_{\mathrm{T}} - N_{\mathrm{T}}^{\mathrm{R}}}{N_{\mathrm{L}} - N_{\mathrm{L}}^{\mathrm{R}}} \tag{5.18}$$

The Factor Factor, F, can be computed by taking the ratio of the number of tight fake leptons $N_{\mathrm{T}}^{\mathrm{F}}$, which equal to $N_{\mathrm{T}} - N_{\mathrm{T}}^{\mathrm{R}}$, where N_{T}, is the number of tight leptons, measured in data, and $N_{\mathrm{T}}^{\mathrm{R}}$ is the number of real tight leptons estimated using simulation, and the denominator, which contains the number of loose leptons, N_{L}, measured in data, and the number of loose real leptons, $N_{\mathrm{L}}^{\mathrm{R}}$, estimated using simulation.

This Fake Factor calculation assumes that the composition between the fake measurement region and the signal region are similar. If there are differences in composition, an uncertainty should be taken into account.

The two lepton case is similar to the one lepton case except there are more terms in the matrix because in the two lepton case, either both leptons are tight, or one is tight, one is loose, or both are loose. The relationship between loose and tight leptons is related to the number of real and fake leptons by this matrix:

$$\begin{pmatrix} N_{\mathrm{TT}} \\ N_{\mathrm{LT}} \\ N_{\mathrm{TL}} \\ N_{\mathrm{LL}} \end{pmatrix} = \begin{pmatrix} \varepsilon_1^{\mathrm{R}}\varepsilon_2^{\mathrm{R}} & \varepsilon_1^{\mathrm{F}}\varepsilon_2^{\mathrm{R}} & \varepsilon_1^{\mathrm{R}}\varepsilon_2^{\mathrm{F}} & \varepsilon_1^{\mathrm{F}}\varepsilon_2^{\mathrm{F}} \\ \bar{\varepsilon}_1^{\mathrm{R}}\varepsilon_2^{\mathrm{R}} & \bar{\varepsilon}_1^{\mathrm{F}}\varepsilon_2^{\mathrm{R}} & \bar{\varepsilon}_1^{\mathrm{R}}\varepsilon_2^{\mathrm{F}} & \bar{\varepsilon}_1^{\mathrm{F}}\varepsilon_2^{\mathrm{F}} \\ \varepsilon_1^{\mathrm{R}}\bar{\varepsilon}_2^{\mathrm{R}} & \varepsilon_1^{\mathrm{F}}\bar{\varepsilon}_2^{\mathrm{R}} & \varepsilon_1^{\mathrm{R}}\bar{\varepsilon}_2^{\mathrm{F}} & \varepsilon_1^{\mathrm{F}}\bar{\varepsilon}_2^{\mathrm{F}} \\ \bar{\varepsilon}_1^{\mathrm{R}}\bar{\varepsilon}_2^{\mathrm{R}} & \bar{\varepsilon}_1^{\mathrm{F}}\bar{\varepsilon}_2^{\mathrm{R}} & \bar{\varepsilon}_1^{\mathrm{R}}\bar{\varepsilon}_2^{\mathrm{F}} & \bar{\varepsilon}_1^{\mathrm{F}}\bar{\varepsilon}_2^{\mathrm{F}} \end{pmatrix} \begin{pmatrix} N^{\mathrm{RR}} \\ N^{\mathrm{FR}} \\ N^{\mathrm{RF}} \\ N^{\mathrm{FF}} \end{pmatrix} \tag{5.19}$$

The row vector used to multiply both sides of the equation is:

$$\begin{pmatrix} 1 & -\dfrac{\varepsilon_1^{\mathrm{F}}}{\bar{\varepsilon}_1^{\mathrm{F}}} & -\dfrac{\varepsilon_2^{\mathrm{F}}}{\bar{\varepsilon}_2^{\mathrm{F}}} & -\dfrac{\varepsilon_1^{\mathrm{F}}\varepsilon_2^{\mathrm{F}}}{\bar{\varepsilon}_1^{\mathrm{F}}\bar{\varepsilon}_2^{\mathrm{F}}} \end{pmatrix} \tag{5.20}$$

After canceling terms, replacing terms of the form $\frac{\varepsilon^{\mathrm{F}}}{\bar{\varepsilon}^{\mathrm{F}}}$ by F, and adding the substitutions, $\bar{\varepsilon}^{\mathrm{R}} N^{\mathrm{R}} = N_{\mathrm{L}}^{\mathrm{R}}$ and $\varepsilon^{\mathrm{R}} N^{\mathrm{R}} = N_{\mathrm{T}}^{\mathrm{R}}$, the number of events containing at least one fake lepton becomes,

$$N_{\mathrm{TT}} - N_{\mathrm{TT}}^{\mathrm{RR}} = F_1(N_{\mathrm{LT}} - N_{\mathrm{LT}}^{\mathrm{RR}}) + F_2(N_{\mathrm{TL}} - N_{\mathrm{TL}}^{\mathrm{RR}}) - F_1 F_2(N_{\mathrm{LL}} - N_{\mathrm{LL}}^{\mathrm{RR}}) \tag{5.21}$$

This estimates the number of events with one fake lepton and two fake leptons simultaneously.

Finally, in the three lepton case used in this analysis, the relationship between loose and tight leptons is related to the number of real and fake leptons by the following matrix:

$$\begin{pmatrix} N_{\mathrm{TTT}} \\ N_{\mathrm{LTT}} \\ N_{\mathrm{TLT}} \\ N_{\mathrm{TTL}} \\ N_{\mathrm{LLT}} \\ N_{\mathrm{LTL}} \\ N_{\mathrm{TLL}} \\ N_{\mathrm{LLL}} \end{pmatrix} = \begin{pmatrix}
\varepsilon_1^{\mathrm{R}}\varepsilon_2^{\mathrm{R}}\varepsilon_3^{\mathrm{R}} & \varepsilon_1^{\mathrm{F}}\varepsilon_2^{\mathrm{R}}\varepsilon_3^{\mathrm{R}} & \varepsilon_1^{\mathrm{R}}\varepsilon_2^{\mathrm{F}}\varepsilon_3^{\mathrm{R}} & \varepsilon_1^{\mathrm{R}}\varepsilon_2^{\mathrm{R}}\varepsilon_3^{\mathrm{F}} & \varepsilon_1^{\mathrm{F}}\varepsilon_2^{\mathrm{F}}\varepsilon_3^{\mathrm{R}} & \varepsilon_1^{\mathrm{F}}\varepsilon_2^{\mathrm{R}}\varepsilon_3^{\mathrm{F}} & \varepsilon_1^{\mathrm{R}}\varepsilon_2^{\mathrm{F}}\varepsilon_3^{\mathrm{F}} & \varepsilon_1^{\mathrm{F}}\varepsilon_2^{\mathrm{F}}\varepsilon_3^{\mathrm{F}} \\
\bar{\varepsilon}_1^{\mathrm{R}}\varepsilon_2^{\mathrm{R}}\varepsilon_3^{\mathrm{R}} & \bar{\varepsilon}_1^{\mathrm{F}}\varepsilon_2^{\mathrm{R}}\varepsilon_3^{\mathrm{R}} & \bar{\varepsilon}_1^{\mathrm{R}}\varepsilon_2^{\mathrm{F}}\varepsilon_3^{\mathrm{R}} & \bar{\varepsilon}_1^{\mathrm{R}}\varepsilon_2^{\mathrm{R}}\varepsilon_3^{\mathrm{F}} & \bar{\varepsilon}_1^{\mathrm{F}}\varepsilon_2^{\mathrm{F}}\varepsilon_3^{\mathrm{R}} & \bar{\varepsilon}_1^{\mathrm{F}}\varepsilon_2^{\mathrm{R}}\varepsilon_3^{\mathrm{F}} & \bar{\varepsilon}_1^{\mathrm{R}}\varepsilon_2^{\mathrm{F}}\varepsilon_3^{\mathrm{F}} & \bar{\varepsilon}_1^{\mathrm{F}}\varepsilon_2^{\mathrm{F}}\varepsilon_3^{\mathrm{F}} \\
\varepsilon_1^{\mathrm{R}}\bar{\varepsilon}_2^{\mathrm{R}}\varepsilon_3^{\mathrm{R}} & \varepsilon_1^{\mathrm{F}}\bar{\varepsilon}_2^{\mathrm{R}}\varepsilon_3^{\mathrm{R}} & \varepsilon_1^{\mathrm{R}}\bar{\varepsilon}_2^{\mathrm{F}}\varepsilon_3^{\mathrm{R}} & \varepsilon_1^{\mathrm{R}}\bar{\varepsilon}_2^{\mathrm{R}}\varepsilon_3^{\mathrm{F}} & \varepsilon_1^{\mathrm{F}}\bar{\varepsilon}_2^{\mathrm{F}}\varepsilon_3^{\mathrm{R}} & \varepsilon_1^{\mathrm{F}}\bar{\varepsilon}_2^{\mathrm{R}}\varepsilon_3^{\mathrm{F}} & \varepsilon_1^{\mathrm{R}}\bar{\varepsilon}_2^{\mathrm{F}}\varepsilon_3^{\mathrm{F}} & \varepsilon_1^{\mathrm{F}}\bar{\varepsilon}_2^{\mathrm{F}}\varepsilon_3^{\mathrm{F}} \\
\varepsilon_1^{\mathrm{R}}\varepsilon_2^{\mathrm{R}}\bar{\varepsilon}_3^{\mathrm{R}} & \varepsilon_1^{\mathrm{F}}\varepsilon_2^{\mathrm{R}}\bar{\varepsilon}_3^{\mathrm{R}} & \varepsilon_1^{\mathrm{R}}\varepsilon_2^{\mathrm{F}}\bar{\varepsilon}_3^{\mathrm{R}} & \varepsilon_1^{\mathrm{R}}\varepsilon_2^{\mathrm{R}}\bar{\varepsilon}_3^{\mathrm{F}} & \varepsilon_1^{\mathrm{F}}\varepsilon_2^{\mathrm{F}}\bar{\varepsilon}_3^{\mathrm{R}} & \varepsilon_1^{\mathrm{F}}\varepsilon_2^{\mathrm{R}}\bar{\varepsilon}_3^{\mathrm{F}} & \varepsilon_1^{\mathrm{R}}\varepsilon_2^{\mathrm{F}}\bar{\varepsilon}_3^{\mathrm{F}} & \varepsilon_1^{\mathrm{F}}\varepsilon_2^{\mathrm{F}}\bar{\varepsilon}_3^{\mathrm{F}} \\
\bar{\varepsilon}_1^{\mathrm{R}}\bar{\varepsilon}_2^{\mathrm{R}}\varepsilon_3^{\mathrm{R}} & \bar{\varepsilon}_1^{\mathrm{F}}\bar{\varepsilon}_2^{\mathrm{R}}\varepsilon_3^{\mathrm{R}} & \bar{\varepsilon}_1^{\mathrm{R}}\bar{\varepsilon}_2^{\mathrm{F}}\varepsilon_3^{\mathrm{R}} & \bar{\varepsilon}_1^{\mathrm{R}}\bar{\varepsilon}_2^{\mathrm{R}}\varepsilon_3^{\mathrm{F}} & \bar{\varepsilon}_1^{\mathrm{F}}\bar{\varepsilon}_2^{\mathrm{F}}\varepsilon_3^{\mathrm{R}} & \bar{\varepsilon}_1^{\mathrm{F}}\bar{\varepsilon}_2^{\mathrm{R}}\varepsilon_3^{\mathrm{F}} & \bar{\varepsilon}_1^{\mathrm{R}}\bar{\varepsilon}_2^{\mathrm{F}}\varepsilon_3^{\mathrm{F}} & \bar{\varepsilon}_1^{\mathrm{F}}\bar{\varepsilon}_2^{\mathrm{F}}\varepsilon_3^{\mathrm{F}} \\
\bar{\varepsilon}_1^{\mathrm{R}}\varepsilon_2^{\mathrm{R}}\bar{\varepsilon}_3^{\mathrm{R}} & \bar{\varepsilon}_1^{\mathrm{F}}\varepsilon_2^{\mathrm{R}}\bar{\varepsilon}_3^{\mathrm{R}} & \bar{\varepsilon}_1^{\mathrm{R}}\varepsilon_2^{\mathrm{F}}\bar{\varepsilon}_3^{\mathrm{R}} & \bar{\varepsilon}_1^{\mathrm{R}}\varepsilon_2^{\mathrm{R}}\bar{\varepsilon}_3^{\mathrm{F}} & \bar{\varepsilon}_1^{\mathrm{F}}\varepsilon_2^{\mathrm{F}}\bar{\varepsilon}_3^{\mathrm{R}} & \bar{\varepsilon}_1^{\mathrm{F}}\varepsilon_2^{\mathrm{R}}\bar{\varepsilon}_3^{\mathrm{F}} & \bar{\varepsilon}_1^{\mathrm{R}}\varepsilon_2^{\mathrm{F}}\bar{\varepsilon}_3^{\mathrm{F}} & \bar{\varepsilon}_1^{\mathrm{F}}\varepsilon_2^{\mathrm{F}}\bar{\varepsilon}_3^{\mathrm{F}} \\
\varepsilon_1^{\mathrm{R}}\bar{\varepsilon}_2^{\mathrm{R}}\bar{\varepsilon}_3^{\mathrm{R}} & \varepsilon_1^{\mathrm{F}}\bar{\varepsilon}_2^{\mathrm{R}}\bar{\varepsilon}_3^{\mathrm{R}} & \varepsilon_1^{\mathrm{R}}\bar{\varepsilon}_2^{\mathrm{F}}\bar{\varepsilon}_3^{\mathrm{R}} & \varepsilon_1^{\mathrm{R}}\bar{\varepsilon}_2^{\mathrm{R}}\bar{\varepsilon}_3^{\mathrm{F}} & \varepsilon_1^{\mathrm{F}}\bar{\varepsilon}_2^{\mathrm{F}}\bar{\varepsilon}_3^{\mathrm{R}} & \varepsilon_1^{\mathrm{F}}\bar{\varepsilon}_2^{\mathrm{R}}\bar{\varepsilon}_3^{\mathrm{F}} & \varepsilon_1^{\mathrm{R}}\bar{\varepsilon}_2^{\mathrm{F}}\bar{\varepsilon}_3^{\mathrm{F}} & \varepsilon_1^{\mathrm{F}}\bar{\varepsilon}_2^{\mathrm{F}}\bar{\varepsilon}_3^{\mathrm{F}} \\
\bar{\varepsilon}_1^{\mathrm{R}}\bar{\varepsilon}_2^{\mathrm{R}}\bar{\varepsilon}_3^{\mathrm{R}} & \bar{\varepsilon}_1^{\mathrm{F}}\bar{\varepsilon}_2^{\mathrm{R}}\bar{\varepsilon}_3^{\mathrm{R}} & \bar{\varepsilon}_1^{\mathrm{R}}\bar{\varepsilon}_2^{\mathrm{F}}\bar{\varepsilon}_3^{\mathrm{R}} & \bar{\varepsilon}_1^{\mathrm{R}}\bar{\varepsilon}_2^{\mathrm{R}}\bar{\varepsilon}_3^{\mathrm{F}} & \bar{\varepsilon}_1^{\mathrm{F}}\bar{\varepsilon}_2^{\mathrm{F}}\bar{\varepsilon}_3^{\mathrm{R}} & \bar{\varepsilon}_1^{\mathrm{F}}\bar{\varepsilon}_2^{\mathrm{R}}\bar{\varepsilon}_3^{\mathrm{F}} & \bar{\varepsilon}_1^{\mathrm{R}}\bar{\varepsilon}_2^{\mathrm{F}}\bar{\varepsilon}_3^{\mathrm{F}} & \bar{\varepsilon}_1^{\mathrm{F}}\bar{\varepsilon}_2^{\mathrm{F}}\bar{\varepsilon}_3^{\mathrm{F}}
\end{pmatrix} \begin{pmatrix} N^{\mathrm{RRR}} \\ N^{\mathrm{FRR}} \\ N^{\mathrm{RFR}} \\ N^{\mathrm{RRF}} \\ N^{\mathrm{FFR}} \\ N^{\mathrm{FRF}} \\ N^{\mathrm{RFF}} \\ N^{\mathrm{FFF}} \end{pmatrix} \tag{5.22}$$

The row vector used to multiply both sides of the equation is:

$$\begin{pmatrix} 1 & -\dfrac{\varepsilon_1^{\mathrm{F}}}{\bar{\varepsilon}_1^{\mathrm{F}}} & -\dfrac{\varepsilon_2^{\mathrm{F}}}{\bar{\varepsilon}_2^{\mathrm{F}}} & -\dfrac{\varepsilon_3^{\mathrm{F}}}{\bar{\varepsilon}_3^{\mathrm{F}}} & \dfrac{\varepsilon_1^{\mathrm{F}}\varepsilon_2^{\mathrm{F}}}{\bar{\varepsilon}_1^{\mathrm{F}}\bar{\varepsilon}_2^{\mathrm{F}}} & \dfrac{\varepsilon_1^{\mathrm{F}}\varepsilon_3^{\mathrm{F}}}{\bar{\varepsilon}_1^{\mathrm{F}}\bar{\varepsilon}_3^{\mathrm{F}}} & \dfrac{\varepsilon_2^{\mathrm{F}}\varepsilon_3^{\mathrm{F}}}{\bar{\varepsilon}_2^{\mathrm{F}}\bar{\varepsilon}_3^{\mathrm{F}}} & -\dfrac{\varepsilon_1^{\mathrm{F}}\varepsilon_2^{\mathrm{F}}\varepsilon_3^{\mathrm{F}}}{\bar{\varepsilon}_1^{\mathrm{F}}\bar{\varepsilon}_2^{\mathrm{F}}\bar{\varepsilon}_3^{\mathrm{F}}} \end{pmatrix} \tag{5.23}$$

After canceling terms, replacing terms of the form $\frac{\varepsilon^{\mathrm{F}}}{\bar{\varepsilon}^{\mathrm{F}}}$ by F, and adding the substitutions, $\bar{\varepsilon}^{\mathrm{R}} N^{\mathrm{R}} = N_{\mathrm{L}}^{\mathrm{R}}$ and $\varepsilon^{\mathrm{R}} N^{\mathrm{R}} = N_{\mathrm{T}}^{\mathrm{R}}$, the number of events containing at least one fake lepton becomes,

$$
\begin{aligned}
N_{\mathrm{TTT}} - N_{\mathrm{TTT}}^{\mathrm{RRR}} &= F_1(N_{\mathrm{LTT}} - N_{\mathrm{LTT}}^{\mathrm{RRR}}) + F_2(N_{\mathrm{TLT}} - N_{\mathrm{TLT}}^{\mathrm{RRR}}) \\
&+ F_3(N_{\mathrm{TTL}} - N_{\mathrm{TTL}}^{\mathrm{RRR}}) - F_1F_2(N_{\mathrm{LLT}} - N_{\mathrm{LLT}}^{\mathrm{RRR}}) \\
&- F_1F_3(N_{\mathrm{LTL}} - N_{\mathrm{LTL}}^{\mathrm{RRR}}) - F_2F_3(N_{\mathrm{TLL}} - N_{\mathrm{TLL}}^{\mathrm{RRR}}) \\
&+ F_1F_2F_3(N_{\mathrm{LLL}} - N_{\mathrm{LLL}}^{\mathrm{RRR}})
\end{aligned} \tag{5.24}
$$

The number of events with two or three fake leptons are small compared to the number of events with one fake lepton.

Equation (5.24) can be rewritten for multiple Fake Factor bins,

$$
\begin{aligned}
N_{\mathrm{TTT}} - N_{\mathrm{TTT}}^{\mathrm{RRR}} &= \sum_i F_1(i)(N_{\mathrm{LTT}} - N_{\mathrm{LTT}}^{\mathrm{RRR}}) \\
&+ \sum_i F_1(i)F_2(i)(N_{\mathrm{TLT}} - N_{\mathrm{TLT}}^{\mathrm{RRR}}) \\
&+ \sum_i F_3(i)(N_{\mathrm{TTL}} - N_{\mathrm{TTL}}^{\mathrm{RRR}})
\end{aligned} \tag{5.25}
$$

This assumes that the events with three prompt leptons or top-like fakes have been subtracted from the data.

5.6.1.2 Fake Factor Measurement and Validation

To summarize, the Z+jet and $Z + \gamma$ estimate in each lepton p_{T} bin is obtained by extrapolating from events in the fake control sample using the following formula:

$$
\begin{aligned}
N_{Z+j/Z+\gamma} &= (N_{\mathrm{LTT}} - N_{\mathrm{LTT}}^{\mathrm{prompt}} - N_{\mathrm{LTT}}^{\mathrm{top}})F_W \\
&+ (N_{\mathrm{TLT}} - N_{\mathrm{TLT}}^{\mathrm{prompt}} - N_{\mathrm{TLT}}^{\mathrm{top}})F_Z \\
&+ (N_{\mathrm{TTL}} - N_{\mathrm{TTL}}^{\mathrm{prompt}} - N_{\mathrm{TTL}}^{\mathrm{top}})F_Z,
\end{aligned} \tag{5.26}
$$

where F_W and F_Z are the Fake Factors for the W and Z leptons, $N_{\mathrm{LTT}}^{\mathrm{prompt}}$, $N_{\mathrm{TLT}}^{\mathrm{prompt}}$, $N_{\mathrm{TTL}}^{\mathrm{prompt}}$ are the MC predictions with three prompt leptons, and $N_{\mathrm{LTT}}^{\mathrm{top}}$, $N_{\mathrm{TLT}}^{\mathrm{top}}$, $N_{\mathrm{TTL}}^{\mathrm{top}}$ are the estimates for the top-like fakes.

The Fake Factor is derived in a region orthogonal to the WZ selection and enriched with Z+jets and $Z + \gamma$ events: $m_{\mathrm{T}}^{W} < 30$ GeV, and $E_{\mathrm{T}}^{\mathrm{miss}} < 40$ GeV, a region that will be referred to as the Fake Factor measurement region. These events must satisfy all requirements from Table 5.4 except for the new requirements on m_{T} and $E_{\mathrm{T}}^{\mathrm{miss}}$. The contribution of Z+jets and $Z + \gamma$ in this region is 75% of all the backgrounds.

Loose events have looser identification criteria than signal events to enhance the statistics of the denominator events. Tables 5.5 and 5.6 show the definition for the baseline identification for loose events as well as the denominator criteria. The anti-ID criteria have either the signal isolation or the identification criteria inverted. For

Table 5.5 Definition for baseline leptons in the Fake Factor measurement region

Criteria	Electrons	Muons
Identification	`VeryLooseLLH`	`Loose`
Lepton p_T	> 7 GeV	> 7 GeV
Lepton η	$\|\eta\| < 2.47$, exclude $1.37 < \|\eta\| < 1.52$	$\|\eta\| < 2.47$

Table 5.6 Definition of the anti-ID criteria for the Fake Factor measurement region

Criteria	Electrons	Muons
Overlap removal	pass OR with muons and jets	no OR applied
Identification	`VeryLooseLLH`	`Medium`
anti-ID criteria	(! `MediumLLH` identification ||	($\|\Delta z_0^{\text{baseline}}\| > 0.5$ ||
	$\|d_0^{\text{baseline}}\| > 5$ ||	$\|d_0^{\text{baseline}}\| > 3$ ||
	! GradientLoose isolation)	! GradientLoose isolation)

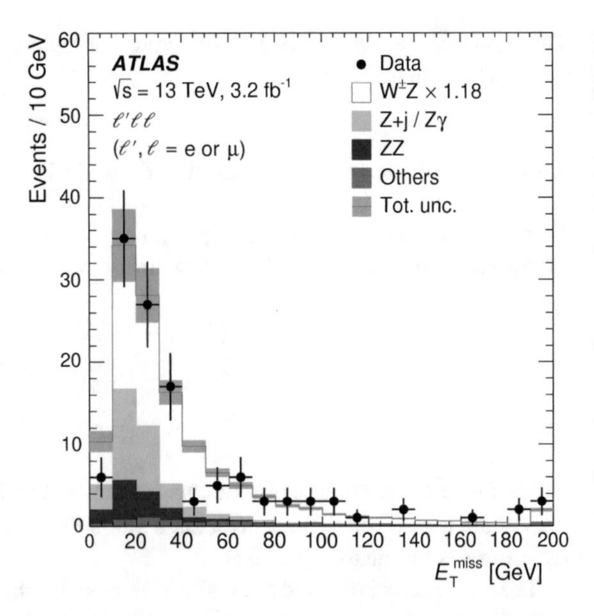

Fig. 5.5 E_T^{miss} distribution in the Fake validation region. The Z+jet and $Z+\gamma$ contribution is estimated using the Fake Factor. A normalization factor is applied to the top Monte Carlo. Backgrounds labeled "Others" consist of $t\bar{t}$, Wt, WW, tZ, ttV, and VVV processes. The WZ process is scaled by 1.18 to match the inclusive cross section

the muon overlap removal criterion, all muons are kept and jets are removed if they are within $\Delta R < 0.4$ of a muon. Trigger matching is applied to the tight leptons.

The Fake Factor estimate of the Z+jet and $Z+\gamma$ background is validated in a subset of the signal region containing events with $30 < m_T^W < 50$ GeV and $E_T^{\text{miss}} < 50$ GeV, which is enriched in background processes. Figure 5.5 shows the good agreement between data and background in the E_T^{miss} distribution in the Fake Factor validation region. The WZ process is scaled by 1.18 to match the inclusive cross section.

There are four main uncertainties associated with the Fake Factor method. There is a statistical uncertainty due to the number of anti-ID data events in the Fake Factor measurement region. Another uncertainty comes from the prompt subtraction, dominated by WZ and ZZ events. The WZ and ZZ yields are varied by 15% and the largest impact on the Fake Factor is taken as an uncertainty. The Fake Factor also has uncertainties associated with the subtraction of top backgrounds; the uncertainty on the top normalization is propagated. Finally, a closure systematic is assessed by calculating a Z+jet/$Z + \gamma$ Monte Carlo Fake Factor in the signal region and compare this with the Z+jet/$Z + \gamma$ Monte Carlo Fake Factor derived in the fake measurement region.

5.6.2 Top-Like Backgrounds

Top-like backgrounds include $t\bar{t}$, Wt, and WW; however, the dominant contribution comes from $t\bar{t}$. To estimate top-like backgrounds, a control region (CR) is constructed with different flavor, opposite charge events (DFOS), meaning that the events are of the form $e^{\pm}e^{\pm}\mu^{\mp}$ or $\mu^{\pm}\mu^{\pm}e^{\mp}$. This guarantees that one of the same flavor leptons is the fake lepton and this region is very pure in events with top-like fakes.

A control region is used to normalize the different background sources in the signal region. This region is built by inverting, or changing, one or more event selection requirements. The goal is to obtain an event sample that is kinematically similar to the signal region but enriched in a particular process.

The normalization factor (NF) extracted from the control region, NF_{bkg} is the ratio of data events, $N^{\text{CR}}_{\text{data}}$, with other background processes ($N^{\text{CR}}_{\text{other}}$) removed over the number of simulated background events for the process, $N^{\text{CR}}_{\text{bkg,MC}}$. All these number of events are determined in the control region. The NF is calculated as

$$NF_{\text{bkg}} = \frac{N^{\text{CR}}_{\text{data}} - N^{\text{CR}}_{\text{other}}}{N^{\text{CR}}_{\text{bkg,MC}}} \tag{5.27}$$

Note the NF provides an overall normalization. The shape is taken directly from the simulated background events in the signal region.

To determine the contribution of the background process in the signal region, $N^{\text{SR}}_{\text{bkg,est.}}$, the normalization is applied to simulated background events in the signal region:

$$N^{\text{SR}}_{\text{bkg, est.}} = NF_{\text{bkg}} \cdot N^{\text{SR}}_{\text{bkg,MC}} \tag{5.28}$$

Systematic uncertainties in the normalization factor arise from different sources. There are statistical uncertainties due to the number of data events available to calculate the NF. There are also experimental and theoretical uncertainties both on the process being normalized and the one being subtracted. To understand the impact of these uncertainties on the normalized background in the signal region, equations (5.27), (5.28) are combined.

$$N^{\text{SR}}_{\text{bkg, est.}} = \frac{N^{\text{CR}}_{\text{data}} - N^{\text{CR}}_{\text{other}}}{N^{\text{CR}}_{\text{bkg,MC}}} \cdot N^{\text{SR}}_{\text{bkg,MC}} = (N^{\text{CR}}_{\text{data}} - N^{\text{CR}}_{\text{other}}) \cdot \alpha_{\text{bkg}}, \tag{5.29}$$

where $\alpha_{\text{bkg}} = \frac{N^{\text{SR}}_{\text{bkg,MC}}}{N^{\text{CR}}_{\text{bkg,MC}}}$ is the transfer factor from the control region to the signal region. Since the transfer factor is the ratio of the simulated yields in the signal region and the simulated yields in the control region, the systematic uncertainties impacting these two yields largely cancel out, especially if both regions are kinematically similar. As a result, normalizing a background using a control region greatly reduces the systematic uncertainties on that background in the signal region.

In this analysis, the top control region applies all the requirements of the signal region, described in Table 5.4, except for the lepton assignment to the parent bosons. DFOS leptons that minimize $|m_{\ell\ell} - m_Z|$ are assigned to the Z boson instead of the SFOS pair. To increase the number of events in the control region, the requirement of $|m_{\ell\ell} - m_Z| < 10$GeV is removed. The third lepton is considered to be the fake lepton. Separate normalization factors are derived for electron and for muon fakes since these factors depend on detector modeling, which may be different between the two lepton flavors.

Figure 5.6 shows the number of jets distribution in the top CR where electron and muon fakes are not plotted separately. The NF derived for the electron fake is $1.41 \pm 0.49 \pm 0.17$ and the NF for the muon fake is $0.54 \pm 0.32 \pm 0.05$, where the

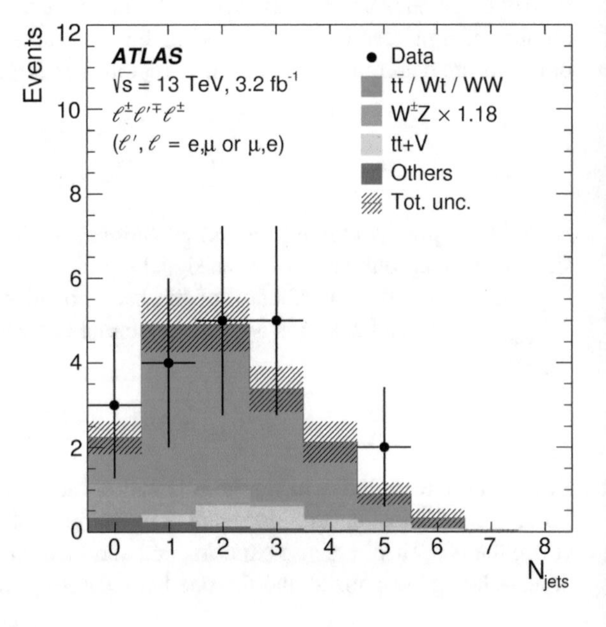

Fig. 5.6 The number of jets distribution in the top control region which uses different flavor, opposite charge events. Backgrounds labeled "Others" consist of ZZ and VVV processes. The WZ process is scaled by 1.18 to match the inclusive cross section

first uncertainty is the statistical uncertainty on the data and the second uncertainty is the statistical uncertainty on the Monte Carlo simulation.

5.6.3 *ZZ Background*

About 70% of the irreducible background is due to ZZ production. Events from ZZ survive the WZ selection either because one lepton falls outside the fiducial volume ($p_{\mathrm{T}} < 7$ GeV, or $|\eta| > 2.5$) or because it falls within the fiducial volume but is not identified. The former is an irreducible source of background, the latter is a reducible background and is suppressed by adding the four-lepton veto, described in Sect. 5.4.

5.6.3.1 Anti-ID Scale Factors

One dominant background contribution comes from $ZZ \rightarrow \ell^{\pm}\ell^{\mp}\ell^{\pm}\ell^{\mp}$, in events where one lepton fails the lepton identification requirements or falls beyond kinematic reach and is thus not identified.

Two methods were used to identify the fourth lepton originating from the second Z boson. In the first method, for each event that passes, we determine which truth lepton corresponds to a reconstructed by ensuring that they have the same flavor and charge, as well as by minimizing the ΔR between the truth and reconstructed leptons. This minimization procedure is verified, as seen in Fig. 5.7.

Next, reconstructed leptons are identified as coming from the Z decay and which are identified as coming from the W boson, referred to in this study as the third lepton. The charge and flavor of the reconstructed lepton coming from the W boson determine the flavor and the charge of the fourth lepton in the ZZ event.

Finally, the invariant mass of the third lepton and a fourth truth lepton is calculated. If there is more than one candidate fourth lepton, minimizing the difference between

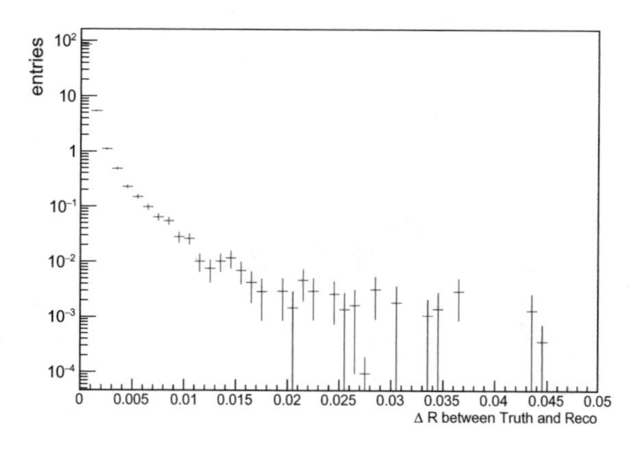

Fig. 5.7 ΔR between the matched truth and reconstructed leptons (raw MC events) for the ZZ background

this calculated invariant mass and the Z boson mass will determine which lepton is the fourth truth lepton coming from the Z decay. Verification of the $m_{\ell\ell}$ minimization calculation is shown in Fig. 5.8.

The second method uses the information from the MCTruthClassifier tool for each container-level lepton of the appropriate flavor in each ZZ event, i.e. a muon (electron) for the $\mu\mu\mu/ee\mu$ ($eee/\mu\mu e$) channels. The container-level lepton must originate from a Z boson decay, according to the MCTruthClassifier tool (and it must not be one of the three already identified leptons). Technically this means:

- Muons of type=6 and origin=13, or type=6 and origin=43
- Electrons of type=2 and origin=13, or type=4 and origin=40, or type=4 and origin=13, or type=2 and origin=43

If such a lepton is found within acceptance, $|\eta| < 2.5$ and $p_T > 7$ GeV, and fails identification, it is used to obtain the anti-ID scale factor that is applied to the ZZ event. To fail identification means to fail the `Loose` requirement for muons, and to fail `LooseLH+BLayer` or d_0 or z_0 selections for electrons.

Table 5.7 presents the number of events after each selection criteria in the studies. ZZ events for which the fourth lepton is outside the acceptance are considered irreducible background. 82% of the muons and 61% of the electrons fall in this category. Leptons that fall within acceptance should have been vetoed since ZZ events have 4 leptons, while the signal region requires exactly 3 leptons. 18% of these leptons are muons, and 39% are electrons.

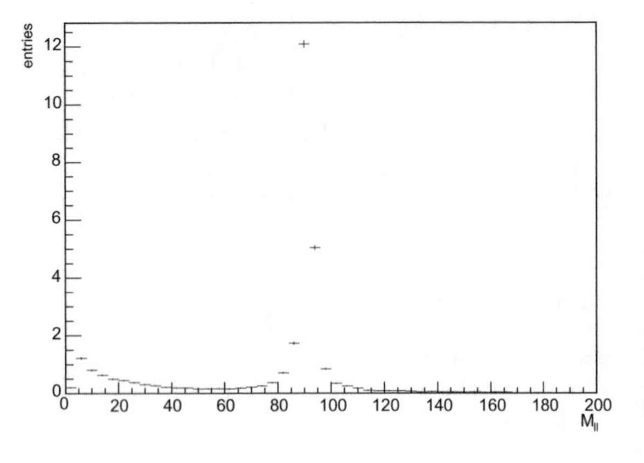

Fig. 5.8 Invariant mass between the third lepton matched with a reconstructed lepton and the fourth truth lepton coming from the Z decay (raw MC events)

Table 5.7 Fraction of ZZ events that falls in the signal region because the fourth lepton is outside the acceptance, or because it fails identification, split by flavor of the fourth lepton

	Outside acceptance	Not identified
Electrons	61%	39%
Muons	82%	18%

The events with an identified lepton are subject to anti-ID scale factors. ID scale factors, SF^{ID}, are used to correct for differences in identification efficiencies between data and MC. They are simply computed as the ratio between the efficiency measured in data and the efficiency measured in MC. Anti-ID scale factors can therefore be computed as shown in Eq. (5.30).

$$SF^{\text{anti-ID}} = \frac{1 - \varepsilon^{\text{data}}}{1 - \varepsilon^{\text{MC}}} = \frac{1 - SF^{\text{ID}} \times \varepsilon^{\text{MC}}}{1 - \varepsilon^{\text{MC}}} \tag{5.30}$$

Anti-ID scale factors for muons have been computed by taking the MC efficiencies and scale factors provided by the Muon CP group. These were provided in three different data-taking periods. The final anti-ID scale factor is thus obtained by taking the luminosity-weighted average of the result for each period. The results are shown in Figs. 5.9 and 5.10, for muons with p_T greater than or smaller than 15 GeV, respectively. For electrons, anti-ID scale factors have been directly provided, and are shown in Fig. 5.11.

These anti-ID scale factors can be significant, and therefore impact the ZZ predicted yield in the signal region. To assess this impact, the anti-ID scale factors are applied to ZZ events, depending on the truth kinematics of the forth non-identified lepton, which are shown in Figs. 5.12 and 5.13, for electrons and muons respectively.

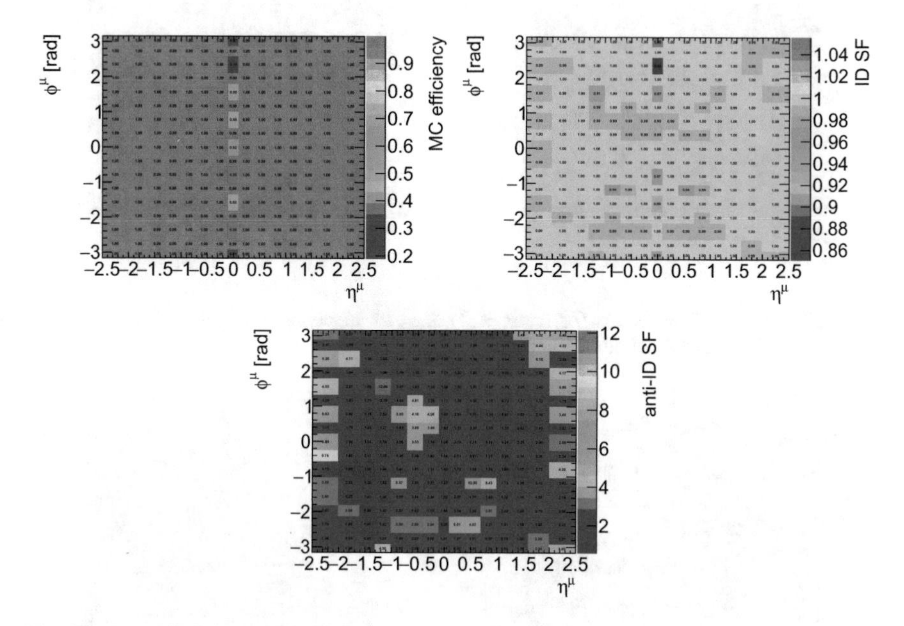

Fig. 5.9 ID scale factors, MC efficiencies and anti-ID scale factors for loose muons with $p_T >$ 15 GeV, as a function of η and ϕ

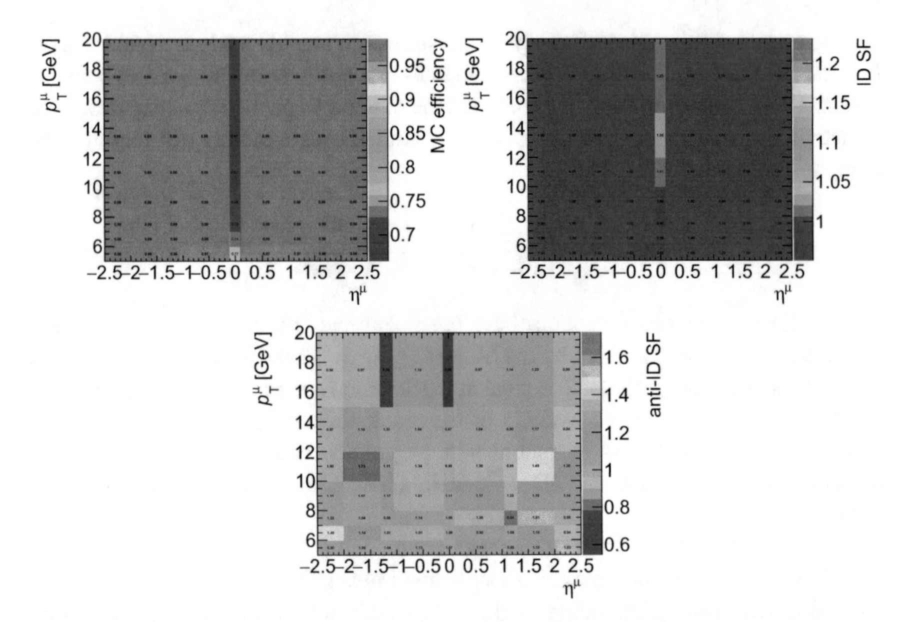

Fig. 5.10 ID scale factors, MC efficiencies and anti-ID scale factors for loose muons with $p_T <$ 15 GeV, as a function of η and p_T

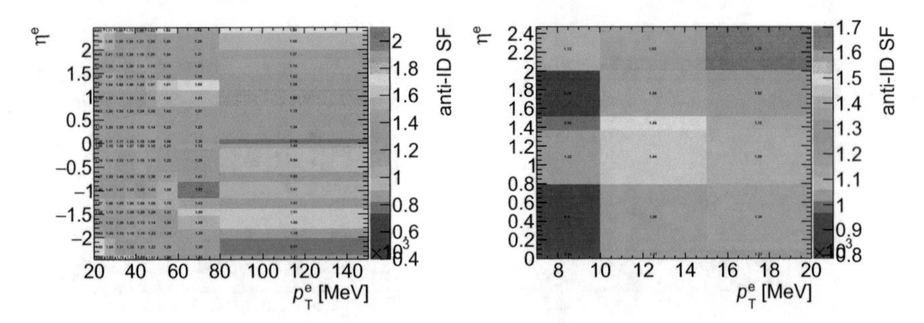

Fig. 5.11 Anti-ID scale factors for `LooseLH+BLayer` electrons with $p_T > 20$ GeV (**a**) and $p_T < 20$ GeV (**b**), as a function of η and p_T

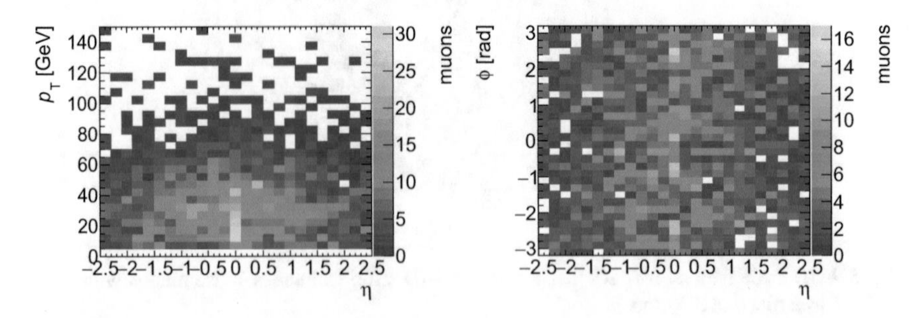

Fig. 5.12 Truth kinematics of the unidentified fourth muons in ZZ events (raw MC events)

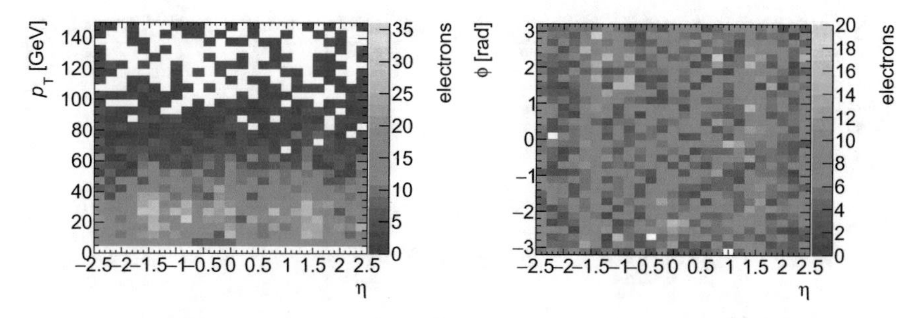

Fig. 5.13 Truth kinematics of the unidentified fourth electrons in ZZ events (raw MC events)

Table 5.8 Predicted ZZ yields after the full WZ selection, with and without anti-ID scale factors applied to unidentified fourth lepton within the acceptance

ZZ prediction	eee	$ee\mu$	$\mu\mu e$	$\mu\mu\mu$	all
no anti-ID SFs	5.38	6.96	7.22	9.57	28.91
with anti-ID SFs	5.87	8.15	7.99	11.13	32.84
relative difference [%]	+9.1	+17.0	+10.1	+16.3	+13.6

The impact of applying anti-ID scale factors on the predicted ZZ yield is summarized in Table 5.8. A 14% increase of the ZZ yield in the signal region is observed when considering the combination of all flavor channels.

The shapes of basic kinematic distributions of ZZ events in the signal region, before and after anti-ID scale factors are applied, are compared in Figs. 5.14, 5.15. As demonstrated by the comparisons, the anti-ID SFs impact mostly the overall ZZ normalization, with no appreciable effect on the shapes of the distributions.

Following these results, the ZZ yield predicted by MC in the signal region is corrected with anti-ID scale factors. The full size of the correction is assigned as systematic uncertainty on the procedure, i.e. $2.3 - 5.9\%$ depending on the flavor channel.

5.6.3.2 ZZ Validation Region

A ZZ validation region (VR) is defined to validate the modeling of the ZZ background. The ZZ validation region is the same as the WZ signal region, defined in Table 5.4 except the ZZ veto requirement is reverted and the criteria of $m_{\mathrm{T}}^{W} > 30$ GeV is removed to increase statistics. Just as in the signal region, the ZZ VR requires two leptons that are same flavor, opposite charge with the invariant mass consistent with the mass of the Z boson, $|m_{\ell\ell} - m_Z| < 10$ GeV. If there are more than one pair of leptons that form a Z candidate, the candidate with the invariant mass closes to the Z mass is taken. Two additional leptons are required in the event with $p_{\mathrm{T}} > 20$

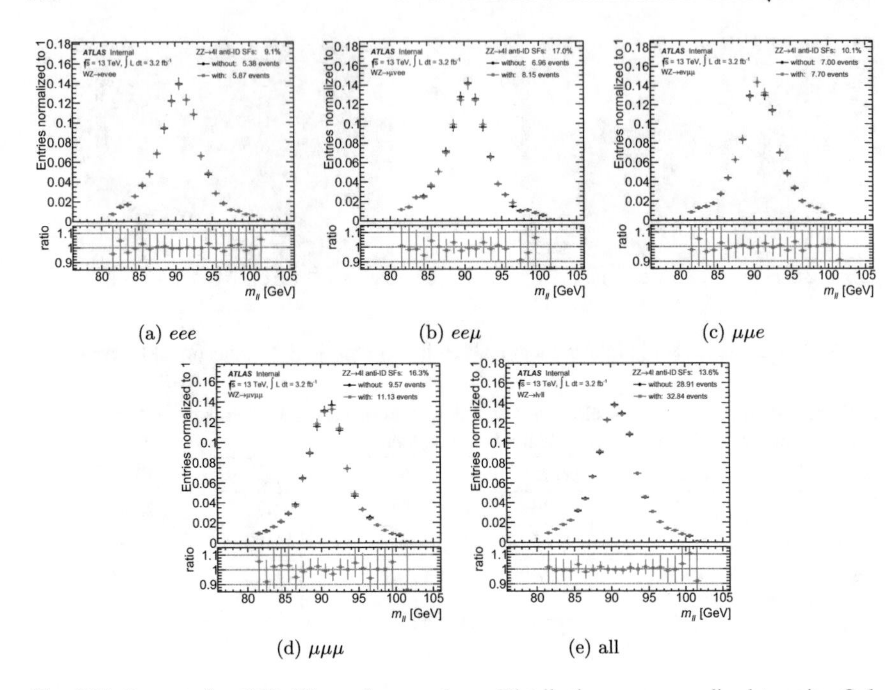

(a) eee (b) $ee\mu$ (c) $\mu\mu e$

(d) $\mu\mu\mu$ (e) all

Fig. 5.14 Impact of anti-ID SFs on the $m_{\ell\ell}$ shape. Distributions are normalized to unity. Only statistical uncertainties are shown

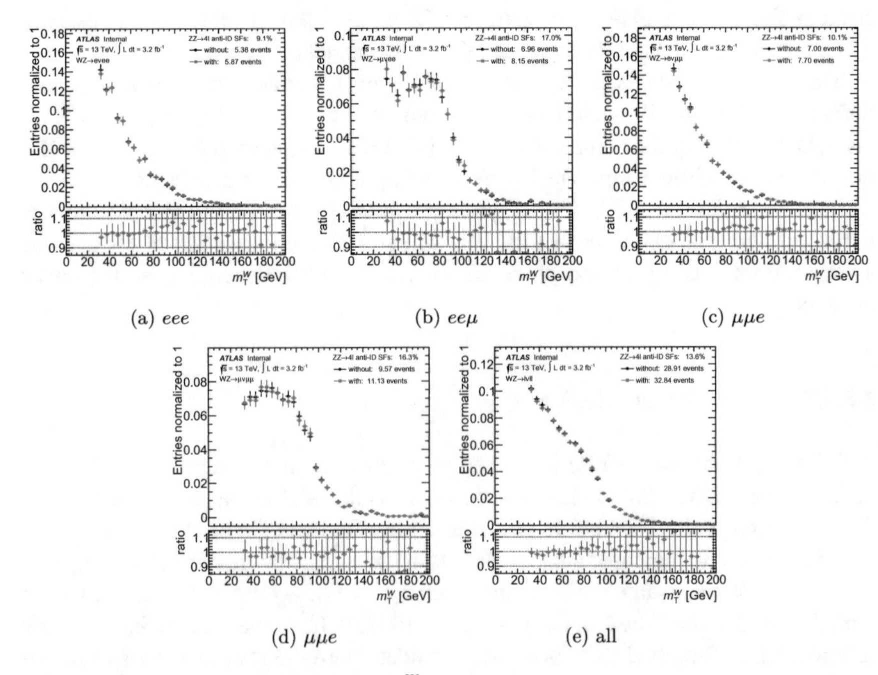

(a) eee (b) $ee\mu$ (c) $\mu\mu e$

(d) $\mu\mu e$ (e) all

Fig. 5.15 Impact of anti-ID SFs on the m_{T}^{W} shape. Distributions are normalized to unity. Only statistical uncertainties are shown

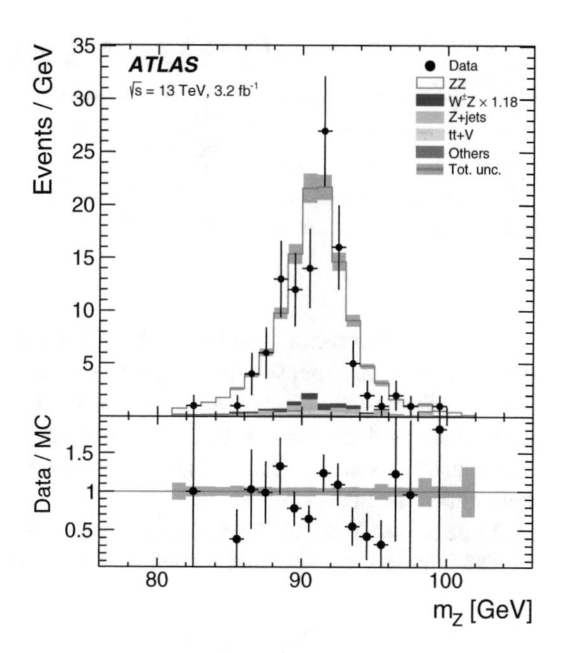

Fig. 5.16 The mass of the Z boson distribution in the ZZ validation region. Z+jets is estimated using Monte Carlo. Backgrounds labeled "Others" consist of $t\bar{t}$, Wt, and VVV processes. The WZ process is scaled by 1.18 to match the inclusive cross section

and $p_{\text{T}} > 7$ GeV, respectively. These additional leptons must satisfy the W lepton criteria, defined in Table 5.3.

The data events passing this selection are compared to the Monte Carlo prediction. Simulation is used to estimate all backgrounds, including the Z+jets/$Z + \gamma$, and top backgrounds. This region is used to check the theoretical prediction of the ZZ process, not the amount of anti-ID leptons. The mass of the Z boson formed by the lepton pair with the invariant mass closest to m_Z is shown in Fig. 5.16. This validation region has 91.8 ± 0.4 ZZ events, with a total simulation prediction of 103.4 ± 1.2 events. The number of observed events is 106. Based on the yields and distribution, the ZZ Monte Carlo is well-modeled.

In addition to the uncertainty from applying anti-ID scale factors, an 8% uncertainty is attributed to the ZZ background which is the theoretical uncertainty on the ZZ cross section.

5.7 Correction for Acceptance and Detector Effects

To calculate the cross section for each channel, a correction factor is calculated for eee, $ee\mu$, $\mu\mu e$, and $\mu\mu\mu$, which extrapolates between the number of reconstructed events to the number of true events in the fiducial region. This process was first described in Sect. 5.1.

$$C_{WZ} = \frac{N^{\text{MC}}_{\text{reco}}}{N^{\text{MC}}_{\text{fiducial}}} \tag{5.31}$$

Table 5.9 C_{WZ} and N_{τ}/N_{all} for each decay channel. Errors are statistical

Channel	C_{WZ}	N_{τ}/N_{all}
eee	0.421 ± 0.003	0.040 ± 0.001
$ee\mu$	$0.553 \pm 0.0.04$	0.038 ± 0.001
$\mu\mu e$	0.552 ± 0.004	0.036 ± 0.001
$\mu\mu\mu$	0.732 ± 0.005	0.040 ± 0.001

The correction factor C_{WZ} is calculated using POWHEG +PYTHIA events, which gives information about whether a lepton originates from a W or Z boson. The kinematics of the particle-level particles are constructed using dressed leptons and lepton assignment is done through the resonant shape algorithm, described in Sect. 5.1. Kinematics cuts are applied that are identical to those applied to reconstructed leptons, shown in Table 5.4.

Table 5.9 summarizes the correction factors derived for each channel.

The fiducial cross section for each channel is calculated as,

$$\sigma^{\text{fid.}}_{W^{\pm}Z\rightarrow\ell'\nu\ell\ell} = \frac{N_{\text{data}} - N_{\text{bkg}}}{\mathcal{L} \cdot C_{WZ}} \times \left(1 - \frac{N_{\tau}}{N_{\text{all}}}\right). \tag{5.32}$$

The final factor is this calculations, $\left(1 - \frac{N_{\tau}}{N_{\text{all}}}\right)$ corrects for the fraction of events where a tau lepton decays leptonically ($\tau \rightarrow \ell\nu$) and enters the fiducial region. These states are indistinguishable from final state leptons from prompt WZ decays and therefore contribute to the signal regions of the four channels used in the measurement region. Thus, the addition of this factors accounts for this. This tau fraction is calculated in each channel, summarized in Table 5.9, and correspond to 4% in each channel.

The fiducial cross section calculated in each channel is combined using a χ^2 minimization method [23]. This procedure assumes that there are common values m_i for each lepton flavor, ℓ, across i channels. The χ^2 function also takes into account the correlated systematics for each channel. It is defined as

$$\chi^2_{\text{exp}}(\boldsymbol{m}, \boldsymbol{b}) = \sum_{\ell}\sum_{i} \frac{\left[m^i - \sum_j \gamma^i_{j,\ell} m^i b_j - \mu^i_{\ell}\right]^2}{\delta^2_{i,\text{stat},\ell}\, \mu^i_{\ell}\left(m^i - \sum_j \gamma^i_{j,\ell} m^i b_j\right) + \left(\delta_{i,\text{uncor},\ell}\, m^i\right)^2} + \sum_j b_j^2. \tag{5.33}$$

where μ^i_{ℓ} is the measured cross section in channel i, $\delta_{i,\text{uncor},\ell}$ is the total relative uncorrelated systematic uncertainty, $\gamma^i_{j,\ell}$ are the relative systematic uncertainties that are correlated across channels, and$\delta^2_{i,\text{stat},\ell}$ is the relative statistical uncertainty. Nuisance parameters, b, are centered at zero and have a standard deviation of one; the term $\sum_j b_j^2$ is the nuisance parameter penalty term.

The total cross section is calculated from the the combined fiducial cross section as

$$\sigma^{\text{tot}}_{WZ} = \frac{\sigma^{\text{fid.}}_{W^{\pm}Z\to\ell'\nu\ell\ell}}{\text{BR}_{W\to\ell\nu}\cdot\text{BR}_{Z\to\ell\ell}\cdot\text{A}_{WZ}} \tag{5.34}$$

A_{WZ} is the acceptance factor, described in Sect. 5.1, and shown in equation (5.5) which calculates at particle level the ratio of the number of events in the fiducial region to the number of events in the total phase space, shown in Table 5.1. The single acceptance factor, $\text{A}_{WZ} = 0.343 \pm 0.02$ (stat.), is estimated using POWHEG + PYTHIA simulation using μee and $e\mu\mu$ events to avoid ambiguity in lepton assignment.

$B_W = 10.86 \pm 0.09$ % and $B_Z = 3.3658 \pm 0.0023$ % are the W and Z boson lepton branching fractions, respectively [24].

5.8 Systematic Uncertainties

The systematic uncertainties in the cross section are due to experimental uncertainties for detector effects, theoretical uncertainties in the acceptance in the fiducial region and extrapolation to the total phase space, and uncertainties in the background estimation.

Experimental systematic uncertainties are obtained by repeating the analysis after applying variations for each systematic source. The largest uncertainty is from the Z+jet/$Z+\gamma$ background; these uncertainties are summarized in Sect. 5.6.1. Experimental uncertainties include uncertainties in the scale and resolution of the electron energy, muon momentum, jet energy, and $E_{\text{T}}^{\text{miss}}$, as well as uncertainties applied to reproduce the trigger, reconstruction, and identification. Jet uncertainties enter into the calculation of the $E_{\text{T}}^{\text{miss}}$, as well as the measurement of the jet multiplicity distribution [25].

The theoretical uncertainties in the A_{WZ} and C_{WZ} factors consist theoretical uncertainties on the WZ prediction which is related to the choice of PDF set, QCD renormalization, μ_R, and factorization, μ_F, scales, and to the parton showering. The uncertainties due to the choice of PDF are computed using the CT10 eigenvectors and the envelope of the difference between the CT10 and CT14 [26], MMHT2014 [27], and NNPDF 3.0 [28] PDF sets, according to the PDF4LHC recommendations [29]. The QCD scale uncertainties are calculated by varying the nominal μ_F and μ_R by a factor of 2 around the nominal scale $m_{WZ}/2$ for all variations satisfying $0.5 \leq \mu_R/\mu_F \leq 2$. Parton shower uncertainties are obtained by showering the Powheg simulation with both PYTHIA and SHERPA and using the difference as an uncertainty; the uncertainty here is taken from simulation studies for the 8 TeV measurement [7]. Theoretical uncertainties are negligible for C_{WZ}; the uncertainty on A_{WZ} is less than 0.5% due to the PDF choice, and less than 0.7% due to QCD scales.

Table 5.10 Summary of the relative uncertainties in the measured fiducial cross section for each channel and for the combination

	eee	μee	$e\mu\mu$	$\mu\mu\mu$	Combined
Relative uncertainties (%)					
e energy scale	0.5	0.2	0.3	< 0.1	0.2
e id. efficiency	1.4	1.1	0.6	–	0.7
μ momentum scale	< 0.1	< 0.1	< 0.1	0.1	< 0.1
μ id. efficiency	–	0.6	1.0	1.4	0.7
$E_{\mathrm{T}}^{\mathrm{miss}}$ and jets	0.3	0.4	0.8	0.7	0.6
Trigger	< 0.1	0.1	0.1	0.2	0.1
Pile-up	0.7	1.1	1.0	0.7	0.9
Misid. lepton background	10	4.6	4.8	3.2	3.6
ZZ background	1.0	0.7	0.6	0.7	0.7
Other backgrounds	0.5	0.5	0.3	0.3	0.4
Uncorrelated	2.2	1.3	1.4	1.7	0.8
Total sys. uncertainty	11	5.1	5.3	4.1	4.1
Luminosity	2.4	2.4	2.3	2.3	2.4
Statistics	14	11	10	8.8	5.1
Total	18	12	11	10	7.0

Theoretical uncertainties on ZZ include the uncertainty on the ZZ cross section of 8% [30–33] and an uncertainty of 3–6% due to the correction applied to ZZ simulation events with unidentified leptons, described in Sect. 5.6.3.1.

The uncertainty due to irreducible background sources is evaluated by propagating the uncertainty on their cross sections: 13% (12% for $t\bar{t}W$ ($t\bar{t}Z$) [34], 20% for VVV [35], and 15% for tZ [7].

A 2.1% uncertainty is applied to the integrated luminosity.

The systematic uncertainties on the measurement are summarized in Table 5.10 and vary between 4% and 10% for the four channels.

5.9 Results

Table 5.11 summarizes the expected and observed number of events. The total uncertainties include experimental, theoretical, and statistical uncertainties.

Figure 5.17 shows detector level distributions . The WZ POWHEG +PYTHIA prediction is scaled by 1.18 to match the measured WZ cross section. These kine-

Table 5.11 Observed and expected number of events after WZ inclusive selection. Uncertainties include statistical, theoretical, and experimental uncertainties

Channel	eee	μee	$e\mu\mu$	$\mu\mu\mu$	All
Data	98	122	166	183	569
Total expected	102 ± 10	118 ± 9	126 ± 11	160 ± 12	506 ± 38
WZ	74 ± 6	96 ± 8	97 ± 8	129 ± 10	396 ± 32
$Z+j$, $Z+\gamma$	16 ± 7	7 ± 5	14 ± 7	9 ± 5	45 ± 17
ZZ	6.7 ± 0.7	8.7 ± 1.0	8.5 ± 0.9	11.7 ± 1.2	36 ± 4
$t\bar{t}$+V	2.7 ± 0.4	3.2 ± 0.4	2.9 ± 0.4	3.4 ± 0.5	12.1 ± 1.6
$t\bar{t}$, Wt, WW+j	1.2 ± 0.8	2.0 ± 0.9	2.4 ± 0.9	3.6 ± 1.5	9.2 ± 3.1
tZ	1.28 ± 0.20	1.65 ± 0.26	1.63 ± 0.26	2.12 ± 0.34	6.7 ± 1.1
VVV	0.24 ± 0.04	0.29 ± 0.05	0.27 ± 0.04	0.34 ± 0.05	1.14 ± 0.18

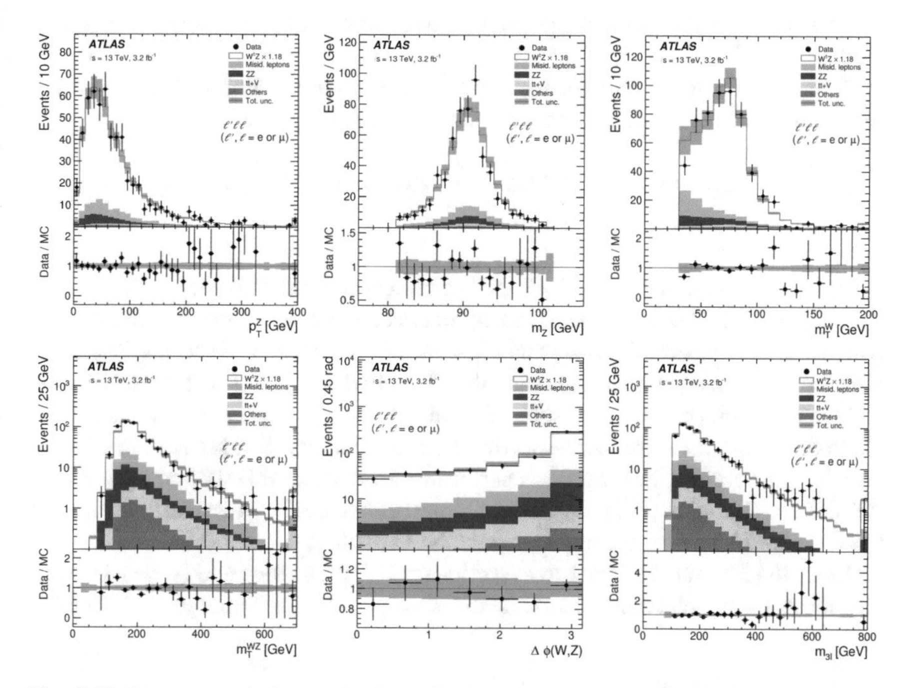

Fig. 5.17 Reconstructed detector-level distributions in the WZ signal region. The POWHEG +PYTHIA8 MC prediction is used for the WZ signal contribution, scaled by a global factor of 1:18 to match the measured inclusive WZ cross section

Table 5.12 Fiducial integrated cross section measured in each of the channels and combined. The uncertainties are given as percentages

Channel	$\sigma^{\text{fid.}}$ (fb)	$\delta_{\text{stat.}}$ (%)	$\delta_{\text{sys.}}$ (%)	$\delta_{\text{lumi.}}$ (%)	$\delta_{\text{tot.}}$ (%)
$\sigma^{\text{fid.}}_{W^{\pm}Z\to\ell'\nu\ell\ell}$					
$e^{\pm}ee$	50.5	14.2	10.6	2.4	17.8
$\mu^{\pm}ee$	55.1	11.1	5.1	2.4	12.4
$e^{\pm}\mu\mu$	75.2	9.5	5.3	2.3	11.1
$\mu^{\pm}\mu\mu$	63.6	8.9	4.1	2.3	10.0
Combined	63.2	5.2	4.1	2.4	7.0
SM prediction	53.4	–	–	–	6.0

matic distributions show good agreement between data and the expected background prediction.

After combining all four channels using the χ^2 minimization method, the cross section for the WZ production in the detector fiducial region is

$$\begin{aligned}\sigma^{\text{fid.}}_{W^{\pm}Z\to\ell'\nu\ell\ell} &= 63.2\,\pm\,3.2\,(\text{stat.})\,\pm\,2.6\,(\text{sys.})\,\pm\,1.5\,(\text{lumi.})\text{ fb}\\ &= 63.2\,\pm\,4.4\text{ fb}\end{aligned} \tag{5.35}$$

By comparison, the SM NLO QCD prediction from POWHEG +PYTHIA is $53.4^{1.6}_{-1.2}(\text{PDF})^{2.1}_{-1.6}(\text{scale})$ fb. The theoretical uncertainties are calculated in the same manner as the ones on the A_{WZ} and C_{WZ} factors described in Sect. 5.8. The theoretical predictions are estimated using the CT10 PDF set and QCD renormalization and factorization scales. The results of the fiducial cross section are summarized in Table 5.12. The measured cross section is larger than the SM prediction, as were also the cross section measurements performed lower center-of-mass energies by the ATLAS collaboration [6, 7]. Figure 5.18 shows the channel-by-channel comparisons between the 13 TeV measurement and the NLO prediction

Using the integrated fiducial cross section for WZ production at $\sqrt{(s)} = 8$TeV [7], the ratio of the fiducial cross sections at different center of mass energies is calculated as

$$\begin{aligned}\frac{\sigma^{\text{fid. PS,13 TeV}}_{W^{\pm}Z}}{\sigma^{\text{fid. PS,8 TeV}}_{W^{\pm}Z}} &= 1.80 \pm 0.10\,(\text{stat.}) \pm 0.08\,(\text{sys.}) \pm 0.06\,(\text{lumi.})\\ &= 1.80 \pm 0.14\,.\end{aligned} \tag{5.36}$$

The uncertainties are treated as uncorrelated between the measurements at the two beam energies. The measured ratio is in good agreement with the SM prediction of 1.78 ± 0.03 from POWHEG.

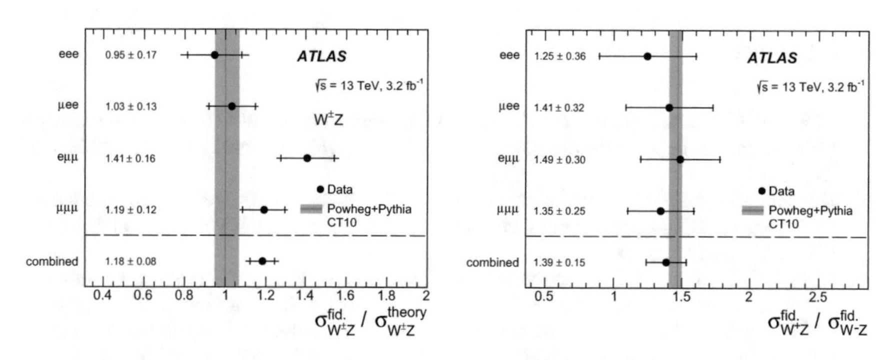

Fig. 5.18 Left: ratio of the measured WZ cross section in the fiducial phase space to the NLO SM prediction from POWHEG +PYTHIA. Right: W^+Z/W^-Z fiducial cross section ratio, compared to the NLO prediction

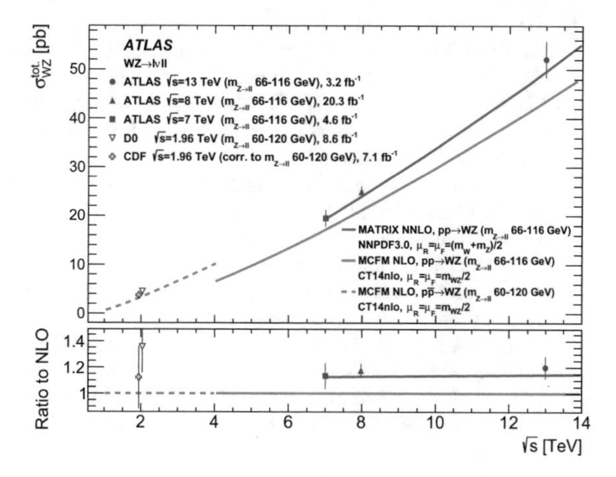

Fig. 5.19 WZ cross section measurements at various center-of-mass energies with Standard Model expectations

The combined fiducial cross section is extrapolated to the total phase space, resulting in

$$\begin{aligned}\sigma^{\text{tot.}}_{W^\pm Z\to\ell'\nu\ell\ell} &= 50.6\ \pm\ 2.6\,(\text{stat.})\ \pm\ 2.0\,(\text{sys.})\ \pm\ 0.9\ \text{th.}\ \pm\ 1.2\,(\text{lumi.})\ \text{fb}\\ &= 50.6\ \pm\ 3.6\ \text{fb}\end{aligned} \tag{5.37}$$

The NLO SM prediction calculated with POWHEG is 42.4 ± 0.8(PDF) ± 1.6(scale) pb. The calculation of the WZ production cross section at NNLO in QCD with MATRIX [36] yields $48.2^{+1.1}_{-1.0}$(scale) pb, which is in better agreement with the measurement. Figure 5.19 shows the comparison between the ATLAS WZ measurements at ps = 7, 8, and 13 TeV, comparing with the NLO predictions in pp and $p\bar{p}$ collisions, and with the newest pp NNLO prediction.

References

1. CDF Collaboration, Aaltonen et al T (2012) Measurement of the WZ cross section and triple gauge couplings in $p\bar{p}$ collisions at $\sqrt{s} = 1.96$ TeV. Phys Rev D86:031104. arXiv:1202.6629 [hep-ex]
2. D0 Collaboration, Abazov VM et al, (2012) A measurement of the WZ and ZZ production cross sections using leptonic final states in 8.6 fb $^{-1}$ of $p\bar{p}$ collisions. Phys Rev D 85:112005. arXiv:1201.5652 [hep-ex]
3. CMS Collaboration (2013) Measurement of the sum of WW and WZ production with $W+$ dijet events in pp collisions at $\sqrt{s} = 7$ TeV. Eur Phys J C73(2):2283. arXiv:1210.7544 [hep-ex]
4. CMS Collaboration (2014) Measurement of WZ and ZZ production in pp collisions at $\sqrt{s} =$ 8 TeV in final states with b-tagged jets. Eur Phys J C74(8):2973. arXiv:1403.3047 [hep-ex]
5. Collaboration CMS (2017) Measurement of the WZ production cross section in pp collisions at $\sqrt{(s)} = 13$ TeV. Phys Lett B 766:268–290 arXiv:1607.06943 [hep-ex]
6. ATLAS Collaboration (2012) Measurement of WZ production in proton-proton collisions at $\sqrt{s} = 7$ TeV with the ATLAS detector. Eur Phys J C72:2173. arXiv:1208.1390 [hep-ex]
7. ATLAS Collaboration (2016) Measurements of $W^{\pm}Z$ production cross sections in pp collisions at $\sqrt{s} = 8$ TeV with the ATLAS detector and limits on anomalous gauge boson self-couplings. Phys Rev D93(9):092004. arXiv:1603.02151 [hep-ex]
8. ATLAS Collaboration (2016) Measurement of the $W^{\pm}Z$ boson pair-production cross section in pp collisions at $\sqrt{s} = 13$ TeV with the ATLAS Detector. Phys Lett B762:1–22. arXiv:1606.04017 [hep-ex]
9. GEANT4 Collaboration, Agostinelli S, et al (2003) GEANT4—a simulation toolkit. Nucl Instrum Meth A 506:250
10. ATLAS Collaboration (2010) The ATLAS simulation infrastructure. Eur Phys J C 70:823. arXiv:1005.4568 [physics.ins-det]
11. Nason P (2004) A New method for combining NLO QCD with shower Monte Carlo algorithms. JHEP 11:040. arXiv:hep-ph/0409146 [hep-ph]
12. Frixione S, Nason P, Oleari C (2007) Matching NLO QCD computations with parton shower simulations: the POWHEG method. JHEP 11:070 arXiv:0709.2092 [hep-ph]
13. Alioli S, Nason P, Oleari C, Re E (2010) A general framework for implementing NLO calculations in shower Monte Carlo programs: the POWHEG BOX. JHEP 06:043 arXiv:1002.2581 [hep-ph]
14. Melia T, Nason P, Rontsch R, Zanderighi G (2011) W+W-, WZ and ZZ production in the POWHEG BOX. JHEP 11:078. arXiv:1107.5051 [hep-ph]
15. Sjöstrand T, Mrenna S, Skands PZ (2008) A brief introduction to PYTHIA 8.1. Comput Phys Commun 178:852–867. arXiv:0710.3820 [hep-ph]
16. Lai H-L, Guzzi M, Huston J, Li Z, Nadolsky PM, Pumplin J, Yuan CP (2010) New parton distributions for collider physics. Phys Rev D 82:074024 arXiv:1007.2241 [hep-ph]
17. Pumplin J et al (2002) New generation of parton distributions with uncertainties from global QCD analysis. JHEP 07:012 arXiv:hep-ph/0201195
18. Gleisberg T, Höche S, Krauss F, Schönherr M, Schumann S, Siegert F, Winter J (2009) Event generation with SHERPA 1.1. JHEP 02:007. arXiv:0811.4622 [hep-ph]
19. Schumann S, Krauss F (2008) A Parton shower algorithm based on Catani-Seymour dipole factorisation. JHEP 03:038 arXiv:0709.1027 [hep-ph]
20. ATLAS Collaboration (2016) Performance of pile-up mitigation techniques for jets in pp collisions at $\sqrt{s} = 8$ TeV using the ATLAS detector. Eur Phys J C76(11):581. arXiv:1510.03823 [hep-ex]
21. ATLAS Collaboration (2012) Measurement of the WW cross section in $\sqrt{s} = 7$ TeV pp collisions with the ATLAS detector and limits on anomalous gauge couplings. Phys Lett B 712:289. arXiv:1203.6232 [hep-ex]
22. Prospects for higgs boson searches using the $H \to WW^{(*)} \to \ell\nu\ell\nu$ decay mode with the ATLAS detector for 10 TeV. Tech Rep ATL-PHYS-PUB-2010-005, CERN, Geneva, Jun, 2010. https://cds.cern.ch/record/1270568

23. H1 Collaboration, Aaron FD, et al (2009) Measurement of the inclusive ep scattering cross section at low Q^2 and x at HERA. Eur Phys J C63:625–678. arXiv:0904.0929 [hep-ex]
24. Olive KA, et al (Particle Data Group) (2014) Review of particle physics. Chinese Phys C 38(9):090001
25. The ATLAS collaboration (2015) Jet calibration and systematic uncertainties for jets reconstructed in the ATLAS detector at $\sqrt{s} = 13$ TeV. Tech Rep ATL-PHYS-PUB-2015-015, CERN, Geneva, July, 2015. https://cds.cern.ch/record/2037613
26. Dulat S, Hou T-J, Gao J, Guzzi M, Huston J, Nadolsky P, Pumplin J, Schmidt C, Stump D, Yuan CP (2016) New parton distribution functions from a global analysis of quantum chromodynamics. Phys Rev D 93(3):033006. arXiv:1506.07443 [hep-ph]
27. Harland-Lang LA, Martin AD, Motylinski P, Thorne RS (2015) Parton distributions in the LHC era: MMHT 2014 PDFs. Eur Phys J C 75(5):204. arXiv:1412.3989 [hep-ph]
28. NNPDF Collaboration, Ball RD et al (2015) Parton distributions for the LHC Run II. JHEP 04:040. arXiv:1410.8849 [hep-ph]
29. Butterworth J et al (2016) PDF4LHC recommendations for LHC Run II. J Phys G 43:023001 arXiv:1510.03865 [hep-ph]
30. Cascioli F, Gehrmann T, Grazzini M, Kallweit S, Maierhfer P, von Manteuffel A, Pozzorini S, Rathlev D, Tancredi L, Weihs E (2014) ZZ production at hadron colliders in NNLO QCD. Phys Lett B 735:311–313 arXiv:1405.2219 [hep-ph]
31. Bierweiler A, Kasprzik T, Khn JH (2013) Vector-boson pair production at the LHC to $\mathcal{O}(\alpha^3)$ accuracy. JHEP 12:071 arXiv:1305.5402 [hep-ph]
32. Baglio J, Ninh LD, Weber MM (2013) Massive gauge boson pair production at the LHC: a next-to-leading order story. Phys Rev D 88:113005. arXiv:1307.4331 [hep-ph]. [Erratum: Phys Rev D 94(9):099902 (2016)]
33. Caola F, Melnikov K, Rntsch R, Tancredi L (2015) QCD corrections to ZZ production in gluon fusion at the LHC. Phys Rev D92(9):094028. arXiv:1509.06734 [hep-ph]
34. Alwall J, Frederix R, Frixione S, Hirschi V, Maltoni F, Mattelaer O, Shao HS, Stelzer T, Torrielli P, Zaro M (2014) The automated computation of tree-level and next-to-leading order differential cross sections, and their matching to parton shower simulations. JHEP 07:079 arXiv:1405.0301 [hep-ph]
35. ATLAS Collaboration (2016) Multi-boson simulation for 13 TeV ATLAS analyses. ATL-PHYS-PUB-2016-002. https://cds.cern.ch/record/2119986
36. Grazzini M, Kallweit S, Rathlev D, Wiesemann M (2016) $W^{\pm}Z$ production at hadron colliders in NNLO QCD. Phys Lett B 761:179–183 arXiv:1604.08576 [hep-ph]

Chapter 6
Searches for Electroweak SUSY: Motivation and Models

6.1 Motivation for Searching for Electroweak SUSY

Even though there is no deviation in the standard model measurement and the theory prediction, as shown in Chap. 5, there are still open questions to the SM that SUSY can answer. In order to search for SUSY, the production cross sections of the various SUSY production modes must be calculated, as shown in Fig. 6.1.

The largest cross-sections are the squark ($\tilde{q}\tilde{q}$, $\tilde{t}\tilde{t}$) and gluino ($\tilde{g}\tilde{g}$) cross-sections [1, 2] and therefore are a natural first place to look for SUSY at the LHC. The ATLAS experiment has conducted many searches for gluino production [3–9] and for stop production [10–15] in Run 1 and with 36.1 fb^{-1} which did not produce any significant excesses. Limits exclude gluino masses up to $\sim$2 TeV and on stop masses up to 1 TeV for massless LSP, as shown in Fig. 6.2.

Thus, the next natural place to look for SUSY is from electroweak (EWK) production, the production of winos, Higgsinos, and sleptons. The gauginos, winos and Higgsinos, cross-sections [16–19] are much larger than the slepton cross sections [17, 18, 20, 21]. The wino cross-section is about four times larger than the Higgsino cross-section. This chapter will focus on motivating searches for both wino and Higgsino production. The electroweakino production at LHC occurs through s-channel production via W and Z bosons, as shown in Fig. 6.3. Searches at LHC focus on the lighter SUSY particles, $\tilde{\chi}_1^{\pm}\tilde{\chi}_1^{\mp}$ and $\tilde{\chi}_1^{\pm}\tilde{\chi}_2^{0}$.

The gaugino mass spectrum will determine if the SUSY particles produced at the LHC are winos and Higgsinos, as shown in Fig. 6.4. As discussed in Sect. 2.3, the neutral Higgsinos ($\tilde{H}_u^0$, $\tilde{H}_d^0$) and the neutral gauginos ($\tilde{B}$, $\tilde{W}^0$) combine to form four mass eigenstates called neutralinos while the charged Higgsinos ($\tilde{H}_u^+$, $\tilde{H}_d^-$) and the charged gauginos ($\tilde{W}^{\pm}$) combine to form four mass eigenstates called charginos [22]. By convention, these are labeled in ascending order by mass, so that $m_{\widetilde{N}_1} < m_{\widetilde{N}_2} < m_{\widetilde{N}_3} < m_{\widetilde{N}_4}$ and $m_{\widetilde{C}_1} < m_{\widetilde{C}_2}$. The lightest neutralino, $\widetilde{N}_1$ is considered the lightest SUSY particle (LSP).

© The Editor(s) (if applicable) and The Author(s), under exclusive license to Springer Nature Switzerland AG 2020

E. Resseguie, *Electroweak Physics at the Large Hadron Collider with the ATLAS Detector*, Springer Theses,
https://doi.org/10.1007/978-3-030-57016-3_6

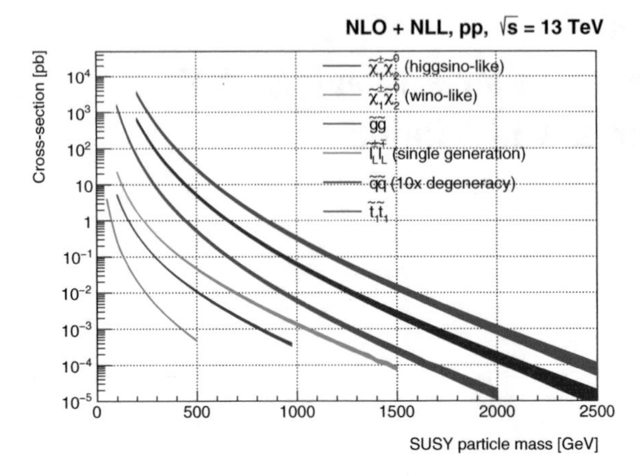

Fig. 6.1 SUSY production cross-sections at $\sqrt{s} = 13$ TeV

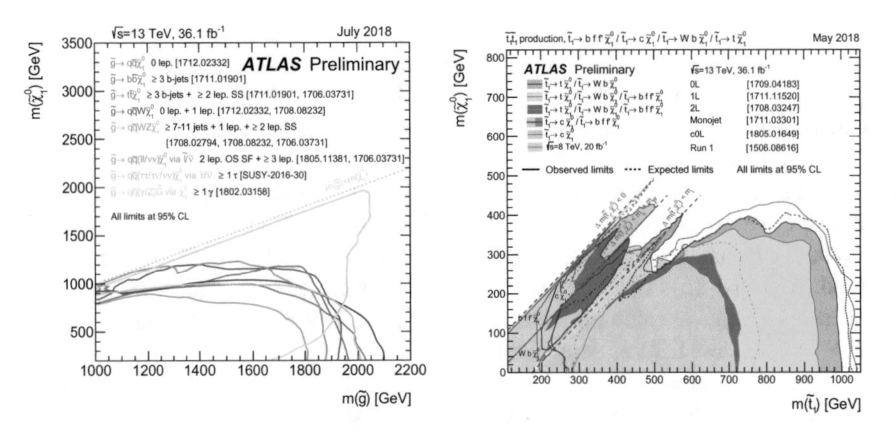

Fig. 6.2 Exclusion limits at 95% CL based on 13 TeV data in the mass of LSP vs. mass of gluino plane (left) and mass of LSP vs. mass of stop (right)

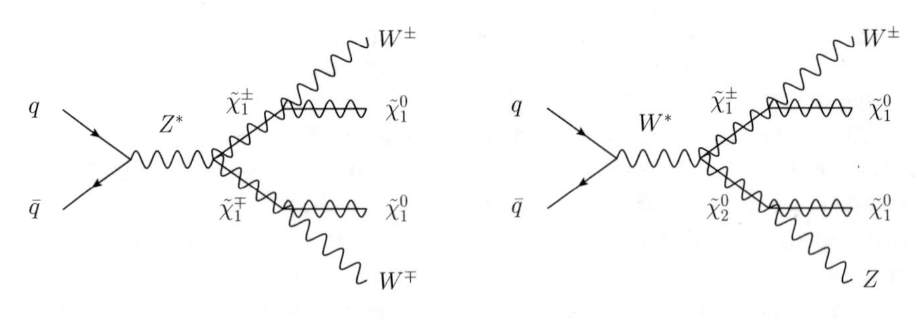

Fig. 6.3 Direct s-channel production and decay of $\tilde{\chi}_1^{\pm}\tilde{\chi}_1^{\mp}$ (left) and $\tilde{\chi}_1^{\pm}\tilde{\chi}_2^0$ (right)

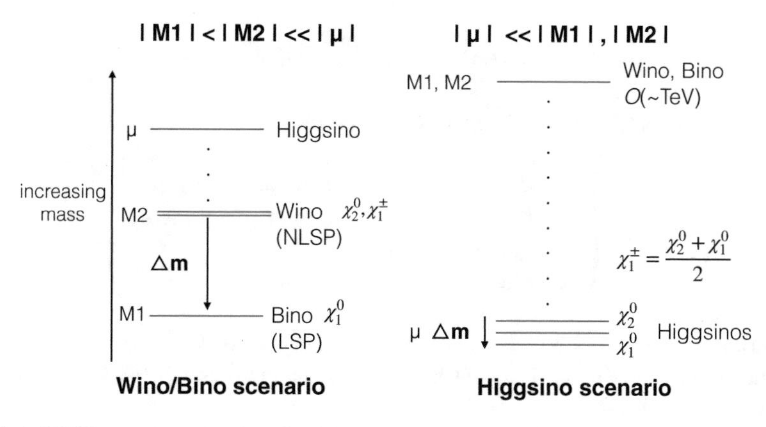

Fig. 6.4 SUSY mass spectrum for wino-bino and Higgsino models

For the wino production, the LSP is a bino and the next-to-lightest SUSY particles are mass-degenerate winos. The mass splitting between the NLSP and LSP is not constrained to be large or small. In the Higgsino cases, the three lightest particles are Higgsinos with the mass of the chargino being in between the masses of the two neutralinos. The mass splitting is small, resulting in a compressed spectrum. Winos and binos are decoupled. In order to understand this further, the derivation of the masses of the charginos and neutralinos must be considered.

In the gaugino basis, $\psi^\pm = \left(\tilde{W}^+, \tilde{H}_u^+, \tilde{W}^-, \tilde{H}_d^-\right)$, the chargino mass terms in the Lagrangian are given by

$$\mathcal{L}_{\text{chargino mass}} = -\frac{1}{2}\left(\psi^\pm\right)^T \mathbf{M}_{\tilde{C}} \psi^\pm + \text{c.c.}, \tag{6.1}$$

where $\mathbf{M}_{\tilde{C}}$, the chargino mass matrix is given by

$$\mathbf{M}_{\tilde{C}} = \begin{pmatrix} 0 & 0 & M_2 & \sqrt{2}m_W\cos(\beta) \\ 0 & 0 & \sqrt{2}m_W\sin(\beta) & \mu \\ M_2 & \sqrt{2}m_W\sin(\beta) & 0 & 0 \\ \sqrt{2}m_W\cos(\beta) & \mu & 0 & 0 \end{pmatrix}. \tag{6.2}$$

The masses of the charginos can be found by diagonalizing Eq. (6.2) with eigenvalues

$$m^2_{\tilde{C}_1}, m^2_{\tilde{C}_2} = \frac{1}{2}\left[|M_2|^2 + |\mu|^2 + 2m_W^2 \mp \sqrt{\left(|M_2|^2 + |\mu|^2 + 2m_W^2\right)^2 - 4\left|\mu M_2 - m_W^2 \sin 2\beta\right|^2}\right] \tag{6.3}$$

Similarly, in the neutral gaugino basis, $\psi^0 = \left(\widetilde{B}, \widetilde{W}^0, \widetilde{H}_d^0, \widetilde{H}_u^0\right)$, the neutralino mass terms in the Lagrangian are given by

$$\mathcal{L}_{\text{neutralino mass}} = -\frac{1}{2}\left(\psi^0\right)^T \mathbf{M}_{\widetilde{N}}\psi^0 +, \text{c.c.} \tag{6.4}$$

where $\mathbf{M}_{\widetilde{N}}$, the neutralino mass matrix is given by

$$\mathbf{M}_{\widetilde{N}} = \begin{pmatrix} M_1 & 0 & -c_\beta s_W m_Z & s_\beta s_W m_Z \\ 0 & M_2 & c_\beta c_W m_Z & -s_\beta c_W m_Z \\ -c_\beta s_W m_Z & c_\beta c_W m_Z & 0 & -\mu \\ s_\beta s_W m_Z & -s_\beta c_W m_Z & -\mu & 0 \end{pmatrix}, \tag{6.5}$$

where $s_\beta = \sin\beta$, $c_\beta = \cos\beta$, $s_W = \sin\theta_W$, and $c_W = \cos\theta_W$.

The masses of the neutralinos can be found by diagonalizing Eq. (6.5) using unitary matrix, N, which satisfies

$$\mathbf{N}^*\mathbf{M}_{\widetilde{N}}\mathbf{N}^{-1} = \begin{pmatrix} m_{\widetilde{N}_1} & 0 & 0 & 0 \\ 0 & m_{\widetilde{N}_2} & 0 & 0 \\ 0 & 0 & m_{\widetilde{N}_3} & 0 \\ 0 & 0 & 0 & m_{\widetilde{N}_4} \end{pmatrix}. \tag{6.6}$$

This diagonalization is much more complicated than for the chargino mass matrix.

6.2 Wino-Bino Interpretation

This scenario assumes that m_Z is small compared to M_1, M_2, and μ, such that $m_Z \ll |\mu \pm M_1|, |\mu \pm M_2|$. Since m_Z is small, terms proportional to m_Z can be set to 0 to leading order with corrections treated using perturbation theory. The neutralino masses from $\mathbf{M}_{\widetilde{N}}$ are then:

$$m_{\widetilde{N}_1} = M_1 \tag{6.7}$$

$$m_{\widetilde{N}_2} = M_2 \tag{6.8}$$

$$m_{\widetilde{N}_3}, m_{\widetilde{N}_4} = |\mu| \tag{6.9}$$

Similarly, in the chargino case,

$$m^2_{\widetilde{C}_1}, m^2_{\widetilde{C}_2} = \frac{1}{2}\left[|M_2|^2 + |\mu|^2 \mp \sqrt{\left(|M_2|^2 + |\mu|^2\right)^2 - 4\,|\mu M_2|^2}\right] \tag{6.10}$$

$$m^2_{\widetilde{C}_1}, m^2_{\widetilde{C}_2} = \frac{1}{2}\left[|M_2|^2 + |\mu|^2 \mp \sqrt{(|M_2|^2)^2 + (|\mu|^2)^2 - 2\,|\mu M_2|^2}\right] \tag{6.11}$$

$$m^2_{\widetilde{C}_1}, m^2_{\widetilde{C}_2} = \frac{1}{2}\left[|M_2|^2 + |\mu|^2 \mp \sqrt{\left||M_2|^2 - |\mu|^2\right|^2}\right] \tag{6.12}$$

$$m^2_{\widetilde{C}_1}, m^2_{\widetilde{C}_2} = \frac{1}{2}\left[|M_2|^2 + |\mu|^2 \mp \left||M_2|^2 - |\mu|^2\right|\right] \tag{6.13}$$

Since $|\mu| > M_2$,

$$m^2_{\widetilde{C}_1}, m^2_{\widetilde{C}_2} = \frac{1}{2}\left[|M_2|^2 + |\mu|^2 \mp (|\mu|^2 - |M_2|^2)\right]. \tag{6.14}$$

The masses of the charginos are

$$m_{\widetilde{C}_1} = M_2 \tag{6.15}$$

$$m_{\widetilde{C}_2} = |\mu| \tag{6.16}$$

Thus, in this case, the mass spectrum is $M_1 < M_2 \ll \mu$, meaning that the LSP is a bino and the next-lightest SUSY particle (NLSP) is a wino with $\widetilde{C}_1$ and $\widetilde{N}_2$ being mass degenerate.

The motivation for a bino LSP is that it gives the correct dark matter relic density if co-annihilations are included [23, 24]. Co-annihilation occurs when the relic abundance of dark matter is not only determined by the annihilation cross-section of the LSP but also by the annihilation cross-section of the heavier particles, which later decay into the LSP. The bino annihilation cross-section is too small leading to an over-production of dark matter. In order to get the correct dark matter relic density, co-annihilation between winos and binos is considered [24–26]. In the early universe, winos and binos were in chemical equilibrium. The mass splittings between the winos and binos should not be too large such that a large amount of winos exist at the freeze-out of the bino. Since the wino annihilation cross section is larger than for binos, the binos are sufficiently suppressed to get the correct dark matter relic density.

Moreover, after the LHC Run 1, a parameter scan was completed using ATLAS and CMS limits [27]. The best-fit point was found to be a bino LSP with co-annihilations controlling the dark matter density. In this case, it was found that $\widetilde{C}_1$, $\widetilde{N}_2$, and $\widetilde{N}_1$ had very small splittings. Thus, searches for compressed wino-bino models are well motivated.

6.3 Higgsino Interpretation

While the wino-bino model provides a candidate for dark matter and the unification of couplings at the GUT scale, the Higgsino model is motivated by naturalness, specifically that the electroweak sector should have minimal fine tuning [28]. This requirement is summarized at tree level in the Minimal Supersymmetric Standard Model (MSSM):

$$-\frac{m_Z^2}{2} = |\mu|^2 + m^2_{H_u} \tag{6.17}$$

For minimal fine tuning, both μ and H_u should be on the order of the weak scale, or m_Z. Thus, the Higgsinos, who have mass parameter μ, should be light and the stop and gluino masses which correct $m^2_{H_u}$ at the one-loop order should also not be too heavy. The other SUSY particles are not important for naturalness and can be much heavier, meaning $\mu \ll M_1, M_2$ [28, 29].

Similar to the wino-bino case, a constraint on the mass parameters M_1, M_2, μ with respect to the mass of the Z boson can be applied, $m_Z \ll |M_1, M_2 \pm \mu|$, which treats electroweak corrections using perturbation theory. The neutralino masses are just as in the wino-bino case except that $\widetilde{N}_3, \widetilde{N}_4$ are the lighter particles:

$$m_{\widetilde{N}_1} = M_1 \tag{6.18}$$

$$m_{\widetilde{N}_2} = M_2 \tag{6.19}$$

$$m_{\widetilde{N}_3}, m_{\widetilde{N}_4} = |\mu|. \tag{6.20}$$

To keep the convention that $m_{\widetilde{N}_1}$ is the LSP, the masses can be rewritten as

$$m_{\widetilde{N}_1}, m_{\widetilde{N}_2} = |\mu| \tag{6.21}$$

$$m_{\widetilde{N}_3}, m_{\widetilde{N}_4} = M_1, M_2. \tag{6.22}$$

In the cases of the charginos, since $|\mu| < M2$, Eq. (6.14) becomes

$$m^2_{\widetilde{C}_1}, m^2_{\widetilde{C}_2} = \frac{1}{2}\left[|M_2|^2 + |\mu|^2 \mp (|M_2|^2 - |\mu|^2)\right]. \tag{6.23}$$

The masses of the charginos become

$$m_{\widetilde{C}_1} = |\mu| \tag{6.24}$$

$$m_{\widetilde{C}_2} = M_2. \tag{6.25}$$

Thus in the Higgsino model, there are three semi-degenerate Higgsino states, $\widetilde{N}_1, \widetilde{N}_2, \widetilde{C}_1$ and decoupled $\widetilde{N}_3, \widetilde{N}_4, \widetilde{C}_2$ which are winos and binos. In the derivation, the Higgsino states all have mass parameter μ; however, realistically, they are not degenerate and have in fact have some splitting. To understand these splittings between the Higgsino state, more realistic assumptions can be used: $M_1 \gg M_2 > |\mu|$ and $M_2 \gg M_1 > |\mu|$ [30].

In the case, $M_1 \gg M_2 > |\mu|$, the heavy bino can be integrated out and when $m_W \ll M_2 \mp \mu$, the splitting between the Higgsino states is [30]

$$\left|m_{\chi_1^\pm}\right| - \left|m_{\chi_1^0}\right| \approx \frac{m_W^2\left(1 \mp s_{2\beta}\right)}{2\left(M_2 + |\mu|\right)} \tag{6.26}$$

$$\left|m_{\chi_2^0}\right| - \left|m_{\chi_1^\pm}\right| \approx \frac{m_W^2\left(1 \pm s_{2\beta}\right)}{2\left(M_2 - |\mu|\right)} \tag{6.27}$$

$$\left|m_{\chi_2^0}\right| - \left|m_{\chi_1^0}\right| \approx \frac{m_W^2\left(\pm|\mu|s_{2\beta} + M_2\right)}{\left(M_2^2 - |\mu|^2\right)}. \tag{6.28}$$

The other case, $M_2 \gg M_1 > |\mu|$, follows the same derivation: the heavy wino can be integrated out and $m_W \ll M_1 \mp \mu$ [30]. The Higgsino splittings are

$$\left|m_{\chi_1^\pm}\right| - \left|m_{\chi_1^0}\right| \approx \frac{m_W^2\left(1 \mp s_{2\beta}\right)}{2\left(M_1 + |\mu|\right)} \tag{6.29}$$

$$\left|m_{\chi_2^0}\right| - \left|m_{\chi_1^\pm}\right| \approx \frac{m_W^2\left(1 \pm s_{2\beta}\right)}{2\left(M_1 - |\mu|\right)} \tag{6.30}$$

$$\left|m_{\chi_2^0}\right| - \left|m_{\chi_1^0}\right| \approx \frac{m_W^2\left(\pm|\mu|s_{2\beta} + M_1\right)}{\left(M_1^2 - |\mu|^2\right)}. \tag{6.31}$$

Thus, the mass splittings for the Higgsino states can be summarized as

$$\left|m_{\chi_1^\pm}\right| - \left|m_{\chi_1^0}\right| \propto \frac{m_W^2}{2 \times \min\left(M_1, M_2\right)} \tag{6.32}$$

$$\left|m_{\chi_2^0}\right| - \left|m_{\chi_1^\pm}\right| \propto \frac{m_W^2}{2 \times \min\left(M_1, M_2\right)} \tag{6.33}$$

$$\left|m_{\chi_2^0}\right| - \left|m_{\chi_1^0}\right| \propto \frac{m_W^2}{\min\left(M_1, M_2\right)}. \tag{6.34}$$

If M_1, M_2 are on the order of 1 TeV, the resulting mass splittings are between 3 and 6 GeV, which gives rise to a naturally compressed mass spectrum with the the chargino being halfway between the two neutralinos.

While a Higgsino LSP provides a solution to the naturalness question, it does not provide a good dark matter candidate. The Higgsino annihilation cross-section is high, predicting an under-abundance of dark matter [31]. Many theories arose to explain how Higgsino could contribute to the dark matter relic density. One such example is that the dark matter relic density can be formed by an admixture of Higgsino and axions. or two particles composing dark matter [32, 33]. Another example is that the SUSY particles are a mixture of wino, bino, and Higgsino [34, 35] where the mixture leads to the correct prediction of dark matter density or dark matter is made up of multiple particles, one of which is the wino/bino/Higgsino mixed neutralino.

References

1. Borschensky C, Kramer M, Kulesza A, Mangano M, Padhi S, Plehn T, Portell X (2014) Squark and gluino production cross sections in pp collisions at $\sqrt{s} = 13$, 14, 33 and 100 TeV. Eur Phys J C 74:3174 arXiv:1407.5066 [hep-ph]
2. Beenakker W, Borschensky C, Krmer M, Kulesza A, Laenen E (2016) NNLL-fast: predictions for coloured supersymmetric particle production at the LHC with threshold and Coulomb resummation. JHEP 12:133 arXiv:1607.07741 [hep-ph]

3. Collaboration ATLAS (2018) Search for squarks and gluinos in final states with jets and missing transverse momentum using 36 fb^{-1} of $\sqrt{s} = 13$ TeV pp collision data with the ATLAS detector. Phys Rev D 97(11):112001 arXiv:1712.02332 [hep-ex]
4. ATLAS Collaboration (2018) Search for supersymmetry in final states with missing transverse momentum and multiple b-jets in proton-proton collisions at $\sqrt{s} = 13$ TeV with the ATLAS detector. JHEP 06, 107. arXiv:1711.01901 [hep-ex]
5. Collaboration ATLAS (2017) Search for supersymmetry in final states with two same-sign or three leptons and jets using 36 fb^{-1} of $\sqrt{s} = 13$ TeV pp collision data with the ATLAS detector. JHEP 09:084 arXiv:1706.03731 [hep-ex]
6. Collaboration ATLAS (2017) Search for squarks and gluinos in events with an isolated lepton, jets, and missing transverse momentum at $\sqrt{s} = 13$ TeV with the ATLAS detector. Phys Rev D 96(11):112010 arXiv:1708.08232 [hep-ex]
7. Collaboration ATLAS (2018) Search for new phenomena using the invariant mass distribution of same-flavour opposite-sign dilepton pairs in events with missing transverse momentum in $\sqrt{s} = 13$ TeV pp collisions with the ATLAS detector. Eur Phys J C78(8):625 arXiv:1805.11381 [hep-ex]
8. Collaboration ATLAS (2019) Search for squarks and gluinos in final states with hadronically decaying τ-leptons, jets, and missing transverse momentum using pp collisions at $\sqrt{s} = 13$ TeV with the ATLAS detector. Phys Rev D 99(1):012009 arXiv:1808.06358 [hep-ex]
9. Collaboration ATLAS (2018) Search for photonic signatures of gauge-mediated supersymmetry in 13 TeV pp collisions with the ATLAS detector. Phys Rev D 97(9):092006 arXiv:1802.03158 [hep-ex]
10. Collaboration ATLAS (2017) Search for a scalar partner of the top quark in the jets plus missing transverse momentum final state at $\sqrt{s}$=13 TeV with the ATLAS detector. JHEP 12:085 arXiv:1709.04183 [hep-ex]
11. Collaboration ATLAS (2018) Search for top-squark pair production in final states with one lepton, jets, and missing transverse momentum using 36 fb^{-1} of $\sqrt{s} = 13$ TeV pp collision data with the ATLAS detector. JHEP 06:108 arXiv:1711.11520 [hep-ex]
12. Collaboration ATLAS (2017) Search for direct top squark pair production in final states with two leptons in $\sqrt{s} = 13$ TeV pp collisions with the ATLAS detector. Eur Phys J C77(12):898 arXiv:1708.03247 [hep-ex]
13. Collaboration ATLAS (2018) Search for dark matter and other new phenomena in events with an energetic jet and large missing transverse momentum using the ATLAS detector. JHEP 01:126 arXiv:1711.03301 [hep-ex]
14. Collaboration ATLAS (2018) Search for supersymmetry in final states with charm jets and missing transverse momentum in 13 TeV pp collisions with the ATLAS detector. JHEP 09:050 arXiv:1805.01649 [hep-ex]
15. ATLAS Collaboration (2015) ATLAS Run 1 searches for direct pair production of third-generation squarks at the Large Hadron Collider. Eur Phys J C75:10, 510. arXiv:1506.08616 [hep-ex]. [Erratum: Eur Phys J C76:3, 153 (2016)]
16. Debove J, Fuks B, Klasen M (2011) Threshold resummation for gaugino pair production at hadron colliders. Nucl Phys B 842:51–85 arXiv:1005.2909 [hep-ph]
17. Fuks B, Klasen M, Lamprea DR, Rothering M (2013) Precision predictions for electroweak superpartner production at hadron colliders with Resummino. Eur Phys J C 73:2480 arXiv:1304.0790 [hep-ph]
18. Fuks B, Klasen M, Lamprea DR, Rothering M (2012) Gaugino production in proton-proton collisions at a center-of-mass energy of 8 TeV. JHEP 10:081 arXiv:1207.2159 [hep-ph]
19. Fiaschi J, Klasen M (2018) Neutralino-chargino pair production at NLO+NLL with resummation-improved parton density functions for LHC Run II. Phys Rev D98:5, 055014 (2018). arXiv:1805.11322 [hep-ph]
20. Bozzi G, Fuks B, Klasen M (2007) Threshold resummation for slepton-pair production at hadron colliders. Nucl Phys B777, 157–181. arXiv:hep-ph/0701202 [hep-ph]
21. Fiaschi J, Klasen M (2018) Slepton pair production at the LHC in NLO+NLL with resummation-improved parton densities. JHEP 03:094 arXiv:1801.10357 [hep-ph]

22. Martin SP, A supersymmetry primer. arXiv:hep-ph/9709356 [hep-ph]. [Adv Ser Direct High Energy Phys 18, 1 (1998)]
23. Griest K, Seckel D (1991) Three exceptions in the calculation of relic abundances. Phys Rev D 43:3191–3203
24. Edsjo J, Gondolo P (1997) Neutralino relic density including coannihilations. Phys Rev D 56:1879–1894 arXiv:hep-ph/9704361
25. Baer H, Krupovnickas T, Mustafayev A, Park E-K, Profumo S, Tata X (2005) Exploring the BWCA (bino-wino co-annihilation) scenario for neutralino dark matter. JHEP 12:011. arXiv:hep-ph/0511034 [hep-ph]
26. Harigaya K, Kaneta K, Matsumoto S (2014) Gaugino coannihilations. Phys Rev D 89(11):115021 arXiv:1403.0715 [hep-ph]
27. de Vries KJ et al (2015) The pMSSM10 after LHC Run 1. Eur Phys J C 75(9):422 arXiv:1504.03260 [hep-ph]
28. Papucci M, Ruderman JT, Weiler A (2012) Natural SUSY endures. JHEP 09:035 arXiv:1110.6926 [hep-ph]
29. Kawamura J, Omura Y (2017) Study of dark matter physics in non-universal gaugino mass scenario. JHEP 08:072 arXiv:1703.10379 [hep-ph]
30. Han Z, Kribs GD, Martin A, Menon A (2014) Hunting quasidegenerate Higgsinos. Phys Rev D 89(7):075007 arXiv:1401.1235 [hep-ph]
31. Baer H, Choi K-Y, Kim JE, Roszkowski L (2015) Dark matter production in the early Universe: beyond the thermal WIMP paradigm. Phys Rep 555:1–60 arXiv:1407.0017 [hep-ph]
32. Choi K-Y, Kim JE, Lee HM, Seto O (2008) Neutralino dark matter from heavy axino decay. Phys Rev D 77:123501 arXiv:0801.0491 [hep-ph]
33. Baer H, Lessa A, Rajagopalan S, Sreethawong W (2011) Mixed axion/neutralino cold dark matter in supersymmetric models. JCAP 1106:031 arXiv:1103.5413 [hep-ph]
34. Arkani-Hamed N, Delgado A, Giudice GF (2006) The well-tempered neutralino. Nucl Phys B741: 108–130. arXiv:hep-ph/0601041 [hep-ph]
35. Profumo S, Stefaniak T, Haskins LS (2017) The not-so-well tempered neutralino. Phys Rev D 96(5):055018 arXiv:1706.08537 [hep-ph]

Chapter 7
Search for Wino-Bino Production Decaying via WZ at $\sqrt{s} = 13$ TeV

The previous chapter described the motivation for EWK SUSY and the search for wino-bino models. The analysis presented in this chapter is a search for the production of bino $\tilde{\chi}_1^{\pm}\tilde{\chi}_2^0$ decaying via W and Z bosons to two or three leptons and missing energy at $\sqrt{s} = 13$ TeV with 36.1 fb^{-1}of data collected with the ATLAS detector. This search will be referred to as the conventional searches. The final state with 2 leptons, 2 jets, and $E_{\mathrm{T}}^{\mathrm{miss}}$, will be referred to as the conventional 2ℓ search, or by its final state 2ℓ+jets, while the final state with three leptons and $E_{\mathrm{T}}^{\mathrm{miss}}$will be referred to as the conventional 3ℓ search, or by its final state 3ℓ.

This chapter will focus mainly on the three lepton final state and briefly go over the two lepton final state because the result presents the combination of the two channels [1]. Similar searches were performed during LHC Run 1 at $\sqrt{s} = 7$ TeV and 8 TeV by the ATLAS [2–4] and the CMS collaborations [5–8]. The searches excluded up to 425 GeV, as shown in Fig. 7.1.

7.1 Signal Signature

Figure 7.2 shows the diagrams for the production of $\tilde{\chi}_1^{\pm}\tilde{\chi}_2^0$ decaying via W and Z bosons in proton-proton collisions. In this search, the W and Z bosons are on-shell.

For these gauge-boson-mediated decays, two distinct final states are considered: three-lepton (where lepton refers to an electron or muon) events where both the W and Z bosons decay leptonically or events with two opposite-sign leptons and two jets where the Z boson decays leptonically and the W boson decays hadronically. Leptonic decays of taus are indistinguishable from promptly produced electrons and muons and therefore contribute to the signal regions. The final state, in addition to the leptons, has missing energy from the LSP, $\tilde{\chi}_1^0$, and a neutrino if the W decays leptonically.

© The Editor(s) (if applicable) and The Author(s), under exclusive license to Springer Nature Switzerland AG 2020

E. Resseguie, *Electroweak Physics at the Large Hadron Collider with the ATLAS Detector*, Springer Theses,
https://doi.org/10.1007/978-3-030-57016-3_7

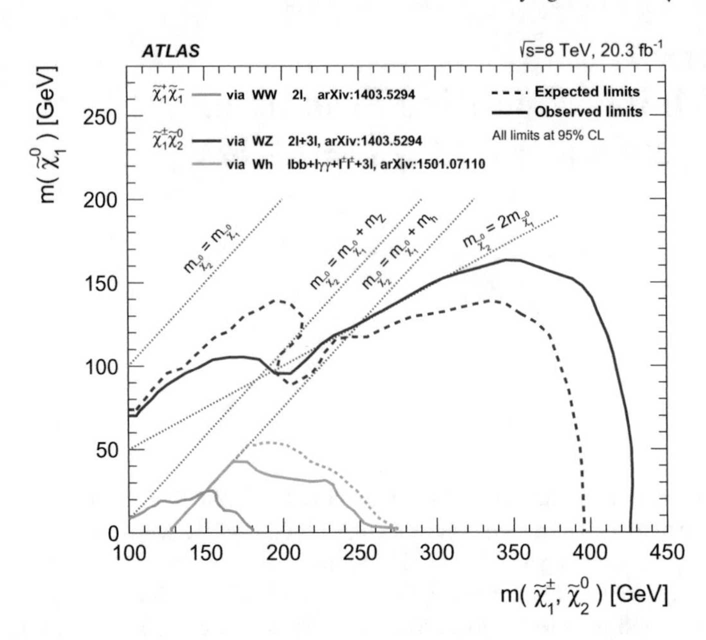

Fig. 7.1 The 95% CL exclusion limits on $\tilde{\chi}_1^+\tilde{\chi}_2^-$, $\tilde{\chi}_1^\pm\tilde{\chi}_2^0$ and $\tilde{\chi}_2^0\tilde{\chi}_3^0$ production with **a** SM-boson-mediated decays and **b** â-mediated decays, as a function of the $\tilde{\chi}_1^\pm\tilde{\chi}_2^0$ and $\tilde{\chi}_1^0$ masses

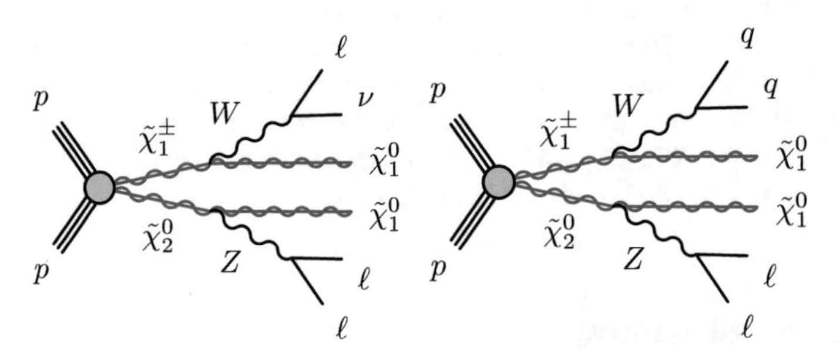

Fig. 7.2 Diagrams for the production of $\tilde{\chi}_1^\pm\tilde{\chi}_2^0$ decaying via W and Z bosons in pp collisions

In this model, the $\tilde{\chi}_1^\pm$ and $\tilde{\chi}_2^0$ are the Next-to-Lightest SUSY particles (NLSP) and mass degenerate winos and the LSP, $\tilde{\chi}_1^0$, is a bino. The mass splitting, Δm, refers to the difference in mass between the wino and the bino. Since the W and Z are on-shell, the mass splitting between the NLSP and the LSP is $\mathcal{O}(m_Z)$.

These signal models are kinematically similar to WZ production discussed in Chap. 5, except for the additional missing energy from the LSP, and, as a result, many techniques in the signal region optimization and background optimization will be the same as for the WZ production cross section measurement.

7.2 Simplified Model Framework

SUSY can produce many models so simplified models [9], in which the masses of the relevant sparticles are the only free parameters, are used for interpretation and to guide the design of the searches.

Simplified models make the assumption that there is no mixing between the SUSY mass parameters. Thus the SUSY particles, $\tilde{\chi}_1^{\pm}$, $\tilde{\chi}_2^0$, and $\tilde{\chi}_1^0$, are 100% wino and bino. There is also 100% branching fraction from sparticle to particle, meaning that the only possible decay modes for the SUSY particles are $\tilde{\chi}_1^{\pm} \rightarrow W^{\pm}\tilde{\chi}_1^0$ and $\tilde{\chi}_2^0 \rightarrow Z\tilde{\chi}_1^0$.

In this model, the mass of the slepton is halfway between the $\tilde{\chi}_1^{\pm}/\tilde{\chi}_2^0$ and the $\tilde{\chi}_1^0$ and sleptons decay all the time to leptons.

Moreover, the model considered assumes R-parity conservation. As a result, the SUSY particles are produced in pairs, and the LSP is stable.

7.3 Overview of Backgrounds

Since the W and Z bosons are on-shell, the main background of this search is WZ production, discussed in Chap. 5. The main difference between the SUSY signal and WZ production is the additional missing energy from the $\tilde{\chi}_1^0$. The main ways to minimize the WZ background is to select events with large missing energy, and with the transverse mass greater than the mass of the W boson. As seen in the WZ cross section measurement, the bulk of WZ events occurs for $m_{\mathrm{T}} < m_W$. The transverse mass used in this search is called $m_{\mathrm{T}}^{\mathrm{min}}$ and will be discussed in Sect. 7.5.2. The WZ simulation is normalized to data in a control region.

The reducible backgrounds are the same as for the WZ cross section measurement and can be split into two groups: Z+jet and $Z+\gamma$, and the top-like backgrounds. The Z+jet and $Z+\gamma$ background appears in the signal region because of the misidentification of a jet as a lepton or photon conversion. The top-like backgrounds, which include WW, Wt, and $t\bar{t}$, have a b-jet or a jet a misidentified as a lepton. A more detailed overview can be found in Sect. 5.5. The Z+jet/$Z+\gamma$ backgrounds are estimated using the Fake Factor method [10, 11] described in Sect. 5.6.1. The top-like backgrounds are normalized using DFOS events in a region close to the signal region, as described in Sect. 5.6.2.

The irreducible backgrounds come from ZZ, Higgs production, VVV, and $t\bar{t}V$, and are estimated using simulation.

7.4 Data Set and MC Samples

The proton-proton collision data corresponds to an integrated luminosity 36.1 fb^{-1} collected at a center-of-mass energy of 13 TeV in 2015 and 2016. The samples include an ATLAS detector simulation [12], based on Geant4 [13], or a fast simulation [12]

Table 7.1 Summary of the signal and background processes with the generator used for the simulation and the order at which the cross section is calculated

Process	Event generator	Parton shower, hadronization	UE tune	PDF	Order α_s
$\tilde{\chi}_1^{\pm}\tilde{\chi}_2^0$	MADGRAPH5_aMC@NLO	PYTHIA	A14	NNPDF2.3	NLO+NLL
WZ, ZZ, WW	SHERPA2.2.1	SHERPA	Default	NNPDF3.0	NLO
V+jet/$V\gamma$	SHERPA2.2.1	SHERPA	Default	NNPDF3.0	NNLO
Wt, s/t-channel	POWHEG	PYTHIA	Default	CT10	NLO + NNLL
$t\bar{t}$	POWHEG	PYTHIA	Default	CT10	NNLO + NNLL
VVV	SHERPA	SHERPA	Default	NNPDF3.0	NLO
$t\bar{t}V$	MADGRAPH5_aMC@NLO	PYTHIA	A14	NNPDF2.3	NLO
Higgs	POWHEG	PYTHIA	A14	NNPDF2.3	NNLO + NNLL

that uses a parametrization of the calorimeter response [14] and Geant4 for the other parts of the detector. The simulated events are reconstructed in the same manner as the data.

Table 7.1 summarizes the Monte Carlo (MC) used specifying the generator used to simulate both background and signal events.

Diboson and triboson processes were simulated with SHERPA 2.2.1 [15, 16] and the cross sections were calculated at NLO.

The $t\bar{t}$ and single top quarks samples in the Wt channel were simulated using POWHEG [17, 18] generator. The $t\bar{t}$ events were normalized using the NNLO+next-to-next-to-leading-logarithm (NNLL) QCD [19] cross-section, while the cross-section for single-top-quark events was calculated at NLO+NNLL [20].

The Z+jets and $Z + \gamma$ background simulation is only used as cross check since this background is estimated using the Fake Factor. They are simulated using the SHERPA generator and the cross section is calculated at NNLO [21].

Higgs boson processes include gluon-gluon fusion, associated VH production, and vector-boson fusion. They were generated using POWHEG [22] and PYTHIA. The cross sections are calculated at NNLO with NNLL accuracy.

The SUSY signal processes were generated from LO matrix elements with up to two extra partons, using the MADGRAPH v2.2.3 generator interfaced to PYTHIA8.186. Signal cross-sections were calculated at NLO with NLL accuracy [23–27]. The nominal cross-section and its uncertainty were taken from an envelope of cross-section predictions using different PDF sets and factorization and renormalization scales, as described in Ref. [28].

Table 7.2 Summary of the baseline and signal levels for electron and muon criteria. Each new level contains the selection of the previous level

Cut	Value/description	
	Baseline electron	Baseline muon
Acceptance	$p_T > 10\,\text{GeV}$, $\lvert\eta^{\text{cluster}}\rvert < 2.47$	$p_T > 10\,\text{GeV}$, $\lvert\eta\rvert < 2.4$
Identification	`LooseAndBLayerLLH`	`Medium`
	Signal electron	Signal muon
Identification	`MediumLH`	`Medium`
Isolation	`GradientLoose`	`GradientLoose`
Impact parameter	$\lvert z_0 \sin\theta\rvert < 0.5$ mm,	$\lvert z_0 \sin\theta\rvert < 0.5$ mm,
	$\lvert d_0/\sigma_{d_0}\rvert < 5$	$\lvert d_0/\sigma_{d_0}\rvert < 3$

7.5 Object and Event Selection

7.5.1 Object Selection

Electrons and muons are identified using identification, isolation, and tracking criteria. Two levels of object selection are used for electrons and muons, summarized in Table 7.2. Each level "baseline" and "signal" applies the selection of the previous levels along with additional criteria. The baseline leptons use the looser identification criteria and lower lepton $p_T$in order to provide a higher efficiency of identifying and removing processes decaying to four prompt leptons. Signal leptons satisfy stricter criteria.

Baseline electrons must have $p_T > 10\,\text{GeV}$ and fall within the inner detector, $|\eta| < 2.47$. The electrons must also satisfy the `LooseAndBLayerLLH` quality criteria. Signal electrons need to pass the impact parameter cuts of $|z_0 \sin\theta| < 0.5$ mm and $|d_0/\sigma_{d_0}| < 5$, designed to suppress fake electrons from pileup jets.They also satisfy tighter identification criteria, `MediumLH` and tighter isolation, `GradientLoose`.

Baseline muons must have $p_T > 10\,\text{GeV}$ and fall within the inner detector, $|\eta| < 2.4$. Signal muons must pass the impact parameter cuts of $|z_0 \sin\theta| < 0.5$ mm and $|d_0/\sigma_{d_0}| < 3$. They must must fulfill a tighter identification criteria, `Medium` and tighter isolation, `GradientLoose`.

Jets are reconstructed from topological clusters using the anti-k_t algorithm with distance parameter $\Delta R = 0.4$. Baseline jets are required to have $p_T > 20$ GeV and fulfill the pseudorapidity requirement of $|\eta| < 4.5$. To suppress jets originating from pileup, jets are further required to pass a JVT cut ($JVT > 0.59$) if the jet p_T is within $20 < p_T < 50$ GeV and it resides within $|\eta| < 2.4$ [29]. Signal jets have the additional requirement of falling within $|\eta| < 2.4$.

Identification of jets containing b-hadrons (b-jets), called b-tagging, is performed with the MV2c10 algorithm, a multivariate discriminant making use of track impact

Table 7.3 Summary of the baseline and signal selection for jets and b-jets

Cut	Value/description
	Baseline jet
Collection	AntiKtEMTopo
Acceptance	$p_T > 20$ GeV, $\|\eta\| < 4.5$
	Signal jet
JVT	$\|JVT\| > 0.59$ for jets with $p_T < 60$ GeV and $\|\eta\| < 2.4$
Acceptance	$p_T > 20$ GeV, $\|\eta\| < 2.4$
	Signal b-jet
b-tagger algorithm	MV2c10, 77% efficiency
Acceptance	$p_T > 20$ GeV, $\|\eta\| < 2.4$

parameters and reconstructed secondary vertices [30, 31]. A requirement is chosen corresponding to a 77% average efficiency obtained for b-jets in simulated $t\bar{t}$ events.

The jet and b-jet selection criteria are summarized in Table 7.3.

Separate algorithms are run in parallel to reconstruct electrons, muons, and jets. A particle can be reconstructed as one or more objects. To resolve these ambiguities, a procedure called "overlap removal" is applied. For electrons, this overlap removal is applied in two steps. At the baseline selection, an electron that shares a track with a muon, and the sub-leading $p_T$electron from two overlapping electrons are removed. The second step removes electrons if they are within $0.2 < \Delta R < 0.4$ of a jet. For muons, overlap removal is applied to baseline muons to separate prompt muons from those originating from the decay of hadrons in a jet. A baseline muon is removed if it is within $\Delta R < 0.4$ of a jet that at least 3 tracks.

The missing transverse momentum, with magnitude $E_{\mathrm{T}}^{\mathrm{miss}}$, is calculated as the negative vector sum of the transverse momenta of the calibrated selected leptons and jets, and the sum of transverse momenta of additional soft objects in the event, which are reconstructed from tracks in the inner detector or calorimeter cell clusters.

7.5.2 *Motivation for $m_{\mathrm{T}}^{\mathrm{min}}$*

In processes decaying via WZ bosons to three leptons and $E_{\mathrm{T}}^{\mathrm{miss}}$, Z leptons are traditionally assigned by finding the same flavor opposite sign pair that minimizes the difference between the invariant mass of the two leptons and the mass of the Z boson; the third remaining lepton is assigned to the W boson decay and used to calculate the transverse mass, as described in Chap. 5. In events with three electrons (eee) or three muons ($\mu\mu\mu$) in the final state, there is ambiguity in assigning the leptons to the W and Z decays, and for higher values of transverse mass, the traditional assignment of leptons is often not correct. Using POWHEG for the WZ production, the truth origin

Table 7.4 Summary of signal points used for optimization. Signal points have either intermediate mass splittings ($\Delta m < 200$) and large mass splittings ($\Delta m \geq 200$)

Δm [GeV]	$(m(\tilde{\chi}_1^{\pm}\tilde{\chi}_2^0), m(\tilde{\chi}_1^0))$ [GeV]
100	(200, 100)
100	(250, 150)
150	(300, 150)
200	(400, 200)
300	(450, 150)
400	(450, 50)

of the leptons is compared to reconstructed origin to determine the fraction of events where the leptons' origin is not identified correctly.

For the production of SUSY particles $\tilde{\chi}_1^{\pm}\tilde{\chi}_2^0$ decaying via W and Z bosons, the transverse mass is the main discriminant between signal and WZ background. It is therefore extremely sensitive large values of the transverse mass due to a wrong choice of the Z leptons. To remedy this, in events with three electrons or three muons, the leptons are assigned by first finding the lepton associated with the W boson that minimizes the value of the transverse mass. The remaining two same flavor opposite sign leptons are associated with the Z boson decay. Using this lepton assignment results a better efficiency to assign the background leptons correctly, and therefore a corresponding improvement in sensitivity.

Four benchmark signal points are used for this study and summarized in Table 7.4: three signal points with large mass splittings, $\Delta m = 200$, 300, and 400 GeV, and three signal points with intermediate mass splittings, Δm = 100 and 150 GeV.

7.5.2.1 Efficiency of Correctly Identifying the Origin of Reconstructed Leptons

The selections for electrons and muons are summarized in Table 7.2. For event cleaning, a few additional cuts are required to select on-shell WZ events and minimize backgrounds. On-shell Z events are also selected, with the invariant mass of the Z leptons within 10 GeV of the mass of the Z boson. A cut on the missing energy of 50 GeV is required to minimize Z+jet/$Z+\gamma$ contribution. A b-jet veto is required to minimize top backgrounds. To minimize ZZ events, a veto on the fourth baseline lepton is applied, similar to what was done in the WZ cross section measurement. Table 7.5 summarizes the preselection cuts.

The lepton mis-identification is only an issue when there is an ambiguity as to which lepton could form the SFOS pair to reconstruct the Z boson therefore, for this study, only eee and $\mu\mu\mu$ events are considered.

After applying the cuts in Table 7.5 and for eee and $\mu\mu\mu$ events, 6.5% of the events have a lepton that has been reconstructed as coming from the Z boson but

Table 7.5 Summary of the cuts used in the $m_{\mathrm{T}}^{\mathrm{min}}$ study

Preselection cuts
Exactly 3 baseline and 3 signal leptons
At least 1 SFOS pair
$\mid m_{\ell\ell} - m_Z \mid \leq 10$ GeV
b-jet veto
$E_{\mathrm{T}}^{\mathrm{miss}} > 50$ GeV

in fact comes from another source according to the truth origin of the lepton. The flavor of the mis-assigned lepton reconstructed as coming from the Z is 52% muon and 48% electron. The parent of the mis-assigned lepton according to the truth MC Classifier Origin is the W boson 95.6 and 4.4% photon conversion. Thus, in most mis-assignment cases, the W lepton is paired with a Z lepton.

To understand the impact of mis-assigning leptons, the main discriminating variable, m_{T}, is defined as

$$m_{\mathrm{T}} = \sqrt{2 p_T^W E_{\mathrm{T}}^{\mathrm{miss}} (1 - cos(\Delta\phi))}, \tag{7.1}$$

where $\Delta\phi$ is the angle between the W lepton and the missing transverse energy vector.

For the SUSY signal, large m_{T} values result from the additional missing energy from the LSP. For the WZ background, large m_{T} values comes from either large p_T of the lepton assigned to the W boson or a large angular separation between the lepton assigned to the W boson and the missing energy vector.

At low $E_{\mathrm{T}}^{\mathrm{miss}}$ and low m_{T}, the efficiency of correctly identifying the leptons originating from the Z boson is very high, above 90%, as seen in Fig. 7.3, which is why the assignment done by minimizing the invariant mass of the leptons with respect to the mass of the Z boson is the appropriate assigning method for the WZ cross section measurement. The assignment becomes problematic at high values of m_{T}, which is the signal region for this search. At large m_{T} values, $m_{\mathrm{T}} > 125$ GeV, the efficiency of assigning Z leptons to the Z boson decreases to 60%. Thus, the mis-classification of leptons can make the WZ background look like signal due to large m_{T} values in the event.

7.5.2.2 Constructing $m_{\mathrm{T}}^{\mathrm{min}}$

The traditional assignment, used in the WZ cross section measurement described in Chap. 5, first assigns Z leptons as the SFOS pair with invariant mass closest to the mass of the Z boson. The remaining lepton is assigned to the W boson.

This new "min" assignment first assigns the lepton that minimizes m_{T} to the W boson. The remaining SFOS pair of leptons is assigned to the Z boson.

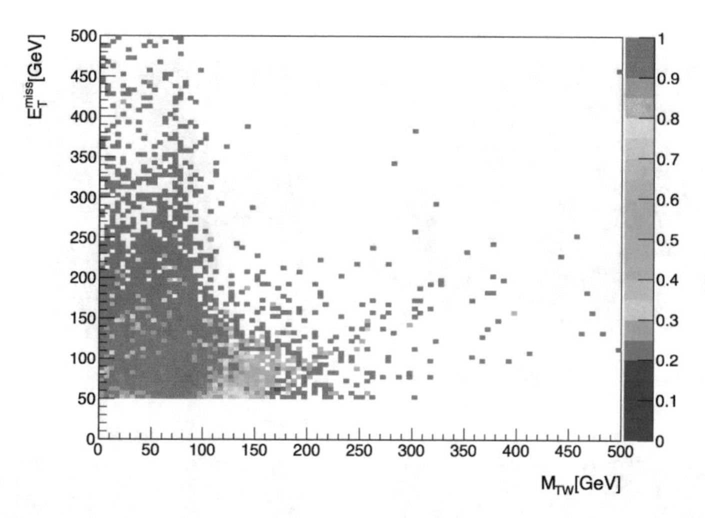

Fig. 7.3 Efficiency of correctly identifying the Z lepton as a function of $E_{\mathrm{T}}^{\mathrm{miss}}$ and m_{T}

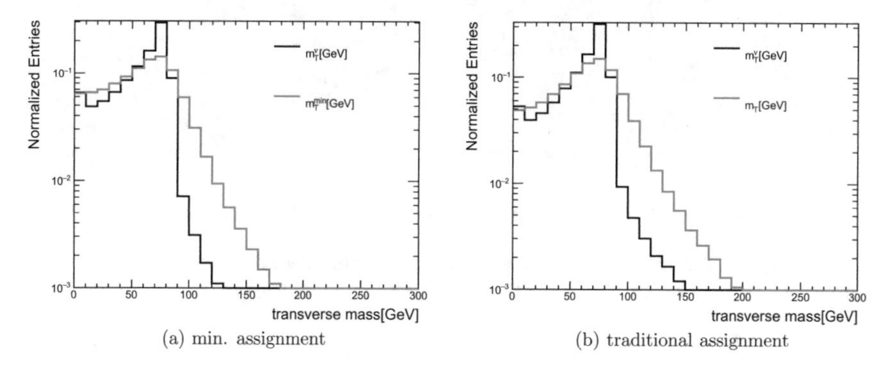

(a) min. assignment

(b) traditional assignment

Fig. 7.4 Comparison of m_{T} calculated with the W lepton and the neutrino (using truth) and $E_{\mathrm{T}}^{\mathrm{miss}}$ in reconstructed events for both min. and traditional assignments

7.5.2.3 Comparison of Performance of m_{T} and $m_{\mathrm{T}}^{\mathrm{min}}$

To determine the benefit of using the min. assignment over the traditional assignment, reconstructed events are compared to truth events. The m_{T} variable is calculated using the reconstructed W lepton and the neutrino, m_{T}^{ν}, as well as using $E_{\mathrm{T}}^{\mathrm{miss}}$ for both the min. and traditional assignments, as shown in Fig. 7.4. Calculating m_{T} with the neutrino as opposed to $E_{\mathrm{T}}^{\mathrm{miss}}$reduces $E_{\mathrm{T}}^{\mathrm{miss}}$mismeasurement effects. When calculating m_{T} with the neutrino, the m_{T} distribution should sharply drop off after $m_W = 80.4\,\mathrm{GeV}$ since $m_{\mathrm{T}} < m_W$. The m_{T} distribution calculated with the neutrino drops more sharply with the min. assignment than with the traditional assignment. Moreover, $m_{\mathrm{T}}^{\mathrm{min}}$ is closer to m_{T}^{ν} than in the traditional assignment since the $m_{\mathrm{T}}^{\mathrm{min}}$ distribution is narrower.

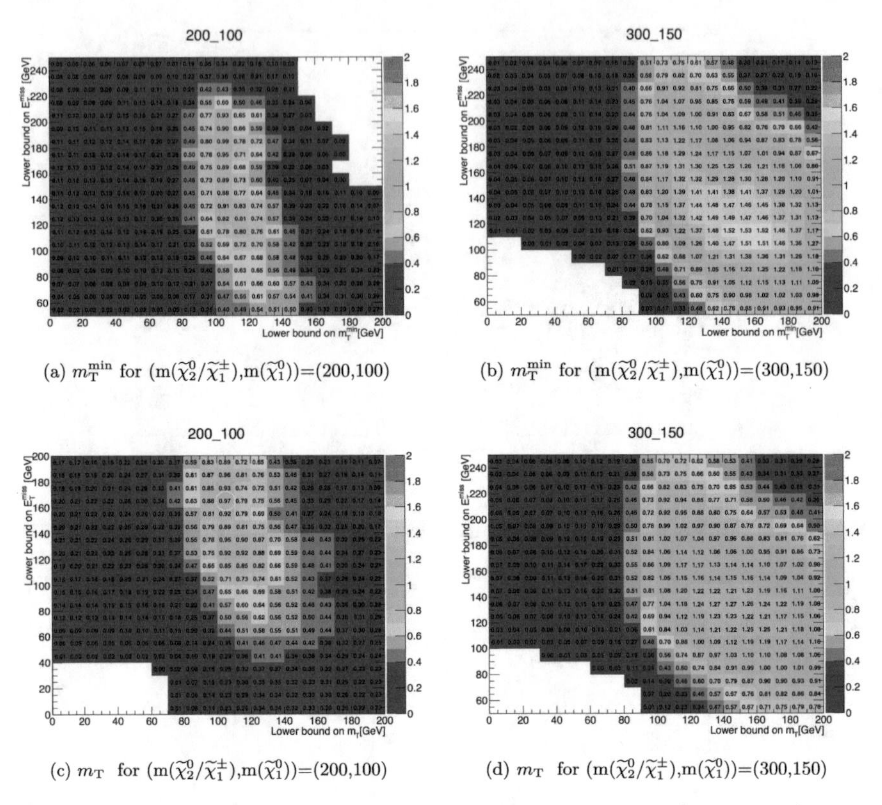

(a) $m_{\mathrm{T}}^{\mathrm{min}}$ for (m($\tilde{\chi}_2^0/\tilde{\chi}_1^{\pm}$),m($\tilde{\chi}_1^0$))=(200,100)

(b) $m_{\mathrm{T}}^{\mathrm{min}}$ for (m($\tilde{\chi}_2^0/\tilde{\chi}_1^{\pm}$),m($\tilde{\chi}_1^0$))=(300,150)

(c) m_{T} for (m($\tilde{\chi}_2^0/\tilde{\chi}_1^{\pm}$),m($\tilde{\chi}_1^0$))=(200,100)

(d) m_{T} for (m($\tilde{\chi}_2^0/\tilde{\chi}_1^{\pm}$),m($\tilde{\chi}_1^0$))=(300,150)

Fig. 7.5 Two dimensional correlation plots showing the significance for $E_{\mathrm{T}}^{\mathrm{miss}}$ versus $m_{\mathrm{T}}^{\mathrm{min}}$ on the top, and $E_{\mathrm{T}}^{\mathrm{miss}}$ versus m_{T} on the bottom for intermediate mass splittings

The performance of m_{T} and $m_{\mathrm{T}}^{\mathrm{min}}$ is also compared by comparing the significance of the signal over the backgrounds. To do this comparison, a loose signal region is defined with first and second lepton $p_T > 25$ GeV to be on plateau for the trigger and third lepton $p_T > 20$ GeV to reduce background contamination. The remaining cuts applied are described in Table 7.5.

Plots of $E_{\mathrm{T}}^{\mathrm{miss}}$ versus $m_{\mathrm{T}}/m_{\mathrm{T}}^{\mathrm{min}}$ are constructed with the significance, called Zn, being the figure of merit. Zn is the significance determined from the p-value and is calculated with 30% flat uncertainty on the background.

Figures 7.5, 7.6 shows two dimensional correlations between $E_{\mathrm{T}}^{\mathrm{miss}}$ and $m_{\mathrm{T}}^{\mathrm{min}}$, and $E_{\mathrm{T}}^{\mathrm{miss}}$ and m_{T}. Using $m_{\mathrm{T}}^{\mathrm{min}}$ instead of m_{T} gives an improvement in significance between 12 and 25%. Therefore, $m_{\mathrm{T}}^{\mathrm{min}}$ does better than m_{T} in terms of improving the significance for all mass points studied.

Figure 7.7 shows the WZ background for both the traditional (in red) and min assignment (in blue) as well as the (450, 50) signal point for the traditional assignment (in pink) and the min assignment (in cyan). The significance, Zn, is the significance calculated for a lower cut of the bin value. There are more WZ background events at

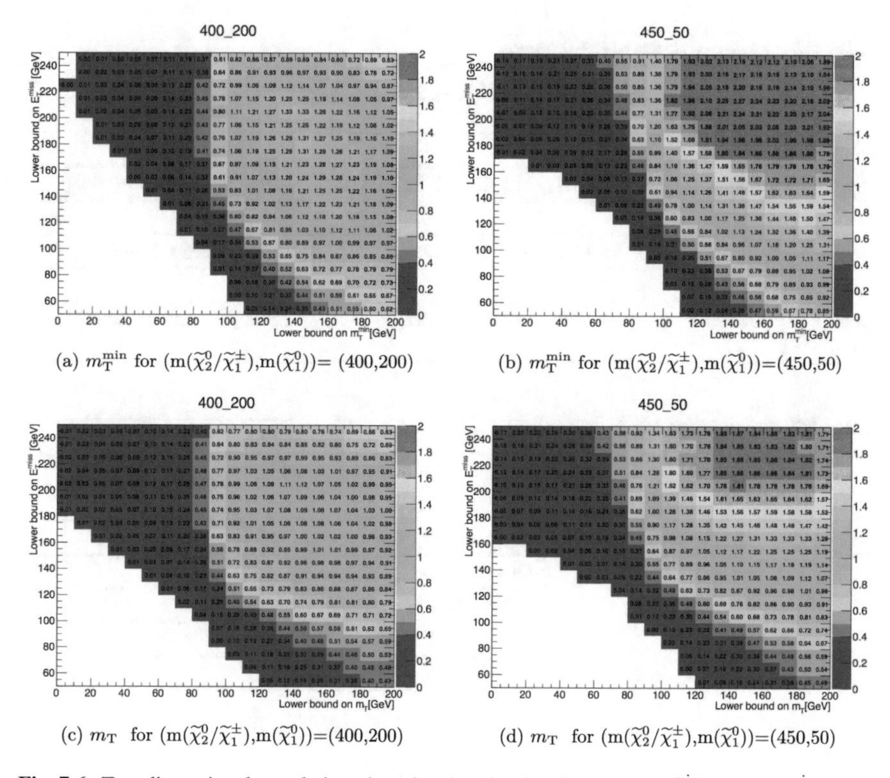

(a) $m_{\mathrm{T}}^{\mathrm{min}}$ for $(\mathrm{m}(\widetilde{\chi}_2^0/\widetilde{\chi}_1^{\pm}),\mathrm{m}(\widetilde{\chi}_1^0))=(400,200)$

(b) $m_{\mathrm{T}}^{\mathrm{min}}$ for $(\mathrm{m}(\widetilde{\chi}_2^0/\widetilde{\chi}_1^{\pm}),\mathrm{m}(\widetilde{\chi}_1^0))=(450,50)$

(c) m_{T} for $(\mathrm{m}(\widetilde{\chi}_2^0/\widetilde{\chi}_1^{\pm}),\mathrm{m}(\widetilde{\chi}_1^0))=(400,200)$

(d) m_{T} for $(\mathrm{m}(\widetilde{\chi}_2^0/\widetilde{\chi}_1^{\pm}),\mathrm{m}(\widetilde{\chi}_1^0))=(450,50)$

Fig. 7.6 Two dimensional correlation plots showing the significance for $E_{\mathrm{T}}^{\mathrm{miss}}$ versus $m_{\mathrm{T}}^{\mathrm{min}}$ on the top, and $E_{\mathrm{T}}^{\mathrm{miss}}$ versus m_{T} on the bottom for large mass splittings

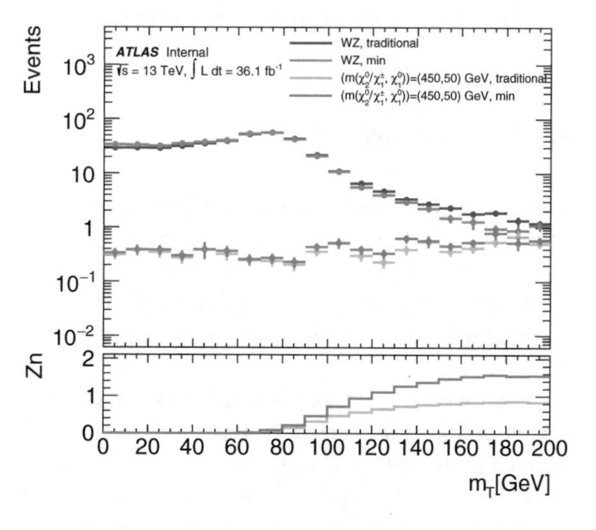

Fig. 7.7 Traditional and min assignment for the WZ background and $(\mathrm{m}(\widetilde{\chi}_2^0/\widetilde{\chi}_1^{\pm}),$ $\mathrm{m}(\widetilde{\chi}_1^0)(450, 50)\,\mathrm{GeV})$ signal point. The significance is the calculated as a lower cut for each bin value

Table 7.6 Summary of the trigger strategy

Lepton p_T	Trigger	
	Data15	Data16
Di-electron channel		
$p_T(e_{1(2)}) > 25$ GeV	HLT_2e12_lhloose_L12EM10VH	HLT_2e17_lhvloose_nod0
Di-muon channel		
$p_T(\mu_{1(2)}) > 25$ GeV	HLT_mu18_mu8noL1	HLT_mu22_mu8noL1
Electron-muon channel		
$p_T(e) > 25$ GeV and $p_T(\mu) > 25$ GeV	HLT_e17_lhloose_mu14	HLT_e17_lhloose_nod0_mu14

Table 7.7 Summary of loose event selection

Preselection cuts
Exactly 3 baseline and 3 signal leptons
At least 1 SFOS pair
$\mid m_{\ell\ell} - m_Z \mid \leq 10$ GeV
b-jet veto
$E_{\mathrm{T}}^{\mathrm{miss}} > 50$ GeV
$p_T^{\ell 1}, p_T^{\ell 2} > 25$ GeV
$p_T^{\ell 3} > 20$ GeV

lower values of m_{T} using the min. assignment than the traditional assignment while the signal remains flat for both assignments. This results in an increase in significance at larger values of m_{T} using the min assignment.

7.5.3 Event Selection

Events considered in the analysis must pass a trigger selection requiring either two electrons, two muons or an electron plus a muon. The trigger-level thresholds on the p_T value of the leptons involved in the trigger decision are in the range 8–22 GeV and are looser than those applied offline to ensure that trigger efficiencies are constant in the relevant phase space. The triggers used in this search are summarized in Table 7.6.

A loose event selection is applied to minimize some of the backgrounds in the search and is summarized in Table 7.7. Section 7.3 summarizes the background sources in the search. Z+jets and $Z + \gamma$ backgrounds do not usually have $E_{\mathrm{T}}^{\mathrm{miss}}$ in their final state so a cut of $E_{\mathrm{T}}^{\mathrm{miss}} > 50$ GeV reduces that background. This background is further minimized by tightening the lepton p_T cuts to 25 GeV for the leading and sub-leading leptons, and 20 GeV for the third lepton.

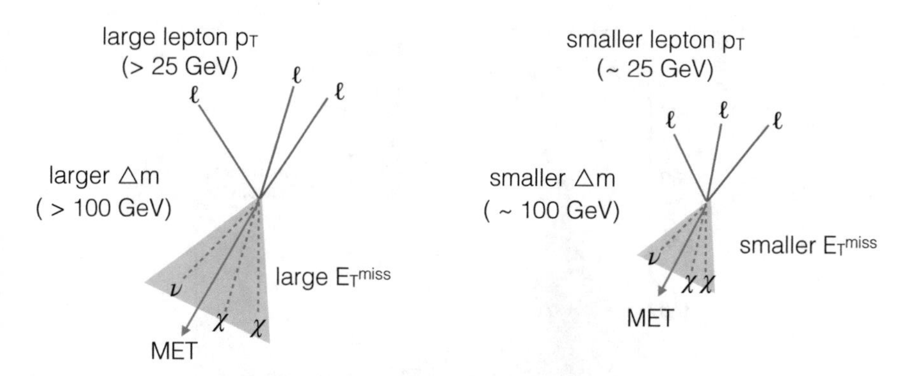

Fig. 7.8 Cartoon showing the difference in kinematics between a signal sample with a large mass splitting and a signal sample with an intermediate mass splitting

Top backgrounds, especially $t\bar{t}$ have a b-jet in their final state so to reduce this background, a b-jet veto is applied. The on-shell Z requirement, $\mid m_{\ell\ell} - m_Z \mid \leq$ 10 GeV, not only helps select signal events, but also reduces top background which do not have a resonance with the mass of a Z boson.

The ZZ background has four leptons in the final state so applying a requirement of exactly 3 leptons after loosening the lepton criteria to baseline instead of signal, defined in Table 7.2, minimizes this background. This fourth lepton veto was also applied in the WZ cross section measurement.

The main background remaining is WZ. The signal differs from the WZ background due to the additional missing energy coming from the LSP, thus $E_{\mathrm{T}}^{\mathrm{miss}}$ is a discriminating variable between signal and background. The amount of $E_{\mathrm{T}}^{\mathrm{miss}}$ in the event is dependent on the mass splitting of the signal. Signals with large mass splittings will have a larger fraction of the final state momentum carried by the LSP than for smaller mass splitting signals, as illustrated in Fig. 7.8. Moreover, larger mass splittings result in larger p_T leptons than smaller mass splittings.

Another discriminating variable between the signal and the WZ process is $m_{\mathrm{T}}^{\mathrm{min}}$. Assigning leptons using $m_{\mathrm{T}}^{\mathrm{min}}$ minimizes the amount of WZ background at large m_{T} values. Moreover, this variable is bound at the mass of the W boson for the WZ process, as shown in Fig. 7.9. The signal with the larger mass splitting, in black, has a small dependence on $m_{\mathrm{T}}^{\mathrm{min}}$. The smaller mass splitting signal point, in red, has a similar dependence on $m_{\mathrm{T}}^{\mathrm{min}}$ as the WZ background, making this signal more difficult to discriminate from the background.

To discriminate signals with larger mass splitting, cuts on large $E_{\mathrm{T}}^{\mathrm{miss}}$ and large $m_{\mathrm{T}}^{\mathrm{min}}$ values is sufficient; however, that is not the case for signal with smaller mass splittings. A different strategy needs to be used in this case.

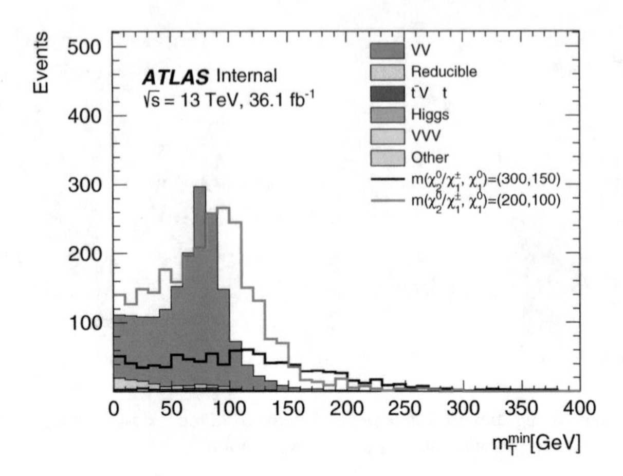

Fig. 7.9 Distribution of $m_{\mathrm{T}}^{\min}$ after loose event selection defined in Table 7.7. Signal is multiplied by 20 to be visible on the plot

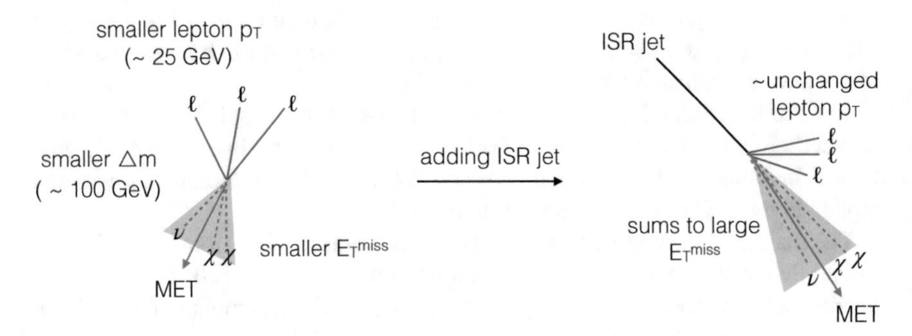

Fig. 7.10 Cartoon illustrating the impact of the addition of an Initial-State-Radiation (ISR) jet on the signal kinematics

7.5.3.1 Jet Veto and ISR Topology

For smaller mass splittings, an Initial-State-Radiation (ISR) jet is required in the event. A signal with a smaller mass splitting has less missing energy and lower p_T lepton. The ISR jet boosts the event, causing the missing energy to sum to large values due to the recoil against the jet. If the mass splitting is small, then the $E_{\mathrm{T}}^{\mathrm{miss}}$ sums up to a larger value than the lepton p_T from the ISR jet boost. This increase in $E_{\mathrm{T}}^{\mathrm{miss}}$ helps distinguish between the signal and the background (Fig. 7.10).

The signal region is therefore split into two regions: one with a jet veto and another with an ISR jet. Jet veto means that there are no signal jets, as defined in Table 7.3, in the signal region, and the ISR region requires the event to have at least one signal jet. The main discriminating variables are $E_{\mathrm{T}}^{\mathrm{miss}}$ and $m_{\mathrm{T}}^{\min}$. In the jet veto region, the cuts on those variables should be lower than in the ISR region.

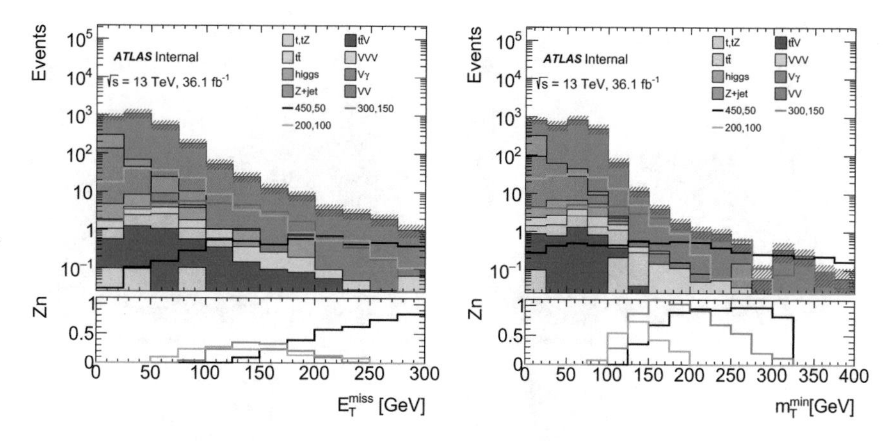

Fig. 7.11 $E_{\mathrm{T}}^{\mathrm{miss}}$ and $m_{\mathrm{T}}^{\mathrm{min}}$ distributions in the jet veto region. The significance is calculated for a lower bound on the bin value

7.5.3.2 Jet Veto Region Optimization

The main discriminating variables, $E_{\mathrm{T}}^{\mathrm{miss}}$ and $m_{\mathrm{T}}^{\mathrm{min}}$, in the jet veto region are shown in Fig. 7.11. Three signal points with different Δm are added to determine optimal signal region cuts: large Δm (450, 50), $\Delta m \simeq m_Z$ (200,100), and a point in between (300, 150). Since the large Δm signal point has larger missing energy, the optimal cut on $E_{\mathrm{T}}^{\mathrm{miss}}$ is larger than for smaller mass splittings, and similarly true for $m_{\mathrm{T}}^{\mathrm{min}}$. Thus, the jet veto signal region can be split into three bins, each targeting a different Δm.

To determine the bins used in the jet veto region, the significance of a lower bound on $E_{\mathrm{T}}^{\mathrm{miss}}$ versus a lower bound on $m_{\mathrm{T}}^{\mathrm{min}}$ for three signal points is studied, as shown in Fig. 7.12. For all signal points, the optimal $m_{\mathrm{T}}^{\mathrm{min}}$ cut is at $m_{\mathrm{T}}^{\mathrm{min}} > 110$, where the WZ background contribution starts decreasing. The optimal cut $E_{\mathrm{T}}^{\mathrm{miss}}$ is dependent on the mass splitting. As the mass splitting increases, the optimal cut on $E_{\mathrm{T}}^{\mathrm{miss}}$ increases. The jet veto signal region with cuts on $E_{\mathrm{T}}^{\mathrm{miss}}$ and $m_{\mathrm{T}}^{\mathrm{min}}$ is summarized in Table 7.8 (Fig. 7.13).

7.5.3.3 ISR Region Optimization

In the ISR region, due to the recoil against the ISR jet, the signal and the background have larger values of $E_{\mathrm{T}}^{\mathrm{miss}}$ and $m_{\mathrm{T}}^{\mathrm{min}}$. As a result, the values where there is an optimal significance for a lower bound of $E_{\mathrm{T}}^{\mathrm{miss}}$ and $m_{\mathrm{T}}^{\mathrm{min}}$ are larger than for the jet veto region. Similar to the jet veto region, three bins in $E_{\mathrm{T}}^{\mathrm{miss}}$ and $m_{\mathrm{T}}^{\mathrm{min}}$ targeting three different signal mass splittings are chosen based on $E_{\mathrm{T}}^{\mathrm{miss}}$ versus $m_{\mathrm{T}}^{\mathrm{min}}$ significance distributions shown in Fig. 7.14.

For smaller mass splittings, the $E_{\mathrm{T}}^{\mathrm{miss}}$ gets a larger contribution from the boost due to the ISR jet than the lepton p_T. In order to study this, an additional variable is defined $p_T^{\ell\ell\ell}$, which is the p_T of the vector sum of the leptons. The leading jet p_T and

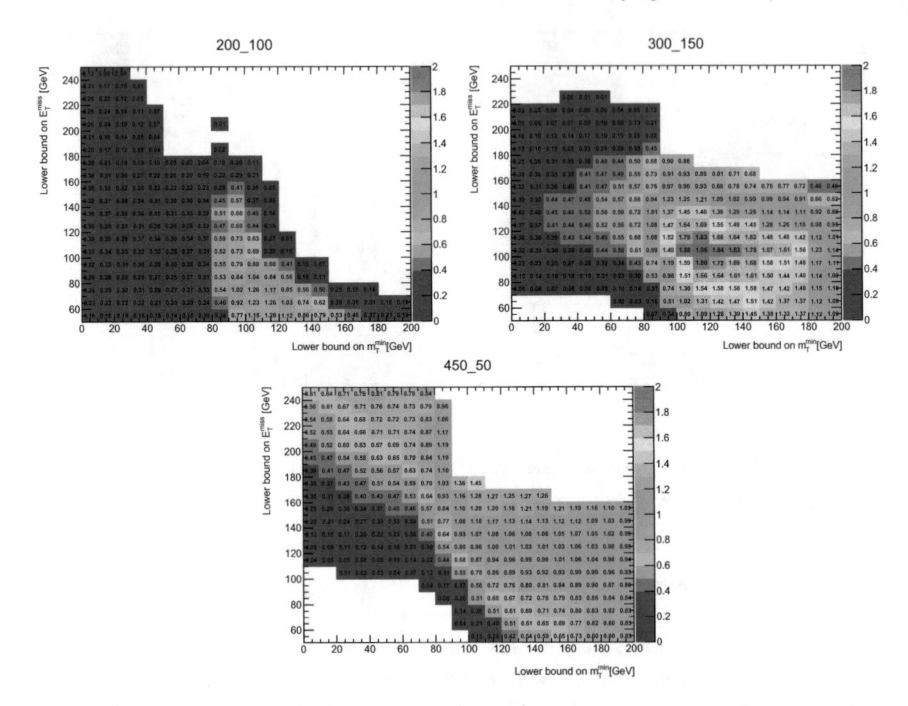

Fig. 7.12 Significance of a lower bound on $E_{\mathrm{T}}^{\mathrm{miss}}$ versus a lower bound on $m_{\mathrm{T}}^{\mathrm{min}}$ in the jet veto region for three signal points with m($\widetilde{\chi}_2^0/\widetilde{\chi}_1^{\pm}$), m($\widetilde{\chi}_1^0$)= (200, 100), (300, 150) and (450, 50)

Table 7.8 Summary of the exclusive signal regions

Bin	Number of jets	$E_{\mathrm{T}}^{\mathrm{miss}}$	$m_{\mathrm{T}}^{\mathrm{min}}$	$p_T^{\ell\ell}$	$p_T^{\mathrm{jet}\ 1}$	$p_T^{3rd\ell}$
SR3-WZ-0Ja	=0	60-120	>110	–	–	–
SR3-WZ-0Jb		120-170	>110	–	–	–
SR3-WZ-0Jc		>170	>110	–	–	–
SR3-WZ-1Ja	>0	120–200	>110	<120	>70	–
SR3-WZ-1Jb		>200	110–160	–	–	–
SR3-WZ-1Jc		>200	>160	–	–	>35

$p_T^{\ell\ell}$ in SR3-WZ-1Ja, defined in Table 7.8, are shown in Fig. 7.15. The significance on $p_T^{\ell\ell}$ is calculated for an upper cut on that variable. The reason an upper bound on $p_T^{\ell\ell}$ is considered is because the smaller mass splittings would have smaller lepton p_T after the boost than background since $E_{\mathrm{T}}^{\mathrm{miss}}$ sums up to a larger value than the lepton p_T. Adding a cut on these two variables, in addition to the cuts on $E_{\mathrm{T}}^{\mathrm{miss}}$ and $m_{\mathrm{T}}^{\mathrm{min}}$ increases the significance for signal with smaller mass splittings.

The significance for a lower cut on the lead jet p_T and an upper cut on $p_T^{\ell\ell}$ is shown in Fig. 7.16.

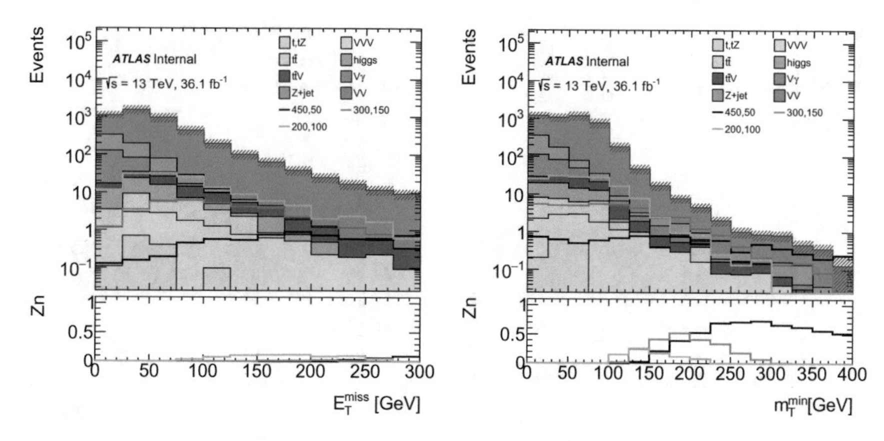

Fig. 7.13 $E_{\mathrm{T}}^{\mathrm{miss}}$ and $m_{\mathrm{T}}^{\mathrm{min}}$ distributions in the jet veto region. The significance is calculated for a lower bound on the bin value

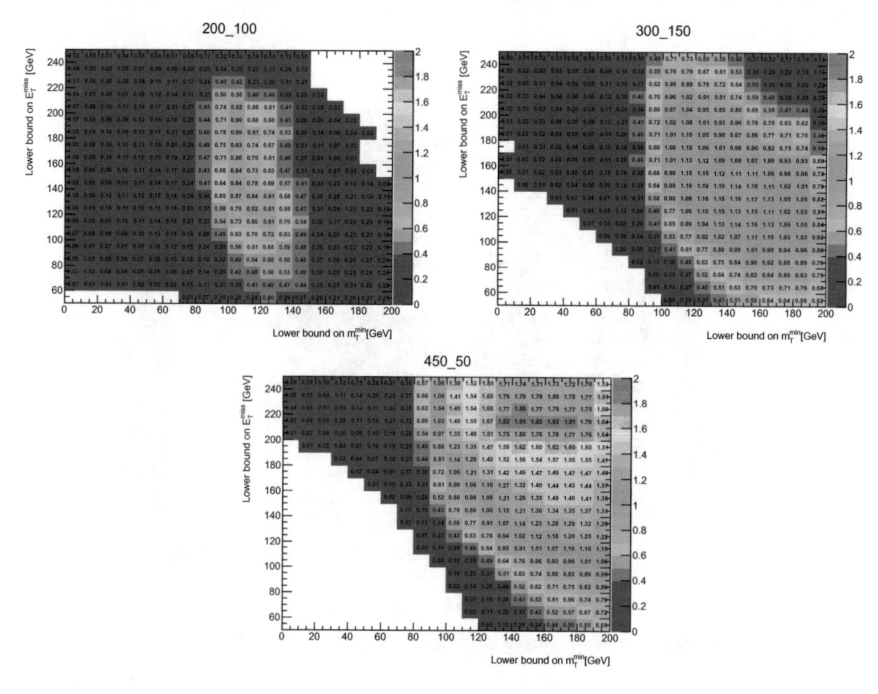

Fig. 7.14 Significance of a lower bound on $E_{\mathrm{T}}^{\mathrm{miss}}$ versus a lower bound on $m_{\mathrm{T}}^{\mathrm{min}}$ in the ISR region for three signal points with $(\mathrm{m}(\widetilde{\chi}_2^0/\widetilde{\chi}_1^{\pm}), \mathrm{m}(\widetilde{\chi}_1^0))$ = (200, 100), (300, 150), and (450, 50)

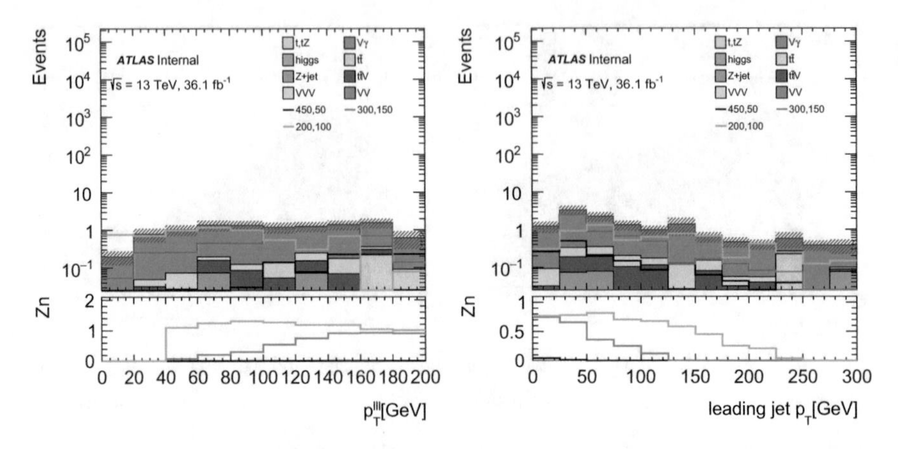

Fig. 7.15 $p_T^{\ell\ell\ell}$ and lead jet p_T distributions in the jet veto region. The significance is calculated for a lower bound on the bin value for the lead jet p_T and for an upper bound on the bin value for $p_T^{\ell\ell\ell}$

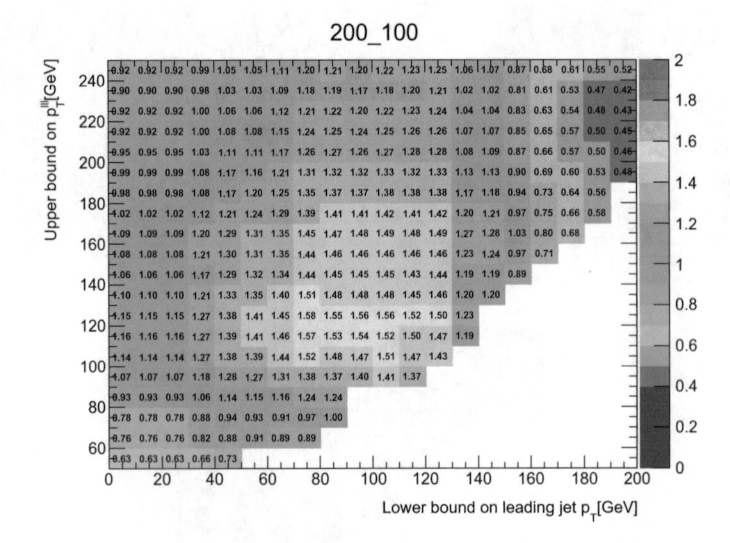

Fig. 7.16 Significance of a lower bound on $E_{\mathrm{T}}^{\mathrm{miss}}$ versus a lower bound on $m_{\mathrm{T}}^{\mathrm{min}}$ in the ISR region for signal with the smallest mass splitting with $(\mathrm{m}(\tilde{\chi}_2^0/\tilde{\chi}_1^{\pm}), \mathrm{m}(\tilde{\chi}_1^0)) = (200, 100)$

Table 7.9 Summary of the estimation methods for each background process in each signal region

Background	Estimation
Z+jets/Z+γ	FF (under "reducible")
$t\bar{t} + Wt + WW$	NF (under "reducible")
WZ	NF (under "VV")
ZZ	MC (under "VV")
Higgs/ VVV/ $t\bar{t}V$	MC

7.5.3.4 Summary of Signal Regions (SR)

The optimal bin selections for the jet veto and the ISR regions is summarized in Table 7.8.

7.6 Background Estimation

The background estimate strategy is summarized in Table 7.9. The dominant background, WZ, is estimated using simulation normalized to data in a control region. The methodology on how to calculate normalization factors can be found in Sect. 5.6.2.

The Z+jets/Z+γ background is estimated using the Fake Factor method, described in Sect. 5.6.1. The top-like background, $t\bar{t}$, Wt, WW, is estimated using a control with different-flavor, opposite-charge events, described in Sect. 5.6.2.

Higgs, VVV, $t\bar{t}V$, and ZZ processes are estimated using simulation.

Table 7.10 summarizes the control regions and validation regions (VR) used in this analysis.

7.6.1 WZ Background

7.6.1.1 Control and Validation Regions

Since the signal regions are separated into a jet veto and an ISR region, the jet multiplicity distribution in the inclusive WZ control region is of particular interest. This is shown in Fig. 7.17. From this distribution is becomes clear that the modeling of the inclusive WZ CR is dominated by events with at least 1 jet, where there is a deficit in data with respect to the MC prediction of the backgrounds. For events with no jets, the data to MC agreement is reasonably good with a slight excess visible in the first bin of the N_{jets} distribution in Fig. 7.17.

Thus, due to difference in modelling between the events with no jets and events with at least one jet, control regions binned in N_{jets} are defined. These CRs are also

Table 7.10 Control and validation region definitions. The m_{SFOS} quantity is the mass of the same-flavor opposite-sign lepton pair and $m_{\ell\ell\ell}$ is the trilepton invariant mass

3ℓ control and validation region definitions

	$p_T^{\ell_3}$ [GeV]	$m_{\ell\ell\ell}$ [GeV]	m_{SFOS} [GeV]	$E_{\text{T}}^{\text{miss}}$[GeV]	$m_{\text{T}}^{\text{min}}$	$n_{\text{non-}b\text{-tagged jets}}$	$n_{b\text{-tagged jets}}$
CR3-WZ-0j	>20	–	81.2–101.2	>60	<110	0	0
CR3-WZ-1j	>20	–	81.2–101.2	>120	<110	>0	0
VR3-offZa	>30	$\notin$ [81.2, 101.2]	$\notin$ [81.2, 101.2]	40–60	–	–	–
VR3-offZb	>20			>40	–	–	>0
VR3-Za-0J	>20	$\notin$ [81.2, 101.2]	81.2–101.2	40–60	–	0	0
VR3-Za-1J				40–60	–	>0	0

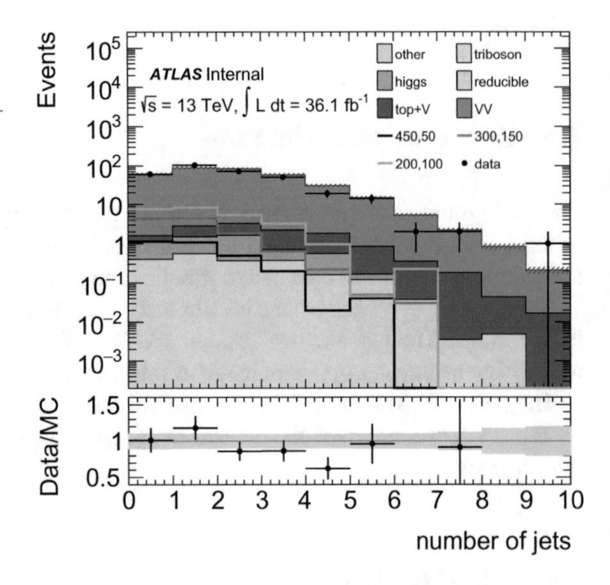

Fig. 7.17 Jet multiplicity distribution in the inclusive WZ control regions. Errors are statistical only, except for the backgrounds derived using data driven estimates, which include systematics as well

kinematically closer to the SRs since the signal regions are also defined with a jet veto or an ISR selection. These control regions are defined to be orthogonal to the signal regions with minimal background contamination with high WZ purity. Different WZ NFs can be extracted in each CR and used to normalize the WZ background in the corresponding SRs. Validation regions (VR) are then used to compare the MC estimates, along with the scale factor, to the observed data (Fig. 7.18).

Both control regions are defined by inverting the $m_{\text{T}}^{\text{min}}$ cut. As shown in Sect. 7.5.2, cutting on $m_{\text{T}}^{\text{min}}$ reduces the WZ background so reverting that cut introduces more WZ background with little signal contamination. The $E_{\text{T}}^{\text{miss}}$ cut applied to the CR matches the $E_{\text{T}}^{\text{miss}}$ cut in the SR. In the validation region, a minimum $E_{\text{T}}^{\text{miss}}$ cut of 40 GeV is introduced to reduce Z+jets/$Z+\gamma$ contamination in this region. The CR and VR are illustrated in Fig. 7.11 and their definitions can be found in Table 7.10.

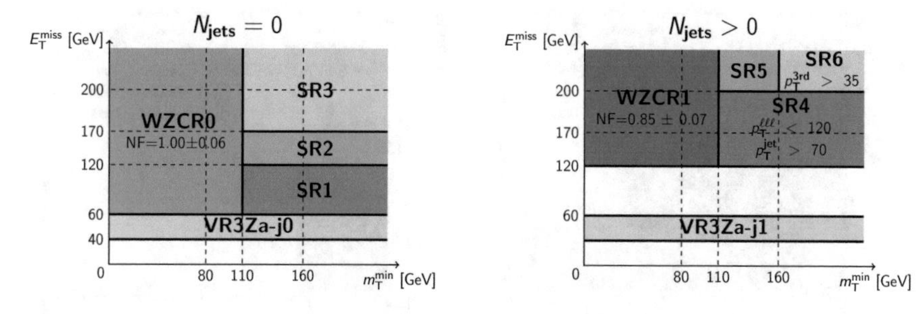

Fig. 7.18 Cartoon defining the SR, CR, and VRs for the jet veto and ISR regions

Table 7.11 Scale factors obtained from the background-only fits

	Scale factor
μ_{WZ0j}	1.075 ± 0.057
μ_{WZ1j}	0.942 ± 0.065

Table 7.12 Yields obtained from the binned control regions from the background-only fit. The errors shown are the statistical plus experimental systematic uncertainties

Table results yields channel	CR3WZ0j	CR3WZ1j
Observed events	486	264
Fitted bkg events	485.79 ± 22.04	263.81 ± 16.23
Fitted WZ0j events	452.81 ± 27.40	0.00 ± 0.00
Fitted WZ1j events	0.00 ± 0.00	241.04 ± 21.52
Fitted ZZ events	12.03 ± 9.98	5.48 ± 4.55
Fitted VVV events	1.58 ± 1.31	1.61 ± 1.33
Fitted ttvNLO events	1.37 ± 1.14	6.85 ± 5.73
Fitted Higgs events	2.51 ± 2.08	2.92 ± 2.42
Fitted FF events	15.48 ± 7.65	5.90 ± 2.92

The normalization factor derived for the jet veto CR is 1.075 ± 0.057, and 0.942 $\pm$ 0.065 for the ISR CR. They are summarized in Table 7.11.

The yields in the control regions are shown in Table 7.12. Figures 7.19, 7.20 show the modeling in the control regions with no normalization factor applied to the WZ background.

The yields in the validation regions are shown in Table 7.13 and Figs. 7.21, 7.22 show the modeling with the normalization factor applied to the WZ background. There is good agreement between the observed data and the expected background.

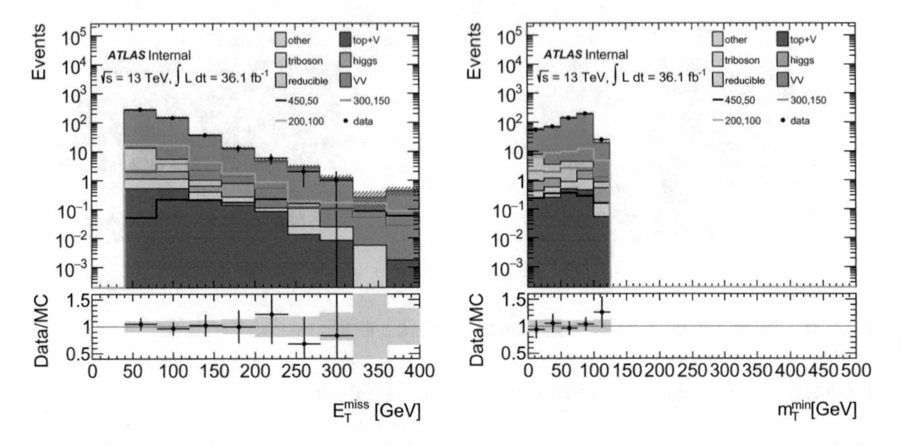

Fig. 7.19 Distributions in the jet veto WZ control region. Errors include a 10% systematic uncertainty for the WZ background, 30% for other MC irreducible backgrounds, and a data driven systematic uncertainty for the fakes

7.6.2 *Reducible Background Estimation*

The reducible background contribution comes from two sources: Z+jets/Z+γ where a jet fakes a lepton or from photon conversion, and top-like backgrounds due to semileptonic hadronic decays. The Z+jets/Z+γ is estimated using the data-driven method, Fake Factor method, first introduced in Sect. 5.6.1. The top-like background is estimated using simulation normalized to data using DFOS events, first described in Sect. 5.6.2. This section shows how these two techniques are applied in this analysis.

7.6.2.1 Z+jets/Z+γ Background

The Z+jets/Z+γ background enters the signal because a jet is mis-identified as a lepton or because of photon conversions. This background cannot be accurately described by simulation; therefore, it is estimated using the Fake Factor method.

The Fake Factor method identifies "ID" leptons whose criteria are identical to signal leptons, described in Table 7.2 and an an "anti-ID" criteria, defined in Table 7.14. The anti-ID criteria is enriched in fake leptons by inverting or relaxing identification and isolation criteria.

The Fake Factor is the ratio of ID to anti-ID leptons and is binned in lepton p_T. Fake Factors are derived for electrons and muons separately. The derivation of the Fake Factor is described in Sect. 5.6.1. In order to properly calculate the Fake Factor, the contribution from backgrounds with three prompt leptons must be subtracted from the data, as shown in Eq. (5.26).

The Fake Factor is derived in a region orthogonal to the signal selection selection and enriched with Z+jets and $Z + \gamma$ events by requiring $E_{\mathrm{T}}^{\mathrm{miss}} < 40$ GeV and $m_{\mathrm{T}}^{\mathrm{min}} <$

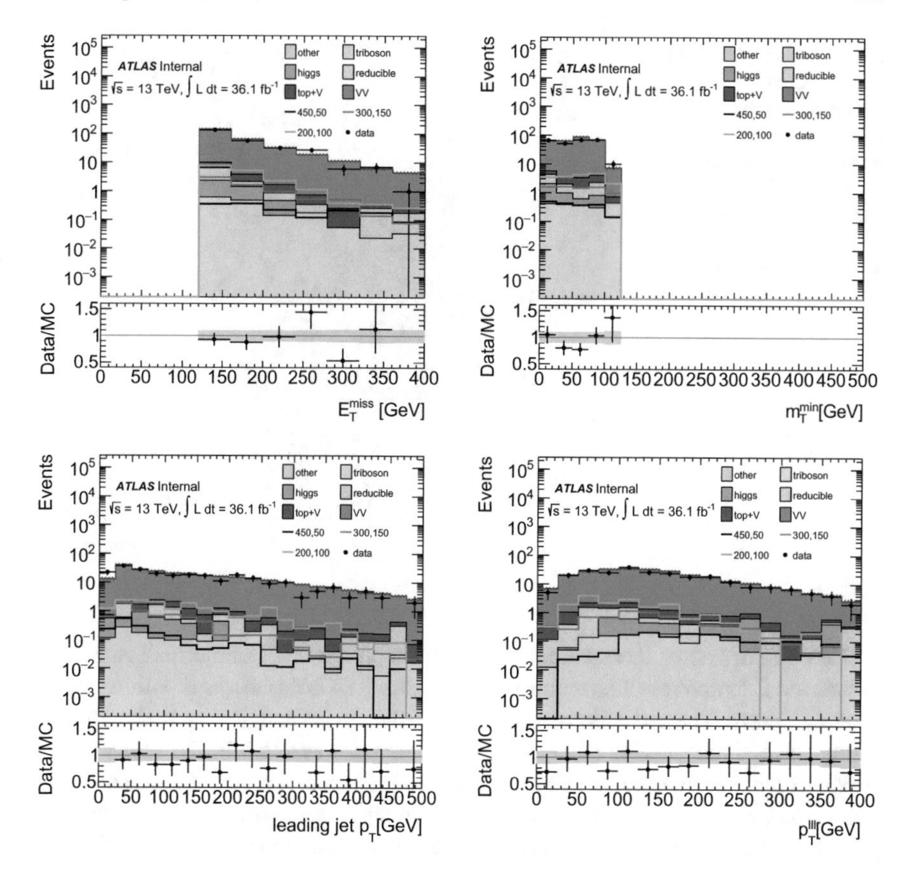

Fig. 7.20 Distributions in the ISR WZ control region. Errors include a 10% systematic uncertainty for the WZ background, 30% for other MC irreducible backgrounds, and a data driven systematic uncertainty for the fakes

Table 7.13 The fitted results shown for the binned validation regions are obtained from the binned control regions using the background-Ont fit. The errors shown are the statistical plus experimental systematic uncertainties

Table results yields channel	VR3Za0j	VR3Za1j
Observed events	618	746
Fitted bkg events	597.85 ± 60.85	809.36 ± 94.68
Fitted WZ0j events	520.34 ± 61.29	0.00 ± 0.00
Fitted WZ1j events	0.00 ± 0.00	620.25 ± 83.54
Fitted ZZ events	23.78 ± 19.72	47.65 ± 39.52
Fitted VVV events	0.93 ± 0.77	1.53 ± 1.27
Fitted ttvNLO events	0.79 ± 0.66	7.27 ± 6.08
Fitted Higgs events	1.21 ± 1.01	5.57 ± 4.62
Fitted FF events	50.79 ± 25.08	127.10 ± 62.77

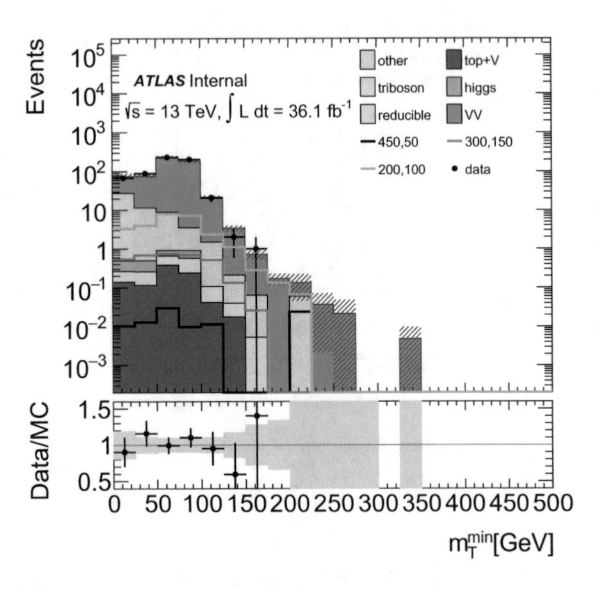

Fig. 7.21 Distributions in the jet veto WZ validation region with NF applied to WZ background. Errors include a 10% systematic uncertainty for the WZ background, 30% for other MC irreducible backgrounds, and a data driven systematic uncertainty for the fakes

30 GeV (Fake Factor measurement region). To enrich this sample in Z+jets/Z+γ events, the Z leptons are also required to reconstruct an invariant mass within 10 GeV of the Z mass. The Fake Factor estimate of the Z+jet and $Z + \gamma$ background is validated in a subset of the signal region containing events with $30 < m_{\mathrm{T}}^{\mathrm{min}} < 50$ GeV and $E_{\mathrm{T}}^{\mathrm{miss}} < 40$ GeV, which is enriched in background processes. The selection for the Fake Factor measurement and validation regions is summarized in Table 7.15 and illustrated in Fig. 7.23 to show how these regions are orthogonal to the WZ control and validation regions.

An additional fake VR is defined with third lepton $p_T > 30$ GeV, dilepton invariant mass outside the Z boson mass window of 10 GeV and with $E_{\mathrm{T}}^{\mathrm{miss}}$ between 40 and 60 GeV. This region is summarized in Table 7.10. Figure 7.24 shows the good agreement between data and background in the $E_{\mathrm{T}}^{\mathrm{miss}}$distribution in this validation region.

The Fake Factors are applied in the signal by regions by requiring that events satisfy the signal region requirements defined in Table 7.8 except that one signal lepton is replaced by an anti-ID lepton. The appropriate Fake Factor derived is applied to that event. The estimate for the number of three lepton events containing at least one fake lepton is shown in Eq. (5.25).

There are several sources of uncertainties for the Fake Factor method. First is the statistical uncertainty on the Fake Factor, which must be accounted for in the final Z+jets/Z+γ estimate.

Second, as the MC samples are used to subtract the diboson contribution from the data, the uncertainty associated to this subtraction must be evaluated. To do so, the MC WZ and ZZ yield is scaled up and down by 15%, and the Fake Factor is recalculated. The largest difference with respect to the nominal Fake Factor is then used as the Fake Factor's uncertainty on the diboson subtraction and assigned as a symmetric uncertainty.

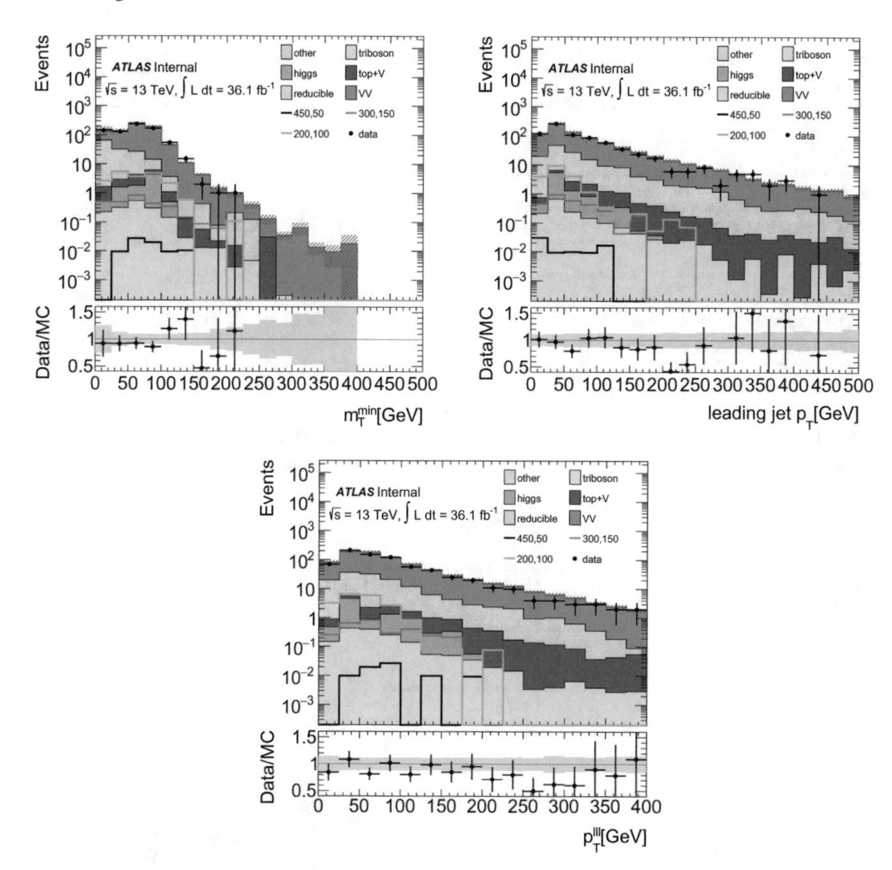

Fig. 7.22 Distributions in the ISR WZ validation region with NF applied to WZ background. Errors include a 10% systematic uncertainty for the WZ background, 30% for other MC irreducible backgrounds, and a data driven systematic uncertainty for the fakes

Third, a closure systematic is assigned to cover kinematic and composition differences between the Fake Factor measurement region and the signal region. To do so, the MC Z+jets/$Z+\gamma$ samples are used, and an MC-based Fake Factor is computed in these two kinematic regions. The difference between these two MC-based Fake Factors is used as the uncertainty on the MC closure. The region used to derive the closure systematic requires three signal leptons, or two signal and one anti-ID leptons, a b-jet veto, a same-flavor, opposite charge, pair with mass within 10 GeV of the Z mass, $m_{\mathrm{T}}^{\mathrm{min}} > 60\,\mathrm{GeV}$, and $E_{\mathrm{T}}^{\mathrm{miss}} > 60\,\mathrm{GeV}$. The looser cut on $m_{\mathrm{T}}^{\mathrm{min}}$, as compared with the signal region requirement of $m_{\mathrm{T}}^{\mathrm{min}} > 110\,\mathrm{GeV}$ is chosen to enhance the MC statistics in this region. The selection for this region is summarized in Table 7.15.

These systematic uncertainties are then added in quadrature to determine a total Fake Factor systematic uncertainty.

Table 7.14 Definition of the anti-ID criteria for the Fake Factor measurement region

Electrons	Muons
$p_T > 20$ GeV	$p_T > 20$ GeV
$\|\eta\| < 2.47$	$\|\eta\| < 2.4$
$\|\Delta z_0 \sin\theta\| < 1.0$	$\|\Delta z_0 \sin\theta\| < 1.0$
Pass `VeryLoose identification`	Pass `Medium` identification
Pass OR requirements	No OR requirements
(`!Medium identification` || $\|d_0$significance$\| > 5$ || `!GradientLoose isolation`)	($\|d_0$significance$\| > 3$ || `!GradientLoose isolation`)

Table 7.15 Summary of measurement and validation regions used for the Z+jets/Z+γ estimates. The closure region is only used with MC events, and is used to derive the MC closure systematic uncertainty on the Fake Factor

	$E_{\mathrm{T}}^{\mathrm{miss}}$ [GeV]	$m_{\mathrm{T}}^{\mathrm{min}}$ [GeV]	$\|m_{\ell\ell} - m_Z\|$ [GeV]	b-jet veto
FF measurement region	<40	<30	<10	N/a
FF validation region	<40	[30 – 50]	<10	N/a
FF closure region	>60	>60	<10	Required

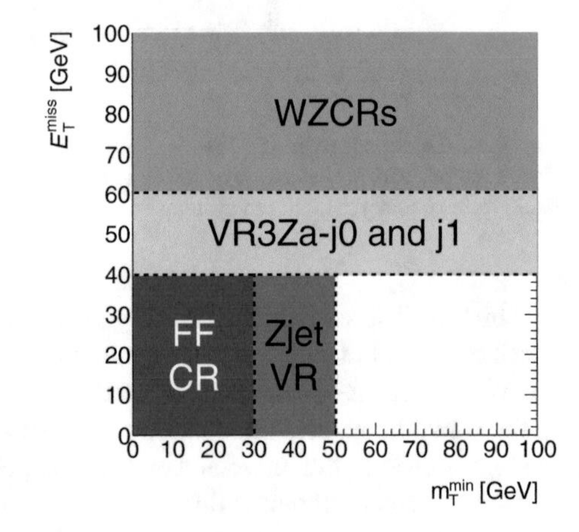

Fig. 7.23 Cartoon demonstrating the selection used to obtain the Fake Factor measurement and validation regions. The WZ CRs, and VRs, VR3Za-j0 and VR3Za-j1, are also displayed

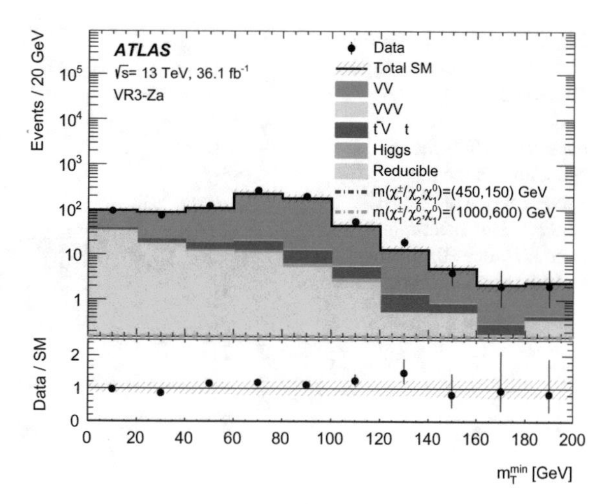

Fig. 7.24 $E_{\mathrm{T}}^{\mathrm{miss}}$distribution in the fake validation region, VR3-offZa. Reducible corresponds to the data-driven fake factor estimate. The uncertainty band includes all statistical and systematic uncertainties

Table 7.16 Selection criteria used to define the control region for the top-like backgrounds

Events with either three signal leptons or two signal leptons and one anti-ID lepton
Only $e^{\pm}e^{\pm}\mu^{\mp}$ and $\mu^{\pm}\mu^{\pm}e^{\mp}$ events
When measuring the normalization factors for events with an anti-ID lepton, the anti-ID lepton must be one of the same-flavor, same-sign leptons
$E_{\mathrm{T}}^{\mathrm{miss}} > 60$ GeV
$m_{\mathrm{T}}^{\mathrm{min}} > 60$ GeV
No requirement on $N_{\text{b-jets}}^{20\ \text{GeV}}$

7.6.2.2 Top-Like Backgrounds

The top-like background contribution in the signal region is estimated using simulation normalized to data in a control region. The top control region is defined using different flavor opposite charge events ($e^{\pm}e^{\pm}\mu^{\mp}$ and $\mu^{\pm}\mu^{\pm}e^{\mp}$) to minimize the WZ contamination and increase top purity. The fake lepton is one of the same flavor leptons. The normalization factor methodology for the top background is further described in Sect. 5.6.2.

The top control region is defined at $E_{\mathrm{T}}^{\mathrm{miss}} > 60$ GeV and $m_{\mathrm{T}}^{\mathrm{min}} > 60$ GeV, no requirement is applied on the invariant mass of the different flavor, opposite charge pair of leptons, and no b-jet veto is applied to increase the statistics in this region. The signal regions are binned in the number of jets; however a normalization factor inclusive in the number of jets is derived since the scale factor is similar between both regions. The selection of the top control region is summarized in Table 7.16.

A normalization factor is derived for electrons and muons separately. Moreover, there are two regions where the top NF is applied: in the signal region, and in the prompt background subtraction in the Fake Factor estimation. Both regions are defined with the same kinematic cuts except that NF derived for the signal regions

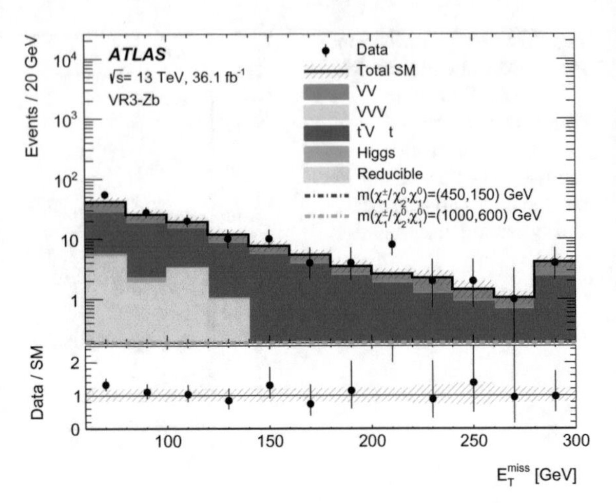

Fig. 7.25 $E_{\mathrm{T}}^{\mathrm{miss}}$distribution in the fake validation region, VR3-offZb. Reducible corresponds to the data-driven fake factor estimate and the top background with the NF applied. The uncertainty band includes all statistical and systematic uncertainties

uses three signal leptons, and the one for the prompt background subtraction is measured using two signal leptons and one anti-ID lepton.

The top NF factor associated with the Z+jets/Z+γ anti-ID control region, an electron NF of 1.04 ± 0.09 is obtained, along with a muon NF of 1.05 ± 0.03. The top NF associated with the three signal lepton top control region is 0.99 ± 0.42 for electrons and 2.37 ± 0.89 for muons.

The modeling of the top background is verified in a validation region, VR3-offZb, defined in Table 7.10. The validation region is defined with same-flavor, opposite sign events, just like the signal regions, with the third lepton $p_T > 20$ GeV and $E_{\mathrm{T}}^{\mathrm{miss}} > 40$ GeV. To minimize WZ contamination, this region vetoes invariant masses of the dilepton pair within 10 GeV of the mass of the Z boson. To increase top background statistics, at least one b-jet is required. Figure 7.25 shows the modeling of the $E_{\mathrm{T}}^{\mathrm{miss}}$ distribution in the top VR. There is good agreement between the data observed and the expected background.

The statistical uncertainty on the normalization factors is propagated to the final estimate, and is used as the systematic uncertainty on the top background.

7.7 Uncertainties

There are several sources of experimental and theoretical uncertainties. The largest sources of systematic uncertainties in the signal regions come from jet energy scale (JES) and resolution (JER), modeling of the soft term of the missing energy, theoretical uncertainties, and uncertainties associated with the calculation of the WZ NF.

The JES and JER uncertainties are derived as a function of jet p_T and η, as well as the jet flavor composition. They are derived using data and simulation using dijet, Z+jet, and γ+jet samples [32, 33].

The systematic uncertainties related to the $E_{\mathrm{T}}^{\mathrm{miss}}$ modeling in the simulation are estimated by propagating the uncertainties in the energy or momentum scale of each of the physics objects, as well as the uncertainties in the soft term's resolution and scale [34].

Other systematics are derived on the muon (electron) momentum (energy) resolution, momentum (energy) scale, reconstruction, and isolation efficiencies. Uncertainties due to the trigger efficiency, and b-tagging efficiency were also calculated. These uncertainties were found to be negligible.

Theoretical uncertainties on the WZ background are the choice of PDF set, QCD renormalization (μ_R) and factorization (μ_F) scales, and the choice of the strong coupling constant (α_s). Further discussion about the calculation of these systematics can be found in Sect. 5.8.

Uncertainties from the Fake Factor method and for the top background are summarized in Sects. 7.6.2.1–7.6.2.2.

A 2.1 % uncertainty is applied to the integrated luminosity.

7.8 Two Leptons and Jets Final State

The signal model in this analysis can also produce two leptons and jets in the final if the W boson decays hadronically instead of leptonically. Figure 7.2 shows the diagram for the production of $\tilde{\chi}_1^{\pm}\tilde{\chi}_2^0$ decaying via W and Z bosons to two leptons, jets, and $E_{\mathrm{T}}^{\mathrm{miss}}$. The results from the two lepton with and jets and the three lepton final states are combined in the limit so the search strategy for the two lepton and jets final state and its results will be introduced in this section.

7.8.1 Event Selection

The electrons, muons and jets are defined using the same criteria as for the three lepton final state. The identification and isolation criteria for electrons and muons is shown in Table 7.2, and, for jets, in Table 7.3.

Candidate events are required to have two leptons that form a same flavor opposite charge pair with invariant mass close to the mass of the Z boson. At least two jets are required which form an invariant mass consistent with the mass of the W boson.

SR2-int and SR2-high target signals with larger mass splittings. in these regions, the W boson is reconstructed from jets with the leading and sub-leading p_T. The only difference in their selection is their requirement on $E_{\mathrm{T}}^{\mathrm{miss}}$ because signal with intermediate mass splittings have less $E_{\mathrm{T}}^{\mathrm{miss}}$ in their final state than signals with large mass splittings.

Table 7.17 Signal region definitions used in the conventional 2ℓ search. W and Z refers to the reconstructed W and Z bosons in the final state. The Z boson is reconstructed using the two leptons. The W boson is reconstructing from jets with the leading and sub-leading p_T for the SR2-int and SR2-high regions. In SR2-low-3J, defined with 3–5 jets, the W boson is reconstructed from minimizing the $\Delta\phi$ between two jets and the Z+$E_{\mathrm{T}}^{\mathrm{miss}}$ system

2ℓ+jets signal region definitions

	SR2-int	SR2-high	SR2-low-2J	SR2-low-3J
$n_{\text{non-}b\text{-tagged jets}}$	≥ 2		2	3–5
$m_{\ell\ell}$ [GeV]	81–101		81–101	86–96
m_{jj} [GeV]	70–100		70–90	70–90
$E_{\mathrm{T}}^{\mathrm{miss}}$[GeV]	>150	>250	>100	>100
p_{T}^{Z} [GeV]	>80		>60	>40
p_{T}^{W} [GeV]	>100			
m_{T2} [GeV]	>100			
$\Delta R_{(jj)}$	<1.5			<2.2
$\Delta R_{(\ell\ell)}$	<1.8			
$\Delta\phi_{(\mathbf{p}_{\mathrm{T}}^{\mathrm{miss}},\mathbf{Z})}$			<0.8	
$\Delta\phi_{(\mathbf{p}_{\mathrm{T}}^{\mathrm{miss}},\mathbf{W})}$	0.5–3.0		>1.5	<2.2
$E_{\mathrm{T}}^{\mathrm{miss}}/p_{\mathrm{T}}^{Z}$			0.6−−1.6	
$E_{\mathrm{T}}^{\mathrm{miss}}/p_{\mathrm{T}}^{W}$			<0.8	
$\Delta\phi_{(\mathbf{p}_{\mathrm{T}}^{\mathrm{miss}},\mathrm{ISR})}$				>2.4
$\Delta\phi_{(\mathbf{p}_{\mathrm{T}}^{\mathrm{miss}},\mathrm{jet1})}$				>2.6
$E_{\mathrm{T}}^{\mathrm{miss}}/p_{\mathrm{T}}^{\mathrm{ISR}}$				0.4–0.8
$\lvert\eta(Z)\rvert$				<1.6
$p_{\mathrm{T}}^{\mathrm{jet3}}$ [GeV]				>30

To suppress $t\bar{t}$, a cut on the stransverse mass, m_{T2}, is required. The stransverse mass is defined in the following section, Sect. 7.8.1.1.

SR2-low regions target signals with mass splittings similar to the mass of the Z boson, where the signal becomes kinematically similar to the WZ background. Just like in the three lepton final state case, the signal regions are separated into "jet veto" and "ISR" regions. In this case, jet veto means that there are no additional jets in the events besides those that reconstruct the W boson. In the jet veto region, SR2-low-2J, the W boson is reconstructed from the leading and sub-leading jet. In the ISR region, SR2-low-3J, defined with 3–5 jets, the W boson is reconstructed from minimizing the $\Delta\phi$ between two jets and the Z+$E_{\mathrm{T}}^{\mathrm{miss}}$ system. Additional jets are considered part of the ISR system. Angular variables select the topology where the W boson recoils against the Z+$E_{\mathrm{T}}^{\mathrm{miss}}$ system in SR2-low-2J, and in SR2-low-2J, the topology selected is where the W+Z+$E_{\mathrm{T}}^{\mathrm{miss}}$ recoils against the ISR jets.

The selections for the two leptons and jets signal regions are summarized in Table 7.17.

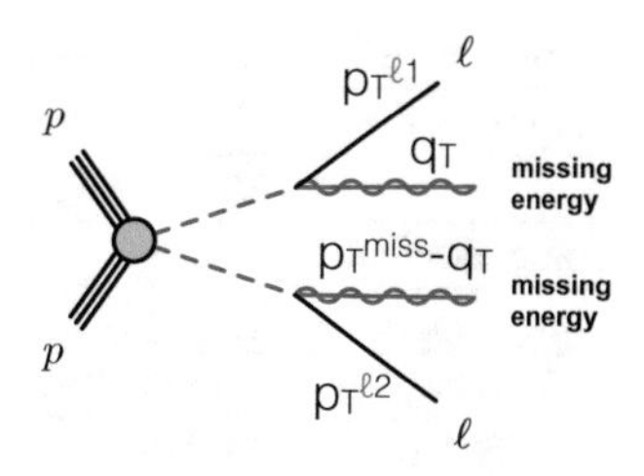

Fig. 7.26 Cartoon illustrating the calculation of m_{T2}

7.8.1.1 Defining m_{T2}

The stransverse mass, m_{T2}, is an extension of m_{T} where instead of one lepton and one source of $E_{\mathrm{T}}^{\mathrm{miss}}$ (neutrino or LSP), now there are two legs each with one lepton and one source of $E_{\mathrm{T}}^{\mathrm{miss}}$, as shown in Fig. 7.26.

Since $E_{\mathrm{T}}^{\mathrm{miss}}$ is calculated in the transverse plane, the mass of the lepton in Fig. 7.26 is:

$$m_{\ell} \geq m_{\mathrm{T}}(\mathbf{p}_{\mathrm{T}}, \mathbf{p}_{\mathrm{T}}^{\mathrm{miss}}), \tag{7.2}$$

where $\mathbf{p}_{\mathrm{T}}$ is the lepton vector in the transverse plane and $\mathbf{p}_{\mathrm{T}}^{\mathrm{miss}}$ is the missing energy vector.

If the splitting of the missing energy is determined as $\mathbf{p}_{\mathrm{T}}^{\mathrm{miss}} = \mathbf{q}_{\mathrm{T}}^{1} + \mathbf{q}_{\mathrm{T}}^{2}$, where $\mathbf{q}_{\mathrm{T}}^{1}$ is the amount of the missing energy on the leg with $p_{T}^{\ell 1}$, and $\mathbf{q}_{\mathrm{T}}^{2}$ is the amount of the missing energy on the leg with $p_{T}^{\ell 2}$, the mass of the lepton becomes:

$$m_{\ell} \geq \max\left(m_{\mathrm{T}}(\mathbf{p}_{\mathrm{T}}^{\ell 1}, \mathbf{q}_{\mathrm{T}}^{1}), m_{\mathrm{T}}(\mathbf{p}_{\mathrm{T}}^{\ell 2}, \mathbf{q}_{\mathrm{T}}^{2})\right) \tag{7.3}$$

However, since the splitting of the missing energy between the two legs is not known, a minimization over all possible splittings must be calculated:

$$m_{\ell}^{2} \geq \min_{\mathbf{q}_{\mathrm{T}}=\mathbf{q}_{\mathrm{T}}^{1}+\mathbf{q}_{\mathrm{T}}^{2}} \left[\max\left(m_{\mathrm{T}}(\mathbf{p}_{\mathrm{T}}^{\ell 1}, \mathbf{q}_{\mathrm{T}}^{1}), m_{\mathrm{T}}(\mathbf{p}_{\mathrm{T}}^{\ell 2}, \mathbf{q}_{\mathrm{T}}^{2})\right)\right] \tag{7.4}$$

Rewriting the equation above with $\mathbf{q}_{\mathrm{T}}^{1}$ as $\mathbf{q}_{\mathrm{T}}$ and $\mathbf{q}_{\mathrm{T}}^{2}$ as $\mathbf{p}_{\mathrm{T}}^{\mathrm{miss}} - \mathbf{q}_{\mathrm{T}}$, the stransverse mass, m_{T2}, [35, 36] is defined as:

$$m_{\mathrm{T2}} = \min_{\mathbf{q}_{\mathrm{T}}} \left[\max\left(m_{\mathrm{T}}(\mathbf{p}_{\mathrm{T}}^{\ell 1}, \mathbf{q}_{\mathrm{T}}), m_{\mathrm{T}}(\mathbf{p}_{\mathrm{T}}^{\ell 2}, \mathbf{p}_{\mathrm{T}}^{\mathrm{miss}} - \mathbf{q}_{\mathrm{T}})\right)\right], \tag{7.5}$$

where $\mathbf{p}_{\mathrm{T}}^{\ell 1}$ and $\mathbf{p}_{\mathrm{T}}^{\ell 2}$ are the transverse momentum vectors of the two leptons, and $\mathbf{q}_{\mathrm{T}}$ is a transverse momentum vector that minimizes the larger of $m_{\mathrm{T}}(\mathbf{p}_{\mathrm{T}}^{\ell 1}, \mathbf{q}_{\mathrm{T}})$ and $m_{\mathrm{T}}(\mathbf{p}_{\mathrm{T}}^{\ell 2}, \mathbf{p}_{\mathrm{T}}^{\mathrm{miss}} - \mathbf{q}_{\mathrm{T}})$.

Table 7.18 Validation region definitions used in the conventional 2ℓ search

2ℓ+jets validation region definitions

	VR2-int(high)	VR2-low-2J(3J)	VR2-VV-int	VR2-VV-low
Loose selection				
$n_{\text{non-}b\text{-tagged jets}}$	≥ 2	2 (3–5)	1	1
$E_{\mathrm{T}}^{\mathrm{miss}}$[GeV]	>150 (>250)	>100	>150	>150
$m_{\ell\ell}$ [GeV]	81–101	81–101 (86–96)		81–101
m_{jj} [GeV]	$\notin$ [60, 100]	$\notin$ [60, 100]		
p_{T}^{Z} [GeV]	>80	>60 (>40)		
p_{T}^{W} [GeV]	>100			
$\lvert\eta(Z)\rvert$		(< 1.6)		
$p_{\mathrm{T}}^{\mathrm{jet3}}$ [GeV]		(>30)		
$\Delta\phi_{(\mathbf{p}_{\mathrm{T}}^{\mathrm{miss}},\mathrm{jet})}$			>0.4	>0.4
m_{T2} [GeV]			>100	
$\Delta R_{(\ell\ell)}$				<0.2
Tight selection				
$\Delta R_{(jj)}$	<1.5	(<2.2)		
$\Delta\phi_{(\mathbf{p}_{\mathrm{T}}^{\mathrm{miss}},W)}$	0.5–3.0	>1.5 (<2.2)		
$\Delta\phi_{(\mathbf{p}_{\mathrm{T}}^{\mathrm{miss}},Z)}$		< 0.8 (–)		
$E_{\mathrm{T}}^{\mathrm{miss}}/p_{\mathrm{T}}^{W}$		< 0.8 (–)		
$E_{\mathrm{T}}^{\mathrm{miss}}/p_{\mathrm{T}}^{Z}$		0.6–1.6 (–)		
$E_{\mathrm{T}}^{\mathrm{miss}}/p_{\mathrm{T}}^{\mathrm{ISR}}$		(0.4–0.8)		
$\Delta\phi_{(\mathbf{p}_{\mathrm{T}}^{\mathrm{miss}},\mathrm{ISR})}$		(> 2.4)		
$\Delta\phi_{(\mathbf{p}_{\mathrm{T}}^{\mathrm{miss}},\mathrm{jet1})}$		(> 2.6)		
m_{T2} [GeV]	>100			
$\Delta R_{(\ell\ell)}$	<1.8			

7.8.2 Background Modeling

The dominant background in this search is diboson processes, WW, ZZ, and WZ. This background is estimated using simulation and the modeling is validated in VRs.

Z+jets events can enter the signal region due to fake $E_{\mathrm{T}}^{\mathrm{miss}}$. Sources of fake $E_{\mathrm{T}}^{\mathrm{miss}}$ are jets reconstructed as missing energy, lepton mis-measurements or neutrinos from semileptonic $b-$ or $c-$ hadron decays. This is difficult to estimate using simulation so a data-driven method, the photon template, is used where γ+jets events estimate the contribution of Z+jets events in the signal region. Similar techniques have been employed by ATLAS [37] and by CMS [38, 39].

The validation regions for the photon template method and for VV are summarized in Table 7.18. Tables 7.19, 7.20 shows the yields in VR2-int, VR2-high, and VR2-low. Table 7.21 shows the yields in the diboson validation regions. The modeling of

Table 7.19 Yields in the validation regions for "loose" and "tight" selections for VR2-int and VR2-high in the conventional 2ℓ search. The Z+jets background is predicted using the data-driven γ+jet method. All systematic and statistical uncertainties are included. The "top" background includes all processes containing one or more top quarks and the "other" backgrounds include all processing containing a Higgs boson and VVV.

	VR2-int (loose)	VR2-int (tight)	VR2-high (loose)	VR2-int (tight)
Observed	246	20	60	6
Total SM	240 ± 26	12.7 ± 1.3	57 ± 5	4.7 ± 0.6
VV	121.8 ± 1.5	11.4 ± 0.6	40.9 ± 0.9	4.7 ± 0.3
Top	42.4 ± 2.8	0.1 ± 0.0	8.5 ± 1.1	–
FNP	27 ± 11	–	6 ± 5	–
Z+jets	49 ± 24	1.2 ± 1.1	1.8 ± 2.0	0.0 ± 0.5

Table 7.20 Yields results for the "loose" and "tight" selections of VR2-low in the conventional 2ℓ search. The Z+jets background is predicted using the data-driven γ+jet method. All systematic and statistical uncertainties are included. The "top" background includes all processes containing one or more top quarks and the "other" backgrounds include all processing containing a Higgs boson and VVV

	VR2-low (loose)	VR2-low (tight)
Observed	919	51
Total SM	980 ± 90	85 ± 12
VV	190 ± 2	15.8 ± 0.7
Top	105 ± 4	11.2 ± 1.6
FNP	41 ± 19	20 ± 8
Z+jets	640 ± 90	39 ± 9

$E_{\mathrm{T}}^{\mathrm{miss}}$ and m_{T2} in the validation regions is shown in Fig. 7.27. There is good agreement between the observed data and the expected background events.

7.8.3 Results

Table 7.22 summarizes the observed data events and the expected background yields in the conventional 2ℓ search SRs. Figure 7.28 shows the $E_{\mathrm{T}}^{\mathrm{miss}}$ distribution in SR2-int and SR2-high, which differ only in the $E_{\mathrm{T}}^{\mathrm{miss}}$ requirement, and in SR2-low of the 2ℓ+jets channel. No significant excess is observed.

Table 7.21 Background results for the diboson validation regions in the conventional 2ℓ search. The Z+jets background is predicted using MC. All systematic and statistical uncertainties are included. The "top" background includes all processes producing one or more top quarks and the "other" backgrounds include all processes producing a Higgs boson or VVV. A "–" symbol indicates that the background contribution is negligible

	VR2-VV-low	VR2-VV-int
Observed	111	114
Total SM	99 ± 4	101 ± 4
VV	89.3 ± 3.5	94 ± 4
Top	7.5 ± 1.1	4.6 ± 1.0
FNP	–	–
Z+jets	2.0 ± 1.5	2.4 ± 1.7

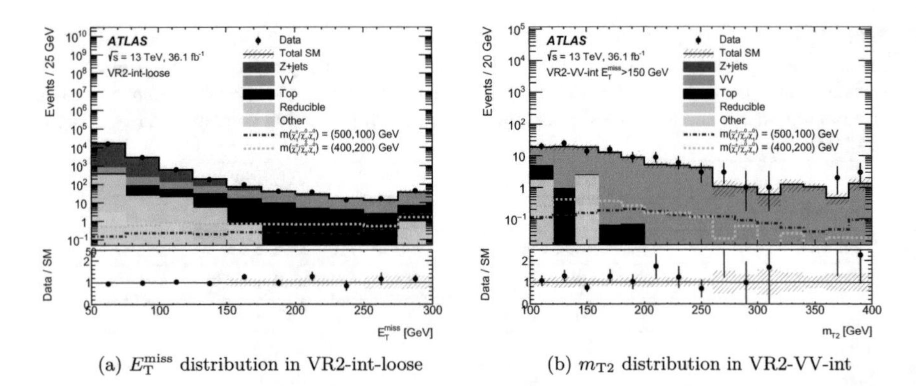

(a) $E_{\mathrm{T}}^{\mathrm{miss}}$ distribution in VR2-int-loose (b) m_{T2} distribution in VR2-VV-int

Fig. 7.27 Distributions of $E_{\mathrm{T}}^{\mathrm{miss}}$ and m_{T2} in the 2ℓ+jets validation region. The "top" background includes $t\bar{t}$, Wt and $t\bar{t}V$, the "other" backgrounds include Higgs bosons and VVV, the "reducible" category corresponds to the data-driven matrix method estimate, and the Z+jets contribution is evaluated with the data-driven γ+jet template method

Table 7.22 Observed events in the conventional 2ℓ search SRs. All systematic and statistical uncertainties are included. The "top" background includes all processes producing one or more top quarks and the "other" backgrounds include all processes producing a Higgs boson or VVV. A "–" symbol indicates that the background contribution is negligible

SR2-	Int	High	Low (combined)
Observed	2	0	11
Total SM	$4.1^{+2.6}_{-1.8}$	$1.6^{+1.6}_{-1.1}$	$4.2^{+3.4}_{-1.6}$
VV	4.0 ± 1.8	1.6 ± 1.1	1.7 ± 1.0
Top	0.15 ± 0.11	0.04 ± 0.03	0.8 ± 0.4
FNP	$0.0^{+0.2}_{-0.0}$	$0.0^{+0.1}_{-0.0}$	$0.7^{+1.8}_{-0.7}$
Z+jets	$0.0^{+1.8}_{-0.0}$	$0.0^{+1.2}_{-0.0}$	$1.0^{+2.7}_{-1.0}$
Other	–	–	–

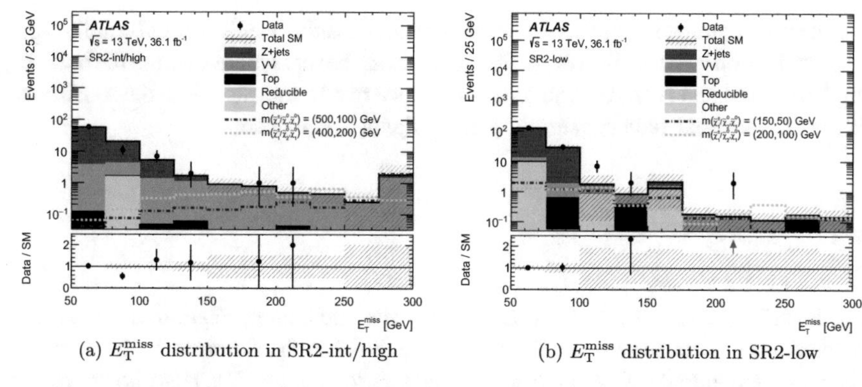

(a) $E_{\mathrm{T}}^{\mathrm{miss}}$ distribution in SR2-int/high (b) $E_{\mathrm{T}}^{\mathrm{miss}}$ distribution in SR2-low

Fig. 7.28 Distributions of $E_{\mathrm{T}}^{\mathrm{miss}}$ in SR2-int/high and SR2-low. The 2 jets and 3–5 jets regions are combined in SR2-low. The "top" background includes $t\bar{t}$, Wt and $t\bar{t}V$, the "other" backgrounds include Higgs bosons and VVV, the "reducible" category corresponds to the data-driven matrix method estimate, and the Z+jets contribution is evaluated with the data-driven γ+jet template method

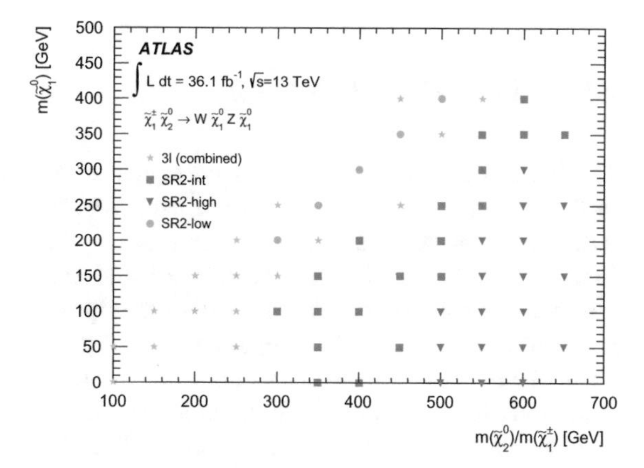

Fig. 7.29 Signal regions contributing to the observed exclusion limit for direct $\tilde{\chi}_1^{\pm}\tilde{\chi}_2^0$ with WZ mediated decays. The markers indicate which result, out of those from SR2-high, SR2-int, SR2-low, and the combination of the three leptons exclusive regions, has the best expected sensitivity

7.8.4 Sensitivity of Each Final State

Since the conventional 2ℓ and 3ℓ searches are orthogonal due to the requirement on the number of leptons, they can be statistically combined to set an exclusion limit on the production of $\tilde{\chi}_1^{\pm}\tilde{\chi}_2^0$ decaying via W and Z bosons. Figure 7.29 shows the sensitivity of each of the signal regions.

The three lepton final state has the greatest sensitivity for the smallest mass splittings, where Δm is close to the mass of the Z boson. The dominant background, WZ is minimized using $m_{\mathrm{T}}^{\mathrm{min}}$ and by making use of both jet veto and ISR regions.

The two lepton and jets final state has greater sensitivity at larger mass splittings. The production cross section of $\tilde{\chi}_1^{\pm}\tilde{\chi}_2^0$ decreases as the mass increases and the branching fraction to jets is larger than to leptons. As a result, this increases the sensitivity of the 2ℓ+jets final state because of greater signal acceptance.

7.9 Results

The HistFitter framework [40] is used for the statistical interpretation of the results. A likelihood is constructed as the product of Poisson distributions with the mean taken as the nominal MC yield in each of the control regions. The HistFitter package constrains the values and uncertainties on the normalization factors. The NFs are used to extrapolate the background prediction into validation regions, where modelling is verified, and the signal regions. Systematic uncertainties are treated as nuisance parameters in the likelihood fit.

7.9.1 Background-Only Fit

The background only fit assumes no signal is present in any region. Only data in the CRs are used to constrain the background only fit. This provides an SR-independent background prediction in all regions. The results are shown in Table 7.23 for SR3-WZ-0Ja to SR3-WZ-0Jc and SR3-WZ-1Ja to SR3-WZ-1Jc. A summary of the observed and expected yields in all of the signal regions considered in this paper is provided in Fig. 7.30. The slepton and 2ℓ+0 jets searches is not discussed in this document but can be found in the publication for this search [1]. No significant excess above the SM expectation is observed in any SR.

Figure 7.31 shows the N-1 $E_{\mathrm{T}}^{\mathrm{miss}}$ distributions in the three lepton signal regions. The signal regions are indicated on the distributions by arrows. Good agreement between data and expectations is observed in all distributions within the uncertainties.

No significant excess is observed in the 2ℓ+jets or three lepton signal regions.

7.9.2 Model Dependent Limits

Since no significant excess is observed, two types of exclusion limits for new physics are calculated using the $\mathrm{CL_s}$ technique [41]: exclusion limits and discovery limits (discussed in Sect. 7.9.3).

Table 7.23 Unblinded yields in the signal regions in the conventional 3ℓ search. The "FF" yields result from the Fake Factor Method. The errors shown are the statistical plus systematic uncertainties

Table results yields channel	SR3L0Ja	SR3L0Jb	SR3L0Jc	SR3L1Ja	SR3L1Jb	SR3L1Jc
Observed events	21	1	2	1	3	4
Fitted bkg events	21.72 ± 1.60	2.68 ± 0.40	1.56 ± 0.30	2.21 ± 0.36	1.82 ± 0.27	1.26 ± 0.39
Fitted WZ0j events	19.46 ± 1.63	2.46 ± 0.37	1.32 ± 0.24	0.00 ± 0.00	0.00 ± 0.00	0.00 ± 0.00
Fitted WZ1j events	0.00 ± 0.00	0.00 ± 0.00	0.00 ± 0.00	1.78 ± 0.30	1.49 ± 0.22	0.92 ± 0.26
Fitted ZZ events	0.82 ± 0.68	0.06 ± 0.05	0.05 ± 0.04	0.05 ± 0.04	0.02 ± 0.01	0.02 ± 0.01
Fitted VVV events	0.32 ± 0.26	0.13 ± 0.11	0.14 ± 0.12	0.12 ± 0.10	0.12 ± 0.10	0.24 ± 0.20
Fitted ttvNLO events	0.04 ± 0.03	0.01 ± 0.01	0.01 ± 0.01	0.14 ± 0.12	0.12 ± 0.10	0.08 ± 0.07
Fitted Higgs events	0.00 ± 0.00	0.00 ± 0.00	0.00 ± 0.00	0.01 ± 0.01	0.00 ± 0.00	0.00 ± 0.00
Fitted FF events	1.09 ± 0.54	0.02 ± 0.01	0.04 ± 0.02	0.11 ± 0.06	0.07 ± 0.04	0.01 ± 0.00

Exclusion limits are set on the masses of the charginos and neutralinos for the simplified models in Fig. 7.2. Figure 7.32 shows the limits from the 3ℓ and 2ℓ+jets channels in the $\tilde{\chi}_1^{\pm}\tilde{\chi}_2^0$ production with decays via W/Z bosons. The 3ℓ limits are calculated using a statistical combination of the six SR3-WZ regions. Since the SRs in the 2ℓ+jets channel are not mutually exclusive, the observed CL_s value is taken from the signal region with the best expected CL_s value. The 3ℓ and 2ℓ+jets channels are then combined, using the channel with the best expected CL_s value for each point in the SUSY particle mass space, shown in Fig. 7.29. $\tilde{\chi}_1^{\pm}$ and $\tilde{\chi}_2^0$ masses up to 580 GeV are excluded for a massless $\tilde{\chi}_1^0$ neutralino.

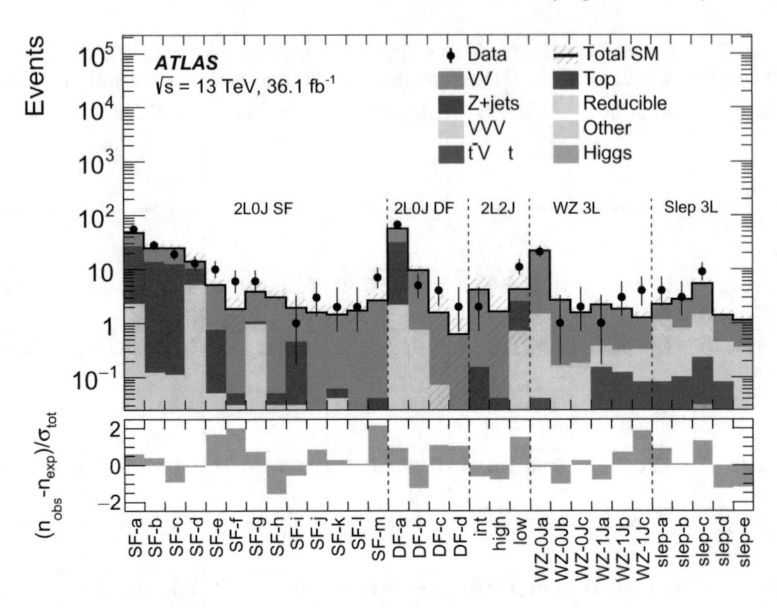

Fig. 7.30 The observed and expected background yields in the signal regions considered in the 2ℓ+0 jets, 2ℓ+jets, and 3ℓ final states. All uncertainties in the background prediction are included in the uncertainty band. The bottom plot shows the difference in standard deviations between the observed and expected yields. n_{obs} and n_{bkg} are the observed data and expected background yields, respectively, $\sigma_{\text{tot}} = \sqrt{n_{\text{bkg}} + \sigma^2_{\text{bkg}} + \sigma^2_{\text{exp}}}$, and σ_{exp} is the total background uncertainty

7.9.3 *Model Independent Limits*

The second type of exclusion limit for new physics is model independent. Upper limits are set on the visible cross-section $\langle\epsilon\sigma^{95}_{obs}\rangle$ as well as on the observed (S^{95}_{obs}) and expected (S^{95}_{exp}) number of events from new physics processes. Dividing S^{95}_{obs} by the integrated luminosity of 36.1 fb^{-1} defines the upper limits on the visible cross-sections, $\langle\epsilon\sigma^{95}_{obs}\rangle$. The p-value and the corresponding significance for the background-only hypothesis is evaluated. These are shown for each of the six 3ℓ signal regions in Table 7.24.

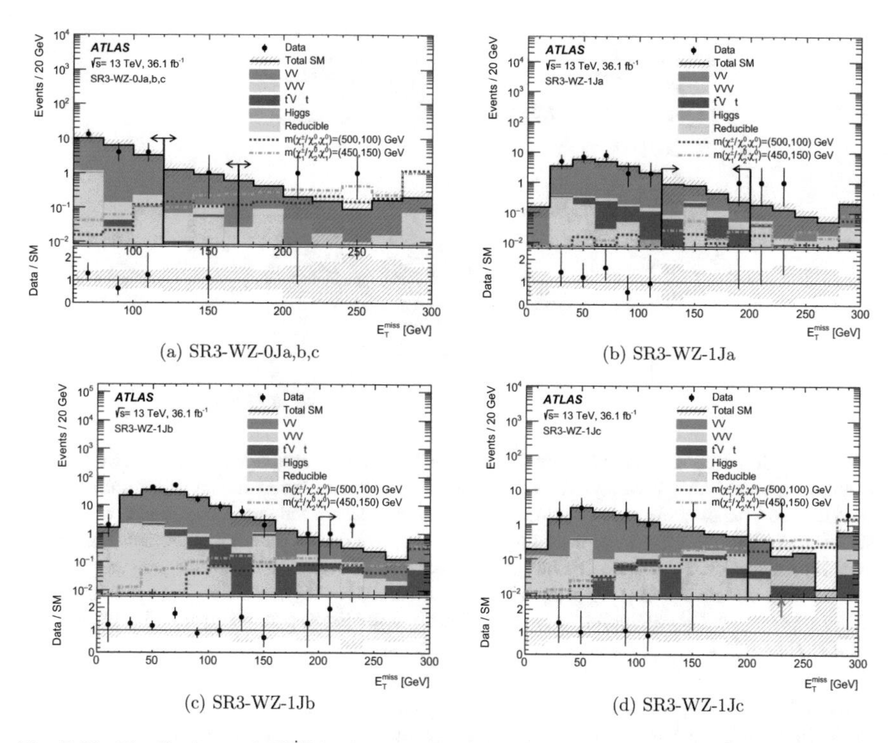

(a) SR3-WZ-0Ja,b,c (b) SR3-WZ-1Ja

(c) SR3-WZ-1Jb (d) SR3-WZ-1Jc

Fig. 7.31 Distributions of $E_{\mathrm{T}}^{\mathrm{miss}}$ in **a** SR3-WZ-0Ja,b,c, **b** SR3-WZ-1Ja, **c** SR3-WZ-1Jb and **d** SR3-WZ-1Jc. The normalization factors extracted from the corresponding CRs are used to rescale the 0-jet and 1-jet WZ background components. The "reducible" category corresponds to the Fake-Factor estimate. The uncertainty bands include all systematic and statistical contributions. Simulated signal models for charginos/neutralinos production are overlayed for comparison

The numbers derived from the model-independent limits can be understood as follows,

- $S^{95}_{exp} = 1.64 \times \sqrt{(\text{exp. error}^2 + \text{obs. data})}$, where 1.64 is the 95% confidence level for a one-sided hypothesis test
- S^{95}_{obs} = (obs. data - exp. data) + S^{95}_{exp}
- $\langle \epsilon\sigma^{95}_{obs} \rangle = \frac{S^{95}_{obs}}{\text{luminosity}}$
- $Z = \frac{\text{(obs. data - exp. data)}}{\sqrt{(\text{exp. error}^2 + \text{obs. data})}}$

These calculations can serve as a cross-check of model-independent limits.

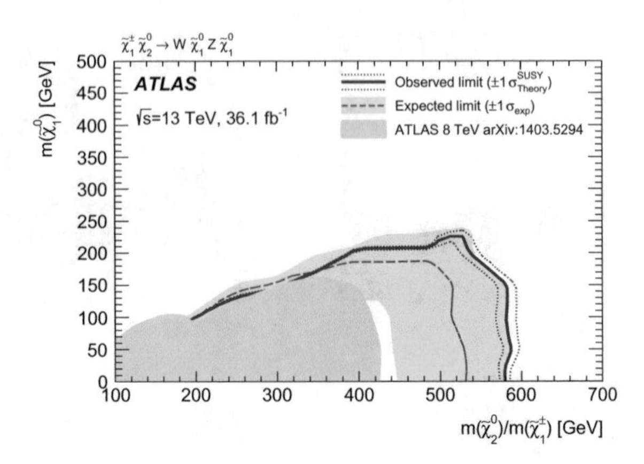

Fig. 7.32 Observed and expected exclusion limits on SUSY simplified models for chargino–neutralino production with decays via W/Z bosons. The observed (solid thick red line) and expected (thin dashed blue line) exclusion contours are indicated. The shaded band corresponds to the $\pm 1\sigma$ variations in the expected limit, including all uncertainties except theoretical uncertainties in the signal cross-section. The dotted lines around the observed limit illustrate the change in the observed limit as the nominal signal cross-section is scaled up and down by the theoretical uncertainty. All limits are computed at 95% confidence level. The observed limits obtained from ATLAS in Run 1 are also shown [42]

Table 7.24 Summary of results and model-independent limits in the inclusive 3ℓ SRs. Signal model-independent upper limits at 95% C.L. on the the visible signal cross-section ($\langle\epsilon\sigma^{95}_{obs}\rangle$), and the observed and expected upper limit on the number of BSM events (S^{95}_{obs} and S^{95}_{exp}, respectively) are also shown. The last two columns show the p-value and the corresponding significance for the background-only hypothesis.

Region	N_{obs}	N_{exp}	$\langle\epsilon\sigma^{95}_{obs}\rangle$[fb]	S^{95}_{obs}	S^{95}_{exp}	$p(s=0)$	Z
WZ-0Ja	21	22 ± 2.9	0.35	12.8	$13.5^{+2.7}_{-5.3}$	0.5	0
WZ-0Jb	1	2.7 ± 0.5	0.10	3.7	$4.6^{+2.1}_{-0.9}$	0.5	0
WZ-0Jc	2	1.6 ± 0.3	0.13	4.8	$4.1^{+1.7}_{-0.7}$	0.28	0.57
WZ-1Ja	1	2.2 ± 0.5	0.09	3.2	$4.5^{+1.6}_{-1.3}$	0.5	0
WZ-1Jb	3	1.8 ± 0.3	0.16	5.6	$4.3^{+1.7}_{-0.9}$	0.18	0.91
WZ-1Jc	4	1.3 ± 0.3	0.20	7.2	$4.2^{+1.7}_{-0.4}$	0.03	1.82

References

1. ATLAS Collaboration (2018) Search for electroweak production of supersymmetric particles in final states with two or three leptons at $\sqrt{s} = 13$ TeV with the ATLAS detector. Eur Phys J C78:12, 995. https://doi.org/10.1140/epjc/s10052-018-6423-7. arXiv:1803.02762 [hepspsex]
2. ATLAS Collaboration (2014) Search for direct production of charginos, neutralinos and sleptons in final states with two leptons and missing transverse momentum in pp collisions at $\sqrt{s} =$ 8 TeV with the ATLAS detector. JHEP 05 071. https://doi.org/10.1007/JHEP05(2014)071. arXiv:1403.5294 [hepspsex]

3. ATLAS Collaboration (2014) Search for direct production of charginos and neutralinos in events with three leptons and missing transverse momentum in $\sqrt{s} =$ 8TeV pp collisions with the ATLAS detector. JHEP 04, 169. https://doi.org/10.1007/JHEP04(2014)169. arXiv:1402.7029 [hepspsex]
4. ATLAS Collaboration (2016) Search for the electroweak production of supersymmetric particles in $\sqrt{s}$=8 TeV pp collisions with the ATLAS detector. Phys Rev D93:5, 052002. https://doi.org/10.1103/PhysRevD.93.052002. arXiv:1509.07152 [hepspsex]
5. CMS Collaboration (2018) Search for electroweak production of charginos and neutralinos in multilepton final states in proton-proton collisions at $\sqrt{s} = 13$ TeV. JHEP 03, 166. https://doi.org/10.1007/JHEP03(2018)166. arXiv:1709.05406 [hepspsex]
6. CMS Collaboration (2012) Search for electroweak production of charginos and neutralinos using leptonic final states in pp collisions at $\sqrt{s} = 7$ TeV. JHEP 11, 147. https://doi.org/10.1007/JHEP11(2012)147. arXiv:1209.6620 [hepspsex]
7. CMS Collaboration (2014) Searches for electroweak neutralino and chargino production in channels with Higgs, Z, and W bosons in pp collisions at 8 TeV. Phys Rev D90:9, 092007. https://doi.org/10.1103/PhysRevD.90.092007. arXiv:1409.3168 [hepspsex]
8. CMS Collaboration (2014) Searches for electroweak production of charginos, neutralinos, and sleptons decaying to leptons and W, Z, and Higgs bosons in pp collisions at 8 TeV, https://doi.org/10.1140/epjc/s10052-014-3036-7Eur Phys J C74:9, 3036, arXiv:1405.7570 [hepspsex]
9. Alwall J, Schuster P, Toro N (2009) Simplified models for a first characterization of new physics at the LHC. Phys Rev D 79, 075020. https://doi.org/10.1103/PhysRevD.79.075020. arXiv:0810.3921 [hepspsph]
10. ATLAS Collaboration (2012) Measurement of the WW cross section in $\sqrt{s} = 7$ TeV pp collisions with the ATLAS detector and limits on anomalous gauge couplings. Phys Lett B 712, 289. https://doi.org/10.1016/j.physletb.2012.05.003. arXiv:1203.6232 [hepspsex]
11. Prospects for Higgs Boson searches using the $H \to WW^{(*)} \to \ell\nu\ell\nu$ decay mode with the ATLAS detector for 10 TeV, Technical report, ATL-PHYS-PUB-2010-005, CERN, Geneva, Jun, 2010. https://cds.cern.ch/record/1270568
12. ATLAS Collaboration (2010) The ATLAS simulation infrastructure. Eur Phys J C 70:823. https://doi.org/10.1140/epjc/s10052-010-1429-9. arXiv:1005.4568 [physics.ins-det]
13. GEANT4 Collaboration, Agostinelli S et al (2003) GEANT4–a simulation toolkit. Nucl Instrum Meth A 506:250. https://doi.org/10.1016/S0168-9002(03)01368-8
14. ATLAS Collaboration (2010) The simulation principle and performance of the ATLAS fast calorimeter simulation FastCaloSim, ATL-PHYS-PUB-2010-013. https://cds.cern.ch/record/1300517
15. Gleisberg T, Höche S, Krauss F, Schönherr M, Schumann S, Siegert F, Winter J (2007) Event generation with SHERPA 1.1. JHEP 02, 007. https://doi.org/10.1088/1126-6708/2009/02/007. arXiv:0811.4622 [hepspsph]
16. ATLAS Collaboration (2016) Multi-boson simulation for 13 TeV ATLAS analyses, ATL-PHYS-PUB-2016-002. https://cds.cern.ch/record/2119986
17. Re E (2011) Single-top Wt-channel production matched with parton showers using the POWHEG method. Eur Phys J C 71, 1547. https://doi.org/10.1140/epjc/s10052-011-1547-z. arXiv:1009.2450 [hepspsph]
18. Frixione S, Nason P, Ridolfi G (2007) A positive-weight next-to-leading-order Monte Carlo for heavy flavour hadroproduction. JHEP 09, 126. https://doi.org/10.1088/1126-6708/2007/09/126. arXiv:0707.3088 [hepspsph]
19. Czakon M, Mitov A (2014) Top++: a program for the calculation of the top-pair cross-section at hadron colliders. Comput Phys Commun 185, 2930. https://doi.org/10.1016/j.cpc.2014.06.021. arXiv:1112.5675 [hepspsph]
20. Kidonakis N (2010) Two-loop soft anomalous dimensions for single top quark associated production with a W^- or H^-. Phys Rev D 82, 054018. https://doi.org/10.1103/PhysRevD.82.054018. arXiv:1005.4451 [hepspsph]
21. Catani S, Cieri L, Ferrera G, de Florian D, Grazzini M (2009) Vector boson production at hadron colliders: a fully exclusive QCD calculation at NNLO. Phys Rev Lett 103, 082001. https://doi.org/10.1103/PhysRevLett.103.082001. arXiv:0903.2120 [hepspsph]

22. Alioli S, Nason P, Oleari C, Re E (2010) A general framework for implementing NLO calculations in shower Monte Carlo programs: the POWHEG BOX. JHEP 06, 043. arXiv:1002.2581 [hepspsph]
23. Beenakker W, Hopker R, Spira M, Zerwas P (1997) Squark and gluino production at hadron colliders. Nucl. Phys. B 492:51. https://doi.org/10.1016/S0550-3213(97)00084-9. arXiv:hep-ph/9610490
24. Kulesza A, Motyka L (2009) Threshold resummation for squark-antisquark and gluino-pair production at the LHC. Phys Rev Lett 102:111802. https://doi.org/10.1103/PhysRevLett.102.111802. arXiv:0807.2405
25. Kulesza A, Motyka L (2009) Soft gluon resummation for the production of gluino-gluino and squark-antisquark pairs at the LHC. Phys Rev D 80, 095004. https://doi.org/10.1103/PhysRevD.80.095004. arXiv:0905.4749 [hepspsph]
26. Beenakker W, Brensing S, Kramer M, Kulesza A, Laenen E et al (2009) Soft-gluon resummation for squark and gluino hadroproduction. JHEP 12, 041. https://doi.org/10.1088/1126-6708/2009/12/041. arXiv:0909.4418 [hepspsph]
27. Beenakker W, Brensing S, Kramer M, Kulesza A, Laenen E, et al (2011) Squark and gluino hadroproduction. Int J Mod Phys A 26, 2637. https://doi.org/10.1142/S0217751X11053560. arXiv:1105.1110 [hepspsph]
28. Borschensky C, Kramer M, Kulesza A, Mangano M, Padhi S, Plehn T, Portell X (2014) Squark and gluino production cross sections in pp collisions at $\sqrt{s} = 13$, 14, 33 and 100 TeV. Eur Phys J C 74, 3174. https://doi.org/10.1140/epjc/s10052-014-3174-y. arXiv:1407.5066 [hepspsph]
29. ATLAS Collaboration (2016) Performance of pile-up mitigation techniques for jets in pp collisions at $\sqrt{s} = 8$ TeV using the ATLAS detector. Eur Phys J C76:11, 581. https://doi.org/10.1140/epjc/s10052-016-4395-z. arXiv:1510.03823 [hepspsex]
30. ATLAS Collaboration (2016) Optimisation of the ATLAS b-tagging performance for the 2016 LHC Run, ATL-PHYS-PUB-2016-012. https://cds.cern.ch/record/2160731
31. ATLAS Collaboration (2016) Performance of b-jet identification in the ATLAS experiment. JINST 11, P04008. https://doi.org/10.1088/1748-0221/11/04/P04008. arXiv:1512.01094 [hepspsex]
32. Jet calibration and systematic uncertainties for jets reconstructed in the ATLAS detector at $\sqrt{s} = 13$ TeV, Technical report, ATL-PHYS-PUB-2015-015, CERN, Geneva, July, 2015. https://cds.cern.ch/record/2037613
33. ATLAS Collaboration (2017) Jet energy scale measurements and their systematic uncertainties in proton-proton collisions at $\sqrt{s} = 13$ TeV with the ATLAS detector. Phys Rev D96:7, 072002. https://doi.org/10.1103/PhysRevD.96.072002. arXiv:1703.09665 [hepspsex]
34. Expected performance of missing transverse momentum reconstruction for the ATLAS detector at $\sqrt{s} = 13$ TeV, Technical report, ATL-PHYS-PUB-2015-023, CERN, Geneva, July, 2015. https://cds.cern.ch/record/2037700
35. Lester CG, Summers DJ (1999) Measuring masses of semi-invisibly decaying particles pair produced at hadron colliders. Phys Lett B 463:99–103. https://doi.org/10.1016/S0370-2693(99)00945-4. arXiv:hep-ph/9906349
36. Barr A, Lester C, Stephens P (2003) A variable for measuring masses at hadron colliders when missing energy is expected; m_{T2}: the truth behind the glamour. J Phys G 29:2343–2363. https://doi.org/10.1088/0954-3899/29/10/304. arXiv:hep-ph/0304226
37. ATLAS Collaboration (2017) Search for new phenomena in events containing a same-flavour opposite-sign dilepton pair, jets, and large missing transverse momentum in $\sqrt{s} = 13$ pp collisions with the ATLAS detector. Eur Phys J C77:3, 144. https://doi.org/10.1140/epjc/s10052-017-4700-5. arXiv:1611.05791 [hepspsex]
38. CMS Collaboration (2012) Search for physics beyond the standard model in events with a Z boson, jets, and missing transverse energy in pp collisions at $\sqrt{s} = 7$ TeV. Phys Lett B716, 260–284. https://doi.org/10.1016/j.physletb.2012.08.026. arXiv:1204.3774 [hepspsex]
39. CMS Collaboration (2015) Search for physics beyond the standard model in events with two leptons, jets, and missing transverse momentum in pp collisions at sqrt(s) = 8 TeV. JHEP 04, 124. https://doi.org/10.1007/JHEP04(2015)124. arXiv:1502.06031 [hepspsex]

40. Baak M, Besjes GJ, Côte D, Koutsman A, Lorenz J (2015) Short D (2015) HistFitter software framework for statistical data analysis. Eur Phys J C 75:153. https://doi.org/10.1140/epjc/s10052-015-3327-7. arXiv:1410.1280 [hepspsex]
41. Read AL (2002) Presentation of search results: the CL(s) technique. J Phys G28:2693–2704. https://doi.org/10.1088/0954-3899/28/10/313
42. ATLAS Collaboration (2014) Search for direct production of charginos, neutralinos and sleptons in final states with two leptons and missing transverse momentum in pp collisions at $\sqrt{s} =$ 8 TeV with the ATLAS detector. JHEP 05, 071. https://doi.org/10.1007/JHEP05(2014)071. arXiv:1403.5294 [hepspsex]

Chapter 8
Studying the Excess in the Recursive Jigsaw Reconstruction (RJR) Search Using the Emulated RJR (eRJR) Technique

Chapter 7 describes the conventional 3ℓ search [2] which does not see an excess of observed events above the background prediction [2] using data collected in 2015 and 2016, corresponding to 36.1 fb^{-1} of data collected. Another search by the ATLAS collaboration using the recursive jigsaw reconstruction (RJR) technique [3, 4] also using 36.1 fb^{-1} of data collected between 2015 and 2016 [1] found excesses of three-lepton events in two regions, one targeting low-mass resonances and another utilizing ISR to target resonances with mass differences with respect to the LSP close to the Z boson mass. The two expected and observed limits are shown in Fig. 8.1. The expected limits around the mass point $\left(m(\tilde{\chi}_1^{\pm}/\tilde{\chi}_2^0), m(\tilde{\chi}_1^0)\right) = (200, 100)$ GeV are the same for both searches; however, the observed limits are different due to the excess of observed data over the background prediction seen by the RJR search.

This chapter will briefly discuss the RJR technique and published result [1] and describe the overlap of events between the RJR and the conventional 3ℓ searches. A new technique, emulated RJR (eRJR) is introduced which explores the intersection between the conventional and RJR approaches to better understand the tension in the exclusion limits produced by the two analyses. This technique emulates the variables used by the RJR technique with conventional laboratory frame discriminating variables, providing a simple set of variables that are easily reproducible. This technique is used to reproduce the three-lepton excesses in the RJR search using 36.1 fb^{-1} of pp collision data collected between 2015 and 2016 by the ATLAS detector at the LHC.

© The Editor(s) (if applicable) and The Author(s), under exclusive license to Springer Nature Switzerland AG 2020

E. Resseguie, *Electroweak Physics at the Large Hadron Collider with the ATLAS Detector*, Springer Theses,
https://doi.org/10.1007/978-3-030-57016-3_8

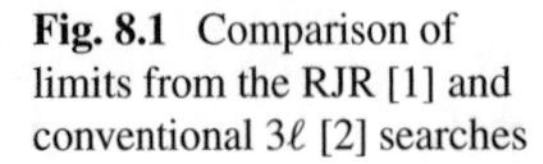

Fig. 8.1 Comparison of limits from the RJR [1] and conventional 3ℓ [2] searches

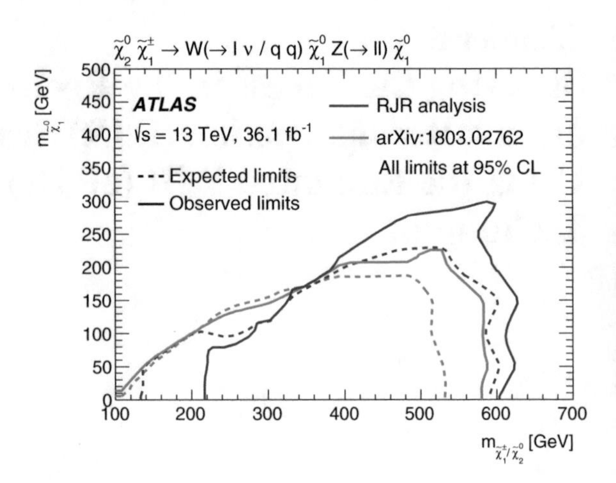

8.1 Recursive Jigsaw Reconstruction (RJR) Search

Figure 8.2 shows the diagrams for the production of $\tilde{\chi}_1^\pm \tilde{\chi}_2^0$ decaying via W and Z bosons in proton-proton collisions. In this search, just like in the search described in Chap. 7, the W and Z bosons are on-shell and decay leptonically via SM branching ratios, leading to a final state with three leptons and missing momentum from two $\tilde{\chi}_1^0$and a neutrino. Signal regions are designed to target the same phase space as the $m_{\mathrm{T}}^{\mathrm{min}}$ search, with SUSY mass splittings $\Delta m = m(\tilde{\chi}_1^\pm/\tilde{\chi}_2^0) - m(\tilde{\chi}_1^0)$ ranging from 100 to 600 GeV. The presence of initial state radiation (ISR) may lead to jets in the final state and boost the $\tilde{\chi}_1^\pm \tilde{\chi}_2^0$ system, enhancing the signature of the missing momentum to target the smaller mass splittings.

The difference between the conventional 3ℓ and the RJR searches is that, while the conventional 3ℓ search makes use of laboratory frame variables only, the RJR search interprets the event using the RJR technique, which provides a way to reconstruct

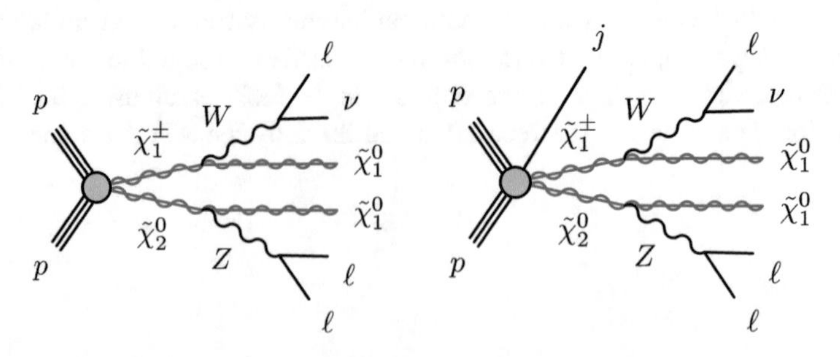

Fig. 8.2 Diagrams for the production of $\tilde{\chi}_1^\pm \tilde{\chi}_2^0$ decaying via W and Z bosons to three leptons and missing transverse energy in pp collisions. The diagram on the right is the production $\tilde{\chi}_1^\pm \tilde{\chi}_2^0$ in association with an initial-state-radiation jet, labelled "j"

the event from the detected particles and in the presence of kinematic unknowns by factorizing missing information according to decays and rest frames of intermediate particles.

8.1.1 Overview of the RJR Technique

In most R-parity conserving SUSY models, the LSP is an invisible particle that rarely, if ever, interacts with matter. It is therefore not directly observed by the ATLAS detector, but manifests itself as missing transverse momentum in an event whose particle transverse momenta would otherwise balance. The relative boost of quarks inside the colliding protons makes it impossible to know the true vector of missing momentum, allowing only for an accurate measurement of the transverse component. For SUSY particles with various decay stages the loss of this information can make it difficult to match the decay products and correctly reconstruct the originally-produced particles, resulting in ambiguities in the reconstruction of the $\tilde{\chi}_1^{\pm}$ and the $\tilde{\chi}_2^0$.

The RJR technique [3, 4] attempts to resolve these ambiguities by analyzing each event starting from the laboratory-frame particles and boosting back to the rest frames of the parent particles. Reconstructed jets, muons, and electrons are used as inputs for the RJR algorithm that determines which leptons come from the chargino or neutralino decays, assuming a specific decay chain. The decay tree, shown in Fig. 8.3a, shows the sparticles produced (PP frame) and each of their decay chains. The final states are separated into visible (V) and invisible (I) objects. The decay tree for the production of $\tilde{\chi}_1^{\pm}\tilde{\chi}_2^0$ decaying via W and Z bosons to three leptons and missing transverse energy is represented in Fig. 8.3b and referred to as the standard tree.

After partitioning the visible objects, the leptons, the remaining unknowns in the events are associated with the two invisible particles, $\tilde{\chi}_{1\text{a}}^0 + \nu_{\text{a}}$ and $\tilde{\chi}_{1\text{b}}^0$: their masses, longitudinal momenta, and how they contribute to the total missing energy. To determine these, the RJR algorithm determines the smallest Lorentz invariant function which results in non-negative mass parameters for the invisible particles [4].

Frame-dependent variables can be constructed using the full four-momenta of the invisible and visible particles, including both longitudinal and transverse components, and are of the form

$$H_{n,m}^{\text{F}} = \sum_{i=1}^{n} |\vec{p}_{\text{vis},\ i}^{\,\text{F}}| + \sum_{j=1}^{m} |\vec{p}_{\text{inv},\ j}^{\,\text{F}}| \tag{8.1}$$

where n represents the number of visible particles, m represents the number of invisible particles, and F is the rest frame in which $H_{n,m}$ is calculated. Variables of the form $H_{n,m}^{\text{F}}$ represent the scalar sum of the visible and invisible particles' four-

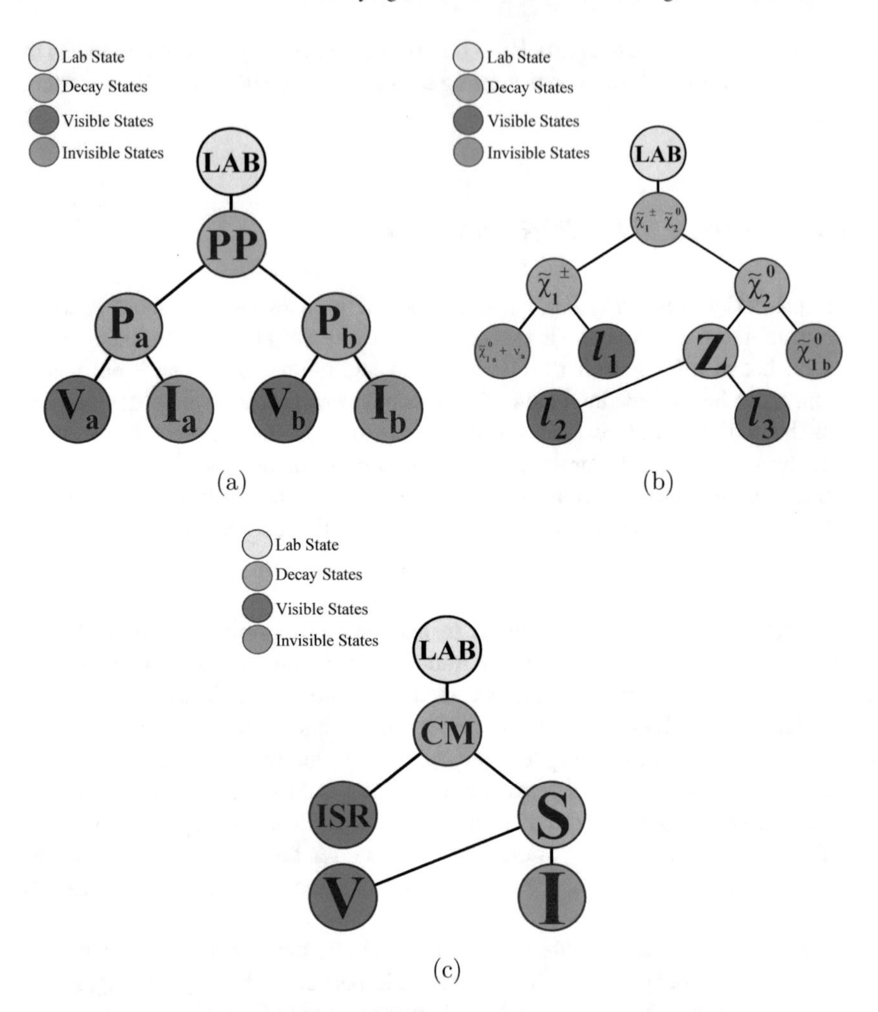

Fig. 8.3 RJR decay trees for **a** generic sparticle production to a final state with one visible and one visible particle per decay leg, **b** standard tree, and **c** ISR tree

momenta, while variables of the form $H_{\mathrm{T}\,n,m}^{\mathrm{F}}$ represent the scalar sum of the visible and invisible particles' transverse momenta.

Variables defined for the standard tree used in the definition of the signal regions are:

- $H_{4,1}^{\mathrm{PP}}$: scale variable described above which is similar to the effective mass, m_{eff} (defined as the scalar sum of the transverse momenta of the visible objects and $E_{\mathrm{T}}^{\mathrm{miss}}$).
- $p_{\mathrm{TPP}}^{\mathrm{lab}}/(p_{\mathrm{TPP}}^{\mathrm{lab}} + H_{\mathrm{T}\,4,1}^{\mathrm{PP}})$: compares the magnitude of the vector sum of the transverse momenta of all objects associated with the PP system in the lab frame ($p_{\mathrm{TPP}}^{\mathrm{lab}}$) to the

overall transverse scale variable considered. For signal events this quantity peaks sharply towards zero while for background processes the distribution is broader.

- $H_{T\,3,1}^{PP}/H_{3,1}^{PP}$: a measure of the fraction of the momentum that lies in the transverse plane.
- $H_{1,1}^{P_b}/H_{2,1}^{P_b}$: for the Z decay frame, this variable compares the scale due to considering both leptons as one object ($H_{1,1}^{P_b}$) as opposed to two visible leptons ($H_{2,1}^{P_b}$).

The decay tree for production of $\tilde{\chi}_1^{\pm}\tilde{\chi}_2^0$ in association with an ISR jet is shown in Fig. 8.3c and referred to as the compressed tree. This decay tree is much simpler and separates the objects in the events as either part of the ISR system or part of the sparticle system (S), which contains the visible and the invisible objects. The ISR jets are selected by minimizing the invariant mass of the system formed by the potential ISR jets and the sparticle system (consisting of leptons and missing energy vector) in the center-of-mass frame (CM). The variables defined for this tree are:

- p_T^{ISR}: the magnitude of the vector sum of the transverse momenta of all jets assigned to the ISR system.
- p_T^{I}: the magnitude of the vector sum of the transverse momenta of the invisible system. This variable is similar to E_T^{miss}.
- p_T^{CM}: the magnitude of the vector-sum of the transverse momenta of the CM system.
- $R_{ISR} \equiv \mathbf{p}_I^{CM} \cdot \hat{p}_{TS}^{CM}/p_{TS}^{CM}$: proxy for $m(\tilde{\chi}_1^0)/m(\tilde{\chi}_2^0/\tilde{\chi}_1^{\pm})$, this variable determines the fraction of the momentum of the sparticle system (S) that is carried by its invisible system (I).
- $\Delta\phi_{ISR,I}$: the azimuthal opening angle between the ISR system and the invisible system in the CM frame.

8.1.2 Data Set and MC Samples

The proton proton collision data corresponds to an integrated luminosity 36.1 fb^{-1} collected at a center-of-mass energy of 13 TeV in 2015 and 2016. The samples include an ATLAS detector simulation [5], based on Geant4 [6], or a fast simulation [5] that uses a parametrization of the calorimeter response [7] and Geant4 for the other parts of the detector. The simulated events are reconstructed in the same manner as the data.

Table 7.1 summarizes the Monte Carlo (MC) used specifying the generator used to simulate both background and signal events. The Monte Carlo used is the same as in the conventional 3ℓ search.

Table 8.1 Preselection and selection criteria for the standard tree regions for the RJR search

Preselection criteria for the standard tree regions

Region	n_{leptons}	n_{jets}	$n_{b\text{-tag}}$	$p_T^{\ell_1}$ [GeV]	$p_T^{\ell_2}$ [GeV]	$p_T^{\ell_3}$ [GeV]
CR3ℓ-VV	=3	<3	= 0	>60	>40	>30
VR3ℓ-VV	=3	<3	= 0	>60	>40	>30
SR3ℓ_Low	= 3	= 0	= 0	>60	>60	>40
SR3ℓ_Int	= 3	<3	= 0	>60	>50	>30
SR3ℓ_High	= 3	<3	= 0	>60	>40	>30

Selection criteria for the standard tree regions

Region	$m_{\ell\ell}$ [GeV]	m_{T} [GeV]	$H_{3,1}^{\text{PP}}$ [GeV]	$\frac{p_{\text{T}}^{\text{PP}}}{p_{\text{T}}^{\text{PP}}+H_{\text{T}\,3,1}^{\text{PP}}}$	$\frac{H_{\text{T}\,3,1}^{\text{PP}}}{H_{3,1}^{\text{PP}}}$	$\frac{H_{1,1}^{P_b}}{H_{2,1}^{P_b}}$
CR3ℓ-VV	∈ (75, 105)	∈ (0, 70)	>250	<0.2	>0.75	–
VR3ℓ-VV	∈ (75, 105)	∈ (70, 100)	>250	<0.2	>0.75	–
SR3ℓ_Low	∈ (75, 105)	>100	>50	<0.05	>0.9	–
SR3ℓ_Int	∈ (75, 105)	>130	>450	<0.15	>0.8	>0.75
SR3ℓ_High	∈ (75, 105)	>150	>550	<0.2	>0.75	>0.8

Table 8.2 Preselection and selection criteria for the 3ℓ_ISR regions for the RJR search

Preselection criteria for 3ℓ_ISR regions

Region	n_{leptons}	n_{jets}	$n_{b\text{-tag}}$	$p_{\text{T}}^{\ell_1}$ [GeV]	$p_{\text{T}}^{\ell_2}$ [GeV]	$p_{\text{T}}^{\ell_3}$ [GeV]
CR3ℓ_ISR-VV	= 3	≥1	= 0	>25	>25	>20
VR3ℓ_ISR-VV	= 3	≥1	= 0	>25	>25	>20
SR3ℓ_ISR	= 3	∈ [1, 3]	= 0	>25	>25	>20

Selection criteria for 3ℓ_ISR regions

Region	$m_{\ell\ell}$ [GeV]	m_{T} [GeV]	$\Delta\phi_{\text{ISR,I}}$	R_{ISR}	$p_{\text{T}}^{\text{ISR}}$ [GeV]	p_{T}^{I} [GeV]	p_{T}^{CM} [GeV]
CR3ℓ_ISR-VV	∈ (75, 105)	<100	>2.0	∈ (0.55, 1.0)	>80	>60	<25
VR3ℓ_ISR-VV	∈ (75, 105)	>60	>2.0	∈ (0.55, 1.0)	>80	>60	>25
SR3ℓ_ISR	∈ (75, 105)	>100	>2.0	∈ (0.55, 1.0)	>100	>80	<25

8.1.3 *Object and Event Selection*

The objects in this search, electrons, muons, jets, $E_{\text{T}}^{\text{miss}}$, have the same definitions as in the $m_{\text{T}}^{\text{min}}$ search and are detailed in Sect. 7.5.1. Tables for the definitions of baseline and signal electrons and muons are shown in Table 7.2 and for jets in Table 7.3.

Table 8.3 Expected and observed yields from the background fit for the CRs and VRs. The normalization factors for VV for the standard and compressed decay trees are different and are extracted from separate fits. The nominal predictions from MC simulation are given for comparison for the VV background. The "Other" category contains the contributions from Higgs boson processes, $t\bar{t}V$ and fake/non-prompt

Region	CR3ℓ-VV	VR3ℓ-VV	CR3ℓ_ISR-VV	VR3ℓ_ISR-VV
Observed events	331	160	98	83
Total (post-fit) SM events	331 ± 18	159 ± 38	98 ± 10	109 ± 24
Other	52 ± 2	5.6 ± 1.2	4.4 ± 1.2	7.1 ± 1.6
Tribosons	1.1 ± 0.1	0.44 ± 0.03	0.22 ± 0.14	0.42 ± 0.04
Fit output, VV	278 ± 18	153 ± 38	93 ± 10	102 ± 24
Fit input, VV	255	140	83	90

Events are selected with three signal and three baseline leptons. The leptons must have at least one same-flavor opposite-charge (SFOS) pair (e^+e^- or $\mu^+\mu^-$) with an invariant mass of the pair $m_{\ell\ell}$ between 75 and 105 GeV, consistent with a Z boson. If there is more than one SFOS pair, the pair chosen is the one that has an invariant mass closest to that of a Z boson. The remaining lepton is assigned to the W boson. The leading source of SM background is WZ production, which, when decaying fully leptonically, has three leptons and $E_{\mathrm{T}}^{\mathrm{miss}}$ from a neutrino in the final state. To reduce the WZ contribution, the transverse mass is calculated from the unpaired third lepton and the $E_{\mathrm{T}}^{\mathrm{miss}}$. It is defined as $m_{\mathrm{T}} = \sqrt{2 p_{\mathrm{T}} E_{\mathrm{T}}^{\mathrm{miss}} \left(1 - \cos(\Delta\phi)\right)}$, where $\Delta\phi$ is the angular separation between the lepton and the missing energy vector, $\mathbf{p}_{\mathrm{T}}^{\mathrm{miss}}$, and will typically be at or below the W boson mass in SM events where the $E_{\mathrm{T}}^{\mathrm{miss}}$ is predominantly from the neutrino of the W decay. The m_{T} calculated in $\tilde{\chi}_1^{\pm}\tilde{\chi}_2^0$ events does not have such a constraint, and the SRs therefore require $m_{\mathrm{T}} \geq 100$ GeV to reduce the SM WZ background. Events containing b-tagged jets are rejected to minimize contributions from the top backgrounds $t\bar{t}$ and Wt.

The search is optimized to target different mass splittings, as shown in Fig. 8.4 by defining four signal regions: ISR, low-mass, intermediate-mass, and high-mass. The low-mass, intermediate-mass, and high-mass signal regions are defined with the standard tree and the ISR signal region is defined with the compressed tree. The only regions that are explicitly orthogonal are SR3ℓ_ISR and SR3ℓ_low because SR3ℓ_ISR requires at least one jet while SR3ℓ_low has a jet veto.

The regions defined with the standard tree require three energetic leptons with $p_{\mathrm{T}} > 60, 40, 30$ GeV for the leading, subleading, and third leptons, respectively. The intermediate-mass signal region tightens this requirement for the second lepton to $p_{\mathrm{T}} > 50$ GeV and in the high-mass region, the second and third lepton p_{T} are 60 GeV and 40 GeV, respectively. These regions have low jet activity with less than 3 jets in all regions except for the low-mass region where a jet veto is applied. Tight selection thresholds on the RJR variables $H_{3,1}^{\mathrm{PP}}$, $\frac{p_{\mathrm{T}}^{\mathrm{PP}}}{p_{\mathrm{T}}^{\mathrm{PP}} + H_{\mathrm{T}\,3,1}^{\mathrm{PP}}}$, $\frac{H_{\mathrm{T}\,3,1}^{\mathrm{PP}}}{H_{3,1}^{\mathrm{PP}}}$, and $\frac{H_{1,1}^{P_b}}{H_{2,1}^{P_b}}$ further reduce the

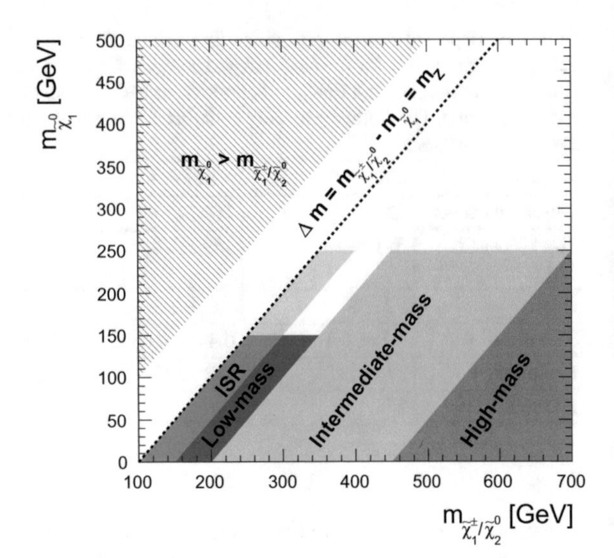

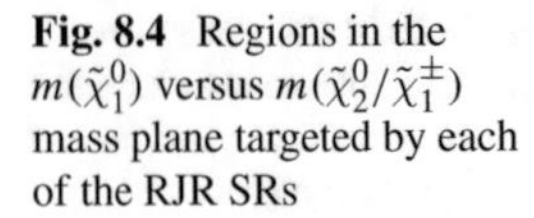
Fig. 8.4 Regions in the $m(\tilde{\chi}_1^0)$ versus $m(\tilde{\chi}_2^0/\tilde{\chi}_1^\pm)$ mass plane targeted by each of the RJR SRs

WZ contribution in the signal region. The signal region cuts are summarized in Table 8.1.

The ISR region has a requirement of $p_T^I > 80$ GeV to reduce the Z+jets background which does not have a source of real E_T^{miss}. The p_T requirement on the three leptons can then be relaxed to be greater than 25, 25, and 20 GeV, ensuring the dilepton triggers are fully efficient. To select the ISR topology in which the system of leptons and E_T^{miss} is recoiling against the ISR jets, the angular separation between the signal jets and $\mathbf{p}_T^{\text{miss}}$, $\Delta\phi_{\text{ISR,I}}$, is required to be greater than 2.0. The ratio between the $\mathbf{p}_T^{\text{miss}}$ and the total transverse momenta of the jets is required to be $0.55 > R_{\text{ISR}} < 1.0$ to ensure the majority of transverse momentum along the jet axis is carried by the invisible particles and not by the high-p_T leptons from the WZ background. A requirement of p_T^{CM} less than 25 GeVfurther reduces background contamination. The selection for the ISR regions are summarized in Table 8.2.

8.1.4 Background Estimation

The backgrounds in this analysis can be classified into two groups: irreducible backgrounds with at least three prompt leptons in the final state, and reducible backgrounds containing at least one fake or non-prompt lepton. The dominant irreducible background is VV (WZ and ZZ) production which is estimated from MC simulation whose yields are normalized to data in CRs. Other irreducible backgrounds include VVV, ttV, and Higgs processes, and are estimated directly from MC simulation due to their small contribution. The reducible backgrounds can be categorized into the top-like $t\bar{t}$, Wt, and WW processes, which mostly consists of non-prompt leptons

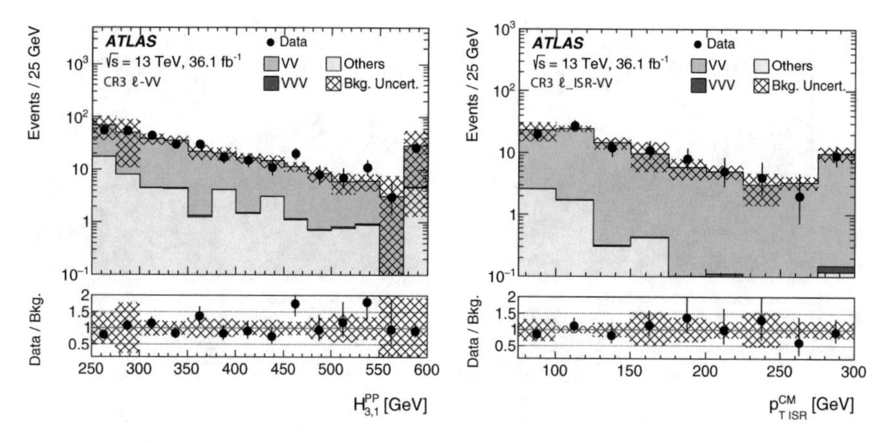

Fig. 8.5 Kinematic distributions in CR-low (left) and CR-ISR (right) after applying all the selection criteria. The histograms show the post-fit MC background predictions. The fake/non-prompt, estimated from a data-driven technique, is included in the category "Others". The last bin includes the overflow. The errors contain statistical and systematic uncertainties

from heavy-flavor hadron decays, and the Z+jets process, which also accounts for the Z+γ process, with fake and non-prompt leptons coming primarily from photon conversions or misidentified jets.

The dominant SM background, VV, is estimated using MC simulation normalized to data in CRs designed to be kinematically similar but orthogonal to SR-low and SR-ISR. The CRs are designed to be enriched in WZ events while keeping the potential signal contamination small, being less than 10% for all signal models. To achieve this an upper bound is placed on the m_{T} of the CRs, targeting events that are likely to have a leptonically decaying W boson and no other sources of $E_{\mathrm{T}}^{\mathrm{miss}}$. Therefore the low-mass CR (CR-low) requires $m_{\mathrm{T}} < 70$ GeV while the ISR CR (CR-ISR) has a slightly looser requirement of $m_{\mathrm{T}} < 100$ GeV, benefiting from the boost of the $E_{\mathrm{T}}^{\mathrm{miss}}$system by the ISR. The other kinematic selections are similar to the corresponding SRs, with some loosened to enhance statistics and reduce signal contamination. Fig. 8.5 shows the background composition in the CR-low and CR-ISR regions, with good agreement seen between data and the background prediction. The normalization factors are found to be 1.09 ± 0.10 for CR-low and 1.13 ± 0.13 for CR-ISR.

The fake/non-prompt background originate from a semileptonic decay of a b- or c-hadrons, decays in flight of light hadrons, misidentification of a light-flavor jet, or photon conversions. These backgrounds include Z+jet and $Z\gamma$, WW, Wt, and $t\bar{t}$, and are estimated using a data-driven method called the Matrix Method [8]. This method uses two types of lepton identification criteria: "signal" and "baseline". The method makes use of the numbers of observed events containing baseline–baseline, baseline–signal, signal–baseline and signal–signal lepton pairs (ordered in p_{T}) in a given SR. The highest-p_{T} electron or muon is taken to be real. Knowing the probabilities for real and FNP leptons satisfying the baseline selection criteria to also satisfy the signal

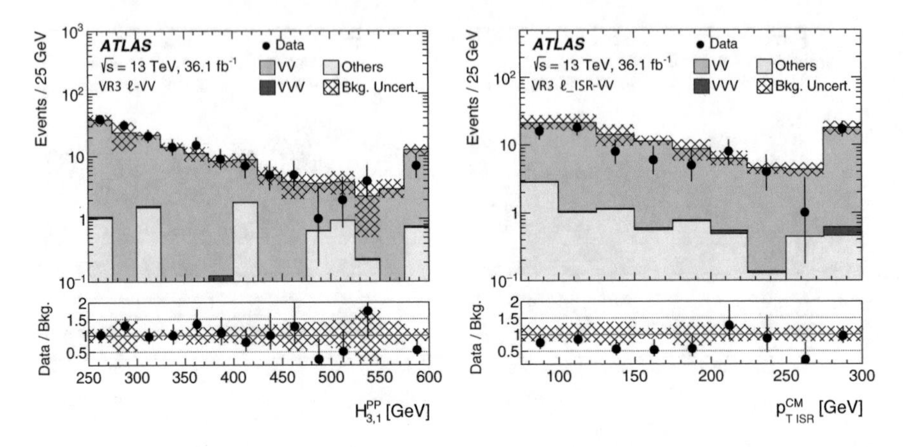

Fig. 8.6 Kinematic distributions in VR-low (left) and VR-ISR (right) after applying all the selection criteria. The histograms show the post-fit MC background predictions. The fake/non-prompt, estimated from a data-driven technique, is included in the category "Others". The last bin includes the overflow. The errors contain statistical and systematic uncertainties

selection, the observed event counts with the different lepton selection criteria can be used to extract a data-driven estimate of the FNP background.

The irreducible backgrounds come from Higgs production, VVV, and $t\bar{t}V$ and are estimated using simulation.

Two validation regions are designed to check the agreement of the background estimation with data in regions kinematically closer to the SRs, typically targeting the extrapolation from CR to SR of a specific variable. The full VR definitions are summarized in Tables 8.1, 8.2. The VR definitions are also chosen to keep signal contamination below 10%. A standard tree VR-low is designed to test the extrapolation over m_{T} between CR-low and the three standard tree signal regions, requiring $70 < m_{\mathrm{T}} < 100$ GeV. The ISR validation region, VR-ISR, inverts the $p_{\mathrm{T}}^{\mathrm{CM}}$ cut to validate the modeling. Figure 8.6 shows distributions in VR-low and VR-ISR for the full background prediction. There is good agreement seen between the expected background prediction and the observed data.

The total yields in the CRs and VRs are shown in Table 8.3 for the standard tree regions and the ISR regions.

8.1.5 Uncertainties

There are several sources of experimental and theoretical uncertainties. The largest sources of systematic uncertainties in the signal regions come from jet energy scale (JES) and resolution (JER), modeling of the soft term of the missing energy, theoretical uncertainties, and uncertainties associated with the calculation of the WZ NF.

The JES and JER uncertainties are derived as a function of jet p_{T} and η, as well as the jet flavor composition. They are derived using data and simulation using dijet, Z+jet, and γ+jet samples [9, 10].

The systematic uncertainties related to the $E_{\mathrm{T}}^{\mathrm{miss}}$ modeling in the simulation are estimated by propagating the uncertainties in the energy or momentum scale of each of the physics objects, as well as the uncertainties in the soft term's resolution and scale [11].

Other systematics are derived on the muon (electron) momentum (energy) resolution, momentum (energy) scale, reconstruction, and isolation efficiencies. Uncertainties due to the trigger efficiency, and b-tagging efficiency were also calculated. These uncertainties were found to be negligible.

Theoretical uncertainties on the WZ background are the choice of PDF set, QCD renormalization (μ_R) and factorization (μ_F) scales, and the choice of the strong coupling constant (α_s). Further discussion about the calculation of these systematics can be found in Sect. 5.8.

Systematic uncertainties are also assigned to the estimated background from fake/non-prompt (FNP) leptons to account for potentially different compositions (heavy flavor, light flavor or conversions) between the signal and control regions. An additional uncertainty is associated with the subtraction of prompt leptons from this CR using simulation.

A 2.1% uncertainty is applied to the integrated luminosity.

The largest systematics are due to the VV theoretical uncertainties and normalization, and statistical uncertainties for both the Monte Carlo and the FNP background. The systematic breakdown in the SRs is summarized in Table 8.4.

8.1.6 Results

The background only fit assumes no signal is present in any region. Only data in the CRs are used to constrain the background only fit. This provides an SR-independent background prediction in all regions. The results are shown in Table 8.5. A summary of the observed and expected yields in all of the signal regions considered in this thesis is provided in Fig. 8.7. No significant excesses above the SM expectation are observed in the SRs targeting intermediate- and high-mass signal models. An excess of events above the background estimate is observed in the low-mass and ISR signal regions.

Distributions in SR-low are shown in Fig. 8.8. There is an overall excess of observed events as compared to the background prediction. Distributions in SR-ISR are shown in Fig. 8.9. The excess appears prominently just at the edge of the m_{T} cut.

To quantify the level of agreement of the observed data with the SM expectations, a model-independent fit is performed separately for each SR, summarized in Table 8.6. The observed data excesses in SR3ℓ_ISR and SR3ℓ_Low have local significances of 3.0 and 2.1, respectively.

Table 8.4 Summary of the systematic uncertainties in each of the SRs. The total systematic uncertainty can be different from the sum in quadrature of individual sources due to the correlations between them resulting from the fit to the data

Signal Region	SR3ℓ_High	SR3ℓ_Int	SR3ℓ_Low	SR3ℓ_ISR
Total uncertainty [%]	44	22	19	26
VV theoretical uncertainties	18	9	12	19
MC statistical uncertainties	37	17	8	10
VV fitted normalization	8	7	9	11
FNP leptons	7	<1	3	5
Jet energy resolution	4	<1	7	3
Jet energy scale	7	<1	2	3
$E_{\mathrm{T}}^{\mathrm{miss}}$ modeling	2	<1	1	4
Lepton reconstruction/identification	3	4	2	2

Table 8.5 Expected and observed yields from the background-only fit for the SRs. The errors shown are the statistical plus systematic uncertainties

Signal region	SR3ℓ_High	SR3ℓ_Int	SR3ℓ_Low	SR3ℓ_ISR
Total observed events	2	1	20	12
Total background events	1.1 ± 0.5	2.3 ± 0.5	10 ± 2	3.9 ± 1.0
Other	$0.03^{+0.07}_{-0.03}$	0.04 ± 0.02	$0.02^{+0.34}_{-0.02}$	$0.06^{+0.19}_{-0.06}$
Triboson	0.19 ± 0.07	0.32 ± 0.06	0.25 ± 0.03	0.08 ± 0.04
Fit output, VV	0.83 ± 0.39	1.9 ± 0.5	10 ± 2	3.8 ± 1.0
Fit input, VV	0.76	1.8	9.2	3.4

Exclusion limits are set on the masses of the charginos, neutralinos and shown in Fig. 8.10. The final state where the Z boson decays leptonically and the W boson decays hadronically, "2L2J", is also considered by this search and is used in combination with the three lepton final state result in the exclusion limit. The 2L2J and three lepton channels that target the same region of phase (i.e. low-mass) are first combined since they are mutually exclusive. The observed $\mathrm{CL_s}$ value is taken from the signal region with the best expected $\mathrm{CL_s}$ value since the high-, intermediate-, and low-mass SRs overlap while the low-mass and ISR SRs are mutually exclusive and can be statistically combined. $\tilde{\chi}_1^{\pm}$ and $\tilde{\chi}_2^0$ masses up to 600 GeV are excluded for a massless $\tilde{\chi}_1^0$ neutralino.

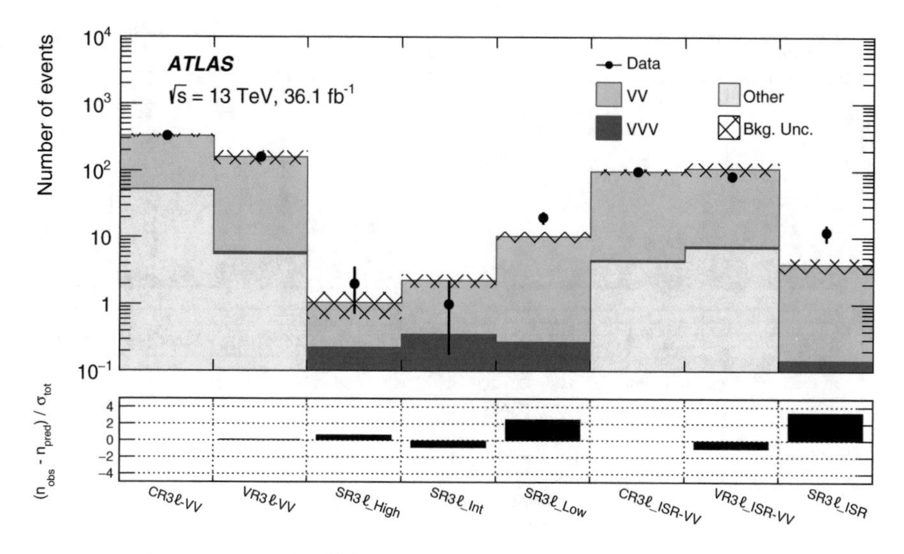

Fig. 8.7 The observed and expected SM background yields in the CRs, VRs and SRs. The statistical uncertainties in the background prediction are included in the uncertainty band, as well as the experimental and theoretical uncertainties. The bottom panel shows the difference in standard deviations between the observed and expected yields

As a result of the excess in observed events above the background expectation, the exclusion curves in Fig. 8.10 show a weaker observed limit compared to the expected limit around $m(\tilde{\chi}_1^{\pm}/\tilde{\chi}_2^0)$ between 100 and 220 GeV.

8.2 Overlap of RJR and Conventional 3ℓ Searches

The conventional 3ℓ and RJR searches target the same phase space and have similar expected exclusions at small mass splittings, $\Delta m = 100$ GeV, as seen in Fig. 8.1. The overlap of data events in the SR is studied to understand how the two searches could have similar expected limits but different observed limits.

One difference between the conventional 3ℓ signal regions and the RJR signal regions is that RJR uses the traditional assignment of leptons to determine m_{T} and $m_{\ell\ell}$, while the conventional 3ℓ search uses the $m_{\mathrm{T}}^{\mathrm{min}}$ assignment. When referring to RJR mimic SRs, the traditional assignment of leptons will be used, denoted as $m_{\ell\ell}$ and m_{T}, while for the conventional 3ℓ search, the variables will be labeled $m_{\ell\ell}^{\mathrm{min}}$ and $m_{\mathrm{T}}^{\mathrm{min}}$. Also, the require on $m_{\ell\ell}^{\mathrm{min}}$ is tighter than in the RJR search with $\mid m_{\ell\ell}^{\mathrm{min}} - m_Z \mid < 10$. The cut on m_{T} is also different in both searches: the RJR search requires $m_{\mathrm{T}} > 100$ GeV while the conventional 3ℓ search requires $m_{\mathrm{T}}^{\mathrm{min}} > 110$ GeV.

To study the events in SR-ISR, only bins SR3-WZ-1Ja, SR3-WZ-1Jb, and SR3-WZ-1Jc of the conventional 3ℓ search, defined in Table 7.8, are considered because this RJR SR requires at least 1 signal jet. After unblinding, there are 12 events in

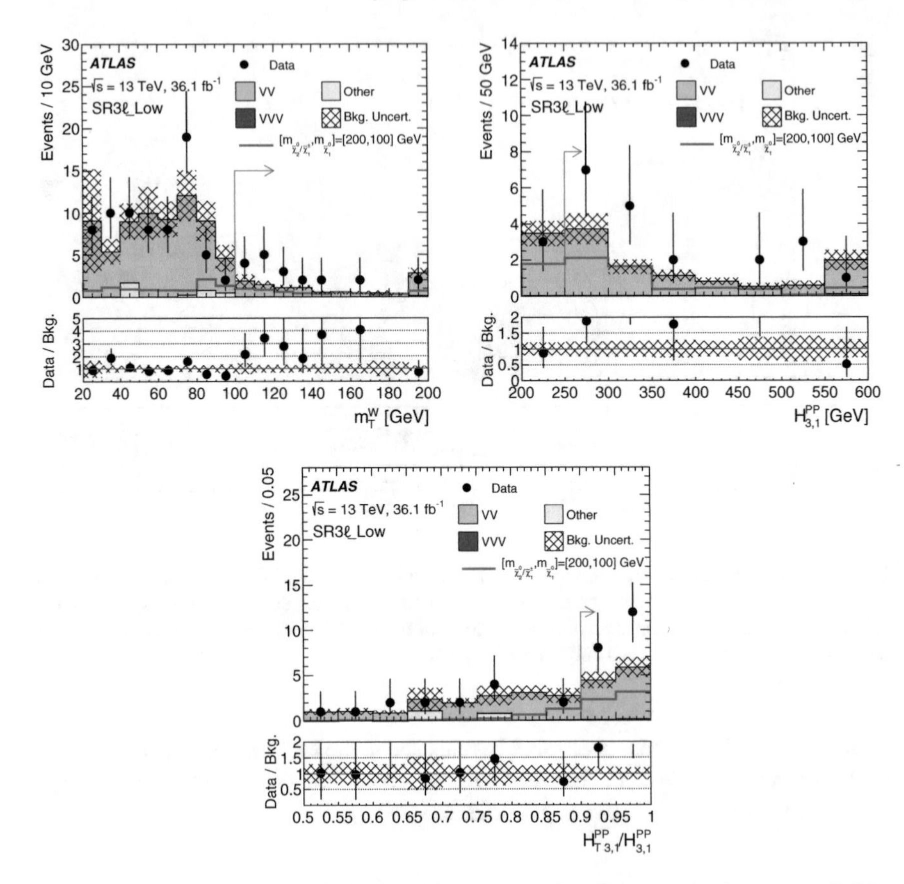

Fig. 8.8 Kinematic distributions in SR-low. The corresponding SR event selections are applied for each distribution except for the variable shown where the selection is indicated by a red arrow. The histograms show the post-fit MC background predictions. The fake/non-prompt, estimated from a data-driven technique, is included in the category "Others". The last bin includes the overflow. The errors contain statistical and systematic uncertainties. A representative signal point is shown with a red dashed line

SR-ISR and 3 of those events are in the conventional 3ℓ SR. Nine out of twelve events are in the RJR SR-ISR but not in the conventional 3ℓ SRs. Those events fail either the $E_{\mathrm{T}}^{\mathrm{miss}}$ and $m_{\mathrm{T}}^{\mathrm{min}}$ cuts of SR3-WZ-1Ja or the $m_{\ell\ell}^{min}$ requirement of being 10 GeV away from the Z-mass. It is important to note that those events do not pass the Z-peak requirement due to the $m_{\mathrm{T}}^{\mathrm{min}}$ assignment of the leptons. If the traditional assignment of the leptons had been used, two of the three events would have entered the $m_{\mathrm{T}}^{\mathrm{min}}$ SRs.

To study the events in SR-low, only bins SR3-WZ-0Ja, SR3-WZ-0Jb, and SR3-WZ-10c of the conventional 3ℓ search, defined in Table 7.8, are considered because this RJR SR requires a jet veto. The three conventional 3ℓ search bins require $E_{\mathrm{T}}^{\mathrm{miss}} >$ 50 GeV to remove Z+jets events while SR-low does not have a $E_{\mathrm{T}}^{\mathrm{miss}}$ cut.

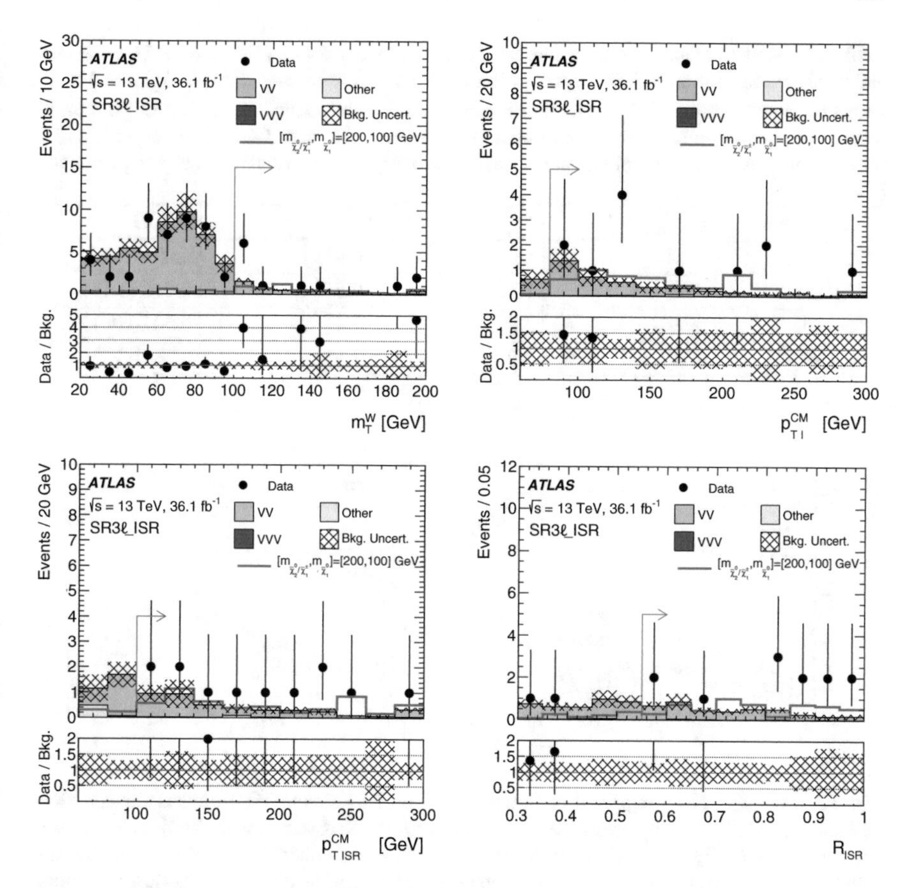

Fig. 8.9 Kinematic distributions in SR-ISR. The corresponding SR event selections are applied for each distribution except for the variable shown where the selection is indicated by a red arrow. The histograms show the post-fit MC background predictions. The fake/non-prompt, estimated from a data-driven technique, is included in the category "Others". The last bin includes the overflow. The errors contain statistical and systematic uncertainties. A representative signal point is shown with a red dashed line

After unblinding, there are 20 events in SR-low and only one event is in the conventional 3ℓ SRs. For the remaining events, they fail either $E_{\mathrm{T}}^{\mathrm{miss}}$ and $m_{\mathrm{T}}^{\mathrm{min}}$ cuts or the $m_{\ell\ell}^{min}$ requirement of being 10 GeV away from the Z-mass. If the traditional assignment had been used in the conventional 3ℓ search, 11 of those events would have been in the conventional 3ℓ SRs.

The kinematics values of the events that do not overlap are shown in Tables A1, A2 in Appendix A.2.

Table 8.6 Model-independent fit results for the SRs. The first column shows the SRs, the second and third columns show the 95% CL upper limits on the visible cross-section ($\langle\epsilon\sigma\rangle^{95}_{\text{obs}}$) and on the number of signal events (S^{95}_{obs}). The fourth column (S^{95}_{exp}) shows the 95% CL upper limit on the number of signal events, given the expected number (and $\pm 1\sigma$ excursions of the expectation) of background events. The last column indicates the discovery p_0-value and its associated significance (Z)

Signal region	$\langle\epsilon\sigma\rangle^{95}_{\text{obs}}$[fb]	S^{95}_{obs}	S^{95}_{exp}	p_0 (Z)
SR3ℓ_ISR	0.42	15.3	$6.9^{+3.1}_{-2.2}$	0.001 (3.02)
SR3ℓ_Low	0.53	19.1	$9.5^{+4.2}_{-1.8}$	0.016 (2.13)
SR3ℓ_Int	0.09	3.3	$4.4^{+2.5}_{-1.5}$	0.50 (0.00)
SR3ℓ_High	0.14	5.0	$3.9^{+2.2}_{-1.3}$	0.23 (0.73)

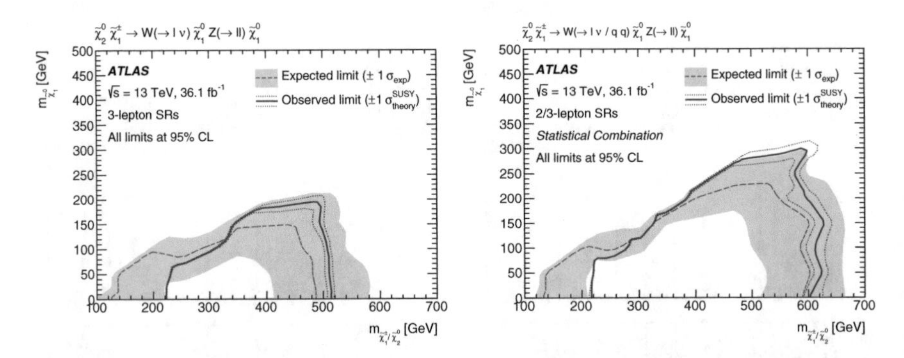

Fig. 8.10 Observed and expected exclusion limits on SUSY simplified models for chargino–neutralino production with decays via W/Z bosons for the RJR search with the W and Z bosons decaying leptonically (left) and with the W boson decaying either hadronically or leptonically (right). The observed (solid thick red line) and expected (thin dashed blue line) exclusion contours are indicated. The shaded band corresponds to the $\pm 1\sigma$ variations in the expected limit, including all uncertainties except theoretical uncertainties in the signal cross-section. The dotted lines around the observed limit illustrate the change in the observed limit as the nominal signal cross-section is scaled up and down by the theoretical uncertainty. All limits are computed at 95% confidence level

8.3 Emulated RJR (eRJR) Technique

Given the excesses that were observed in the RJR search, an emulation of the RJR technique is developed to cross check the RJR result and study the phase space of SR-low and SR-ISR. The intermediate- and high-mass did not see any excesses and are not considered. The emulated Recursive Jigsaw Reconstruction (eRJR) technique emulates the RJR variables using minimal assumptions on the mass of the invisible system and calculates all kinematic variables in the laboratory frame.

8.3.1 Translating RJR Variables into Lab Frame Variables

The eRJR variables, with original RJR variable names from Ref. [1] in parenthesis, used to select the ISR regions are defined as follows:

- $E_{\mathrm{T}}^{\mathrm{miss}}(p_{\mathrm{T}}^{\mathrm{I}})$, the p_{T} of the invisible particles is emulated as the magnitude of the missing transverse momentum.
- $p_{\mathrm{T}}^{\mathrm{jets}}$ $(p_{\mathrm{T}}^{\mathrm{ISR}})$, the p_{T} of the vector sum of the ISR jets' momenta. The ISR system in the eRJR technique is the p_{T} of the vector sum of signal jets' four-momenta in the event.
- $|\Delta\phi\left(E_{\mathrm{T}}^{\mathrm{miss}}, \mathrm{jets}\right)|$ $(\Delta\phi_{\mathrm{ISR},E_{\mathrm{T}}^{\mathrm{miss}}})$, the azimuthal angle between the ISR system and the invisible particles is emulated using the transverse missing momentum, $\mathbf{p}_{\mathrm{T}}^{\mathrm{miss}}$, and the vector sum of signal jets' momenta.
- $R\left(E_{\mathrm{T}}^{\mathrm{miss}}, \mathrm{jets}\right)$ (R_{ISR}), the normalized projection of the invisible system onto the ISR system, representing a ratio of $\mathbf{p}_{\mathrm{T}}^{\mathrm{miss}}$to total jet p_{T}, is emulated as $\frac{|\mathbf{p}_{\mathrm{T}}^{\mathrm{miss}}\cdot\widehat{\mathbf{p}_{\mathrm{T}}^{\mathrm{jets}}}|}{p_{\mathrm{T}}^{\mathrm{jets}}}$, where $\widehat{\mathbf{p}_{\mathrm{T}}^{\mathrm{jets}}}$ is the unit vector of the vector sum of signal jet momenta.
- $p_{\mathrm{T}}^{\mathrm{soft}}$ $(p_{\mathrm{T}}^{\mathrm{CM}})$, the transverse momentum in the frame where the ISR system recoils against the system containing the leptons and the missing energy (CM), is emulated as the magnitude of the p_{T} of the vector sum of the four-momenta of the signal jets, leptons, and $\mathbf{p}_{\mathrm{T}}^{\mathrm{miss}}$, and is highly correlated to the $E_{\mathrm{T}}^{\mathrm{miss}}$soft term, as shown in Appendix A.2.

Similarly, the eRJR variables, with original RJR variable names from Ref. [1] in parenthesis, used in the low-mass regions are defined as follows:

- $p_{\mathrm{T}}^{\mathrm{soft}}$ $(p_{\mathrm{T}}^{\mathrm{PP}})$, the transverse momentum in the center-of mass frame of the protons (PP), is emulated as the magnitude of the p_{T} of the vector sum of the four-momenta of the signal leptons and $\mathbf{p}_{\mathrm{T}}^{\mathrm{miss}}$, being identical to that of the ISR region except for the jet veto applied to the low-mass region.
- $m_{\mathrm{eff}}^{3\ell}$ $(H_{\mathrm{T}\,3,1}^{\mathrm{PP}})$, the scalar sum of the p_{T} of the signal leptons and the invisible system (neutrino and LSPs) in the PP frame, is emulated as the scalar sum of the p_{T} of the signal leptons and $E_{\mathrm{T}}^{\mathrm{miss}}$.
- H^{boost} $(H_{3,1}^{\mathrm{PP}})$, the scalar sum of the magnitude of the momentum of the signal leptons and the invisible system (neutrino and LSPs) in the PP frame, is emulated as the scalar sum of the momentum of the signal leptons and the missing momentum vector (which includes longitudinal and transverse components), $|\mathbf{p}^{\mathrm{miss}}|$, after applying a boost.

To calculate H^{boost}, the longitudinal component of the missing momentum vector, $\mathbf{p}_{||}^{\mathrm{miss}}$, and the boost need to be determined. The $\mathbf{p}_{||}^{\mathrm{miss}}$ variable is calculated as [4]:

$$|\mathbf{p}_{||}^{\mathrm{miss}}| = |\mathbf{p}_{V,||}|\frac{|\mathbf{p}_{\mathrm{T}}^{\mathrm{miss}}|}{\sqrt{(\mathbf{p}_{V,T})^2 + m_V^2}} \tag{8.2}$$

where the $\mathbf{p}_{V,||}$ is the z-component of the vector sum of four-momenta of the three signal leptons, $p_{V,T}$ is the magnitude of the transverse momentum of the vector sum of four-momenta of the three leptons, and m_V is the mass of the three lepton system. The mass of the vector sum of invisible particles are assumed to be zero and do not appear in the equation. The boost of the system can then be calculated as:

$$\beta = \frac{\mathbf{p}}{E} = \frac{\mathbf{p}^V + \mathbf{p}^{\text{miss}}}{E^V + |\mathbf{p}^{\text{miss}}|} \tag{8.3}$$

where $\mathbf{p}^V$ is the vector sum of three-momenta of the three leptons calculated in the laboratory frame. This boost is applied to the three leptons and the $\mathbf{p}^{\text{miss}}$. These new objects are used in the calculation of H^{boost}.

8.3.2 *Validation of the eRJR Technique*

To determine how well the eRJR technique emulates the RJR technique, a few different validations are performed.

8.3.2.1 Comparison of RJR and eRJR Kinematic Variable Shapes at Preselection

The shapes of the eRJR variables are compared with the shapes of the actual RJR variables in both the low and ISR preselection regions. The distributions in the low preselection region, defined in Table 8.1, are shown in Fig. 8.11.There is good agreement between the eRJR and RJR variables in the low preselection. Distributions for kinematic variables in the ISR preselection region, defined in Table 8.2, are shown in Fig. 8.12. The $p_{\text{T}}^{\text{soft}}$ variable is not well emulated and there are more events in the first bin of the $R\left(E_{\text{T}}^{\text{miss}}, \text{jets}\right)$ variable.

To understand the difference in the shape of R_{ISR} and $R\left(E_{\text{T}}^{\text{miss}}, \text{jets}\right)$, the ISR topology needs to be studied. In eRJR, the ISR system is the vector sum of all signal jets while in RJR, the ISR system is comprised only of jets that minimize the center of mass with the leptons and $E_{\text{T}}^{\text{miss}}$. In the ISR topology, the $E_{\text{T}}^{\text{miss}}$ and leptons recoil against the ISR system. For the signal, the $E_{\text{T}}^{\text{miss}}$ is larger than the lepton p_{T}, so the $E_{\text{T}}^{\text{miss}}$ gets a larger boost from the ISR system than the leptons.

To select the topology where the $E_{\text{T}}^{\text{miss}}$ recoils against the jets, two methods are considered:

- Angular separation between $E_{\text{T}}^{\text{miss}}$ and jets: $|\Delta\phi\left(E_{\text{T}}^{\text{miss}}, \text{jets}\right)|$. In the analysis, for all CR, VRs, and SR, this cut is $|\Delta\phi\left(E_{\text{T}}^{\text{miss}}, \text{jets}\right)| > 2.0$, which is greater than $\pm\frac{\pi}{2}$.
- Large projection of $E_{\text{T}}^{\text{miss}}$ onto the jet system in the direction of ISR boost: $R\left(E_{\text{T}}^{\text{miss}}, \text{jets}\right)$. $R\left(E_{\text{T}}^{\text{miss}}, \text{jets}\right)$ should be positive if the MET recoils against the jets.

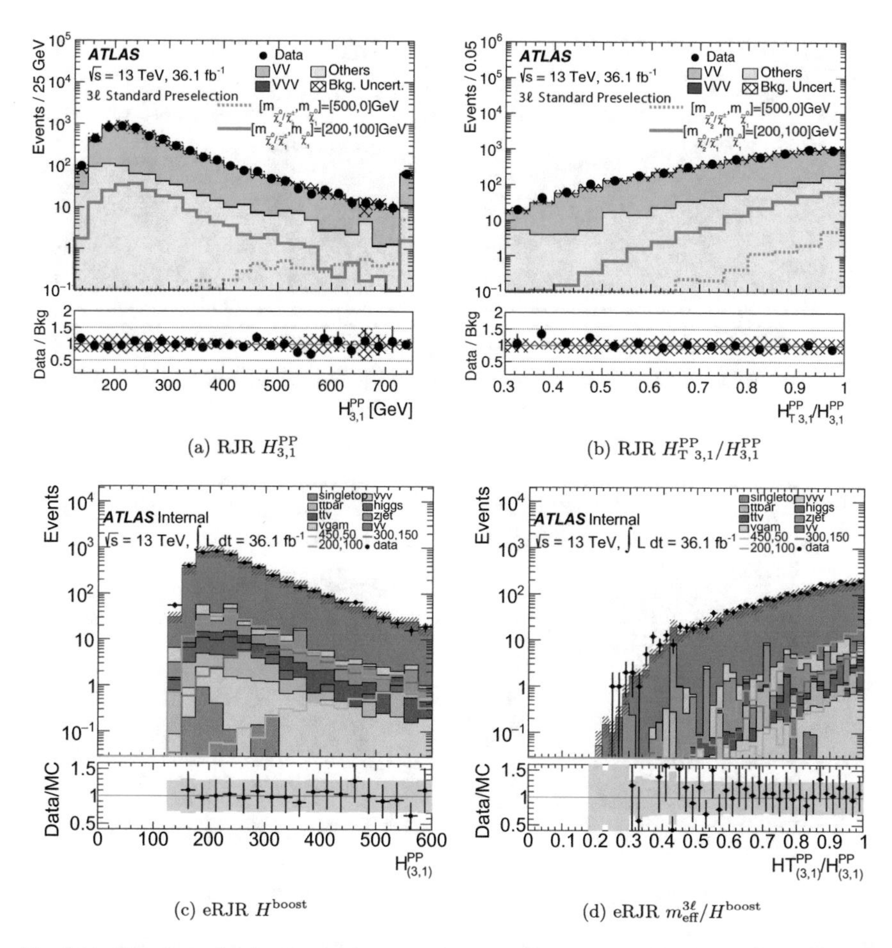

(a) RJR $H_{3,1}^{\mathrm{PP}}$

(b) RJR $H_{\mathrm{T}\,3,1}^{\mathrm{PP}}/H_{3,1}^{\mathrm{PP}}$

(c) eRJR H^{boost}

(d) eRJR $m_{\mathrm{eff}}^{3\ell}/H^{\mathrm{boost}}$

Fig. 8.11 Kinematic distributions in the standard tree preselection regions for RJR and eRJR. The last bin in the RJR distributions contains the overflow and the last bin in the eRJR distributions do not contain the overflow

There is a difference in the definition of eRJR $R\left(E_{\mathrm{T}}^{\mathrm{miss}}, \mathrm{jets}\right)$, as compared with the RJR R_{ISR}:

- eRJR: $R\left(E_{\mathrm{T}}^{\mathrm{miss}}, \mathrm{jets}\right) = \frac{|\mathbf{p}_{\mathrm{T}}^{\mathrm{miss}} \cdot \widehat{\mathbf{p}_{\mathrm{T}}^{\mathrm{jets}}}|}{(p_{\mathrm{T}}^{\mathrm{jets}})^2}$
- RJR: $R_{\mathrm{ISR}} = -\frac{\mathbf{p}_{\mathrm{T}}^{\mathrm{I}} \cdot \widehat{\mathbf{p}_{\mathrm{T}}^{\mathrm{ISR}}}}{(p_{\mathrm{T}}^{\mathrm{ISR}})^2}$

The difference between the two implementations is $|\mathbf{p}_{\mathrm{T}}^{\mathrm{miss}} \cdot \widehat{\mathbf{p}_{\mathrm{T}}^{\mathrm{jets}}}|$ for $R\left(E_{\mathrm{T}}^{\mathrm{miss}}, \mathrm{jets}\right)$ and $-\mathbf{p}_{\mathrm{T}}^{\mathrm{miss}} \cdot \widehat{\mathbf{p}_{\mathrm{T}}^{\mathrm{jets}}}$ for R_{ISR}. This however does not have an impact on the analysis because $|\mathbf{p}_{\mathrm{T}}^{\mathrm{miss}} \cdot \widehat{\mathbf{p}_{\mathrm{T}}^{\mathrm{jets}}}| = -\mathbf{p}_{\mathrm{T}}^{\mathrm{miss}} \cdot \widehat{\mathbf{p}_{\mathrm{T}}^{\mathrm{jets}}}$ for all $\Delta\phi > |\frac{\pi}{2}|$ and the analysis phase space is $|\Delta\phi| > 2.0$, which is greater than $\frac{\pi}{2}$.

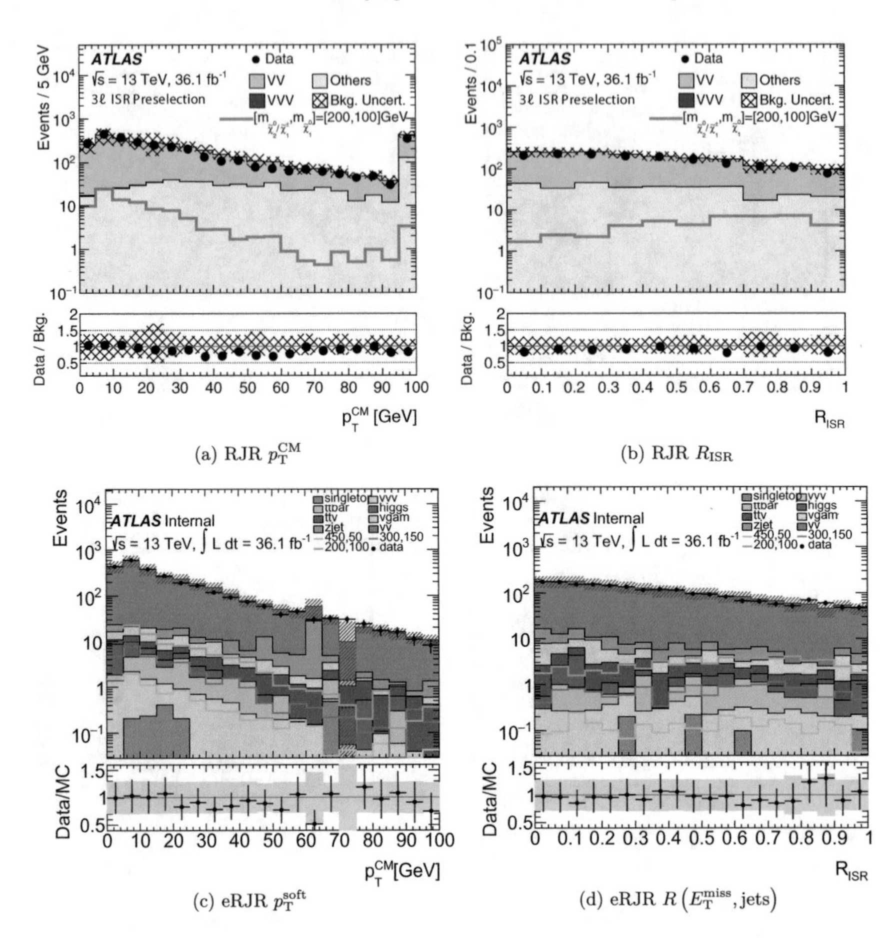

(a) RJR $p_{\mathrm{T}}^{\mathrm{CM}}$

(b) RJR R_{ISR}

(c) eRJR $p_{\mathrm{T}}^{\mathrm{soft}}$

(d) eRJR $R\left(E_{\mathrm{T}}^{\mathrm{miss}}, \mathrm{jets}\right)$

Fig. 8.12 Kinematic distributions in the compressed tree preselection regions for RJR and eRJR. The last bin in the RJR distributions contains the overflow and the last bin in the eRJR distributions do not contain the overflow

8.3.2.2 Correlations RJR and eRJR Kinematic Variables

The eRJR variables are compared with the RJR variables which were evaluated using the implementation at truth level from HepData [12]. The comparisons are done in the preselection regions defined in Tables 8.1, 8.2.

Correlations for the signal point of $\left(m(\tilde{\chi}_1^{\pm}/\tilde{\chi}_2^0), m(\tilde{\chi}_1^0)\right) = (200, 100)$ GeV are shown in the low-mass preselection regions inclusive in jets in Fig. 8.13 and with a jet veto applied in Fig. 8.14. There is a good correlation between the RJR and the eRJR variables; however, the correlation improves when a jet veto is applied, just as in SR-low. The $\frac{m_{\mathrm{eff}}^{3\ell}}{H^{\mathrm{boost}}}$ variable can have values greater than 1 in the eRJR technique because the $m_{\mathrm{eff}}^{3\ell}$ and H^{boost} variables are calculated in different frames;

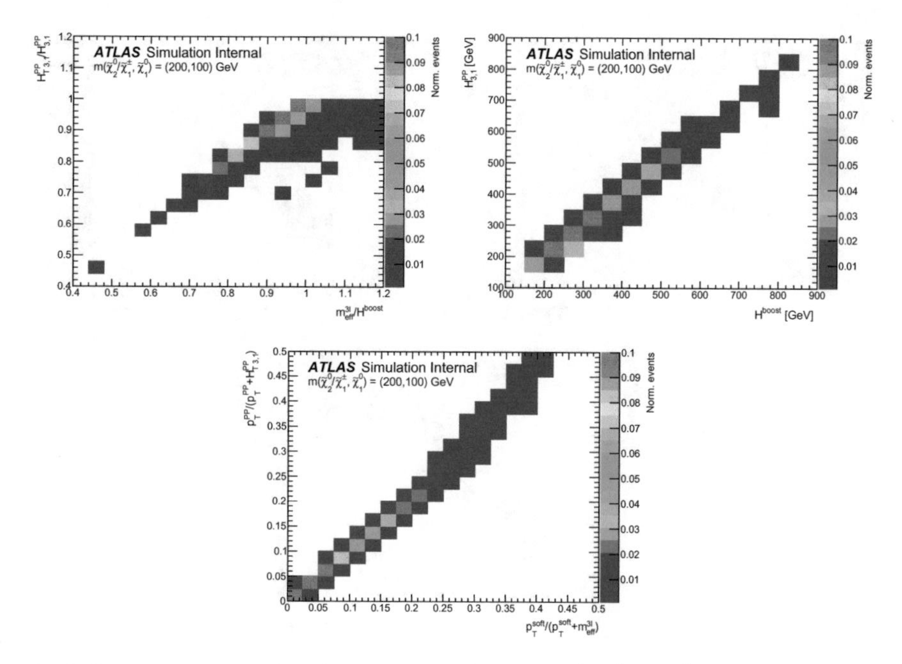

Fig. 8.13 Correlations of RJR and eRJR variables for the $m(\tilde{\chi}_1^{\pm}/\tilde{\chi}_2^0)$, $m(\tilde{\chi}_1^0) = (200, 100)$ signal point in low-mass preselection events inclusive in jets

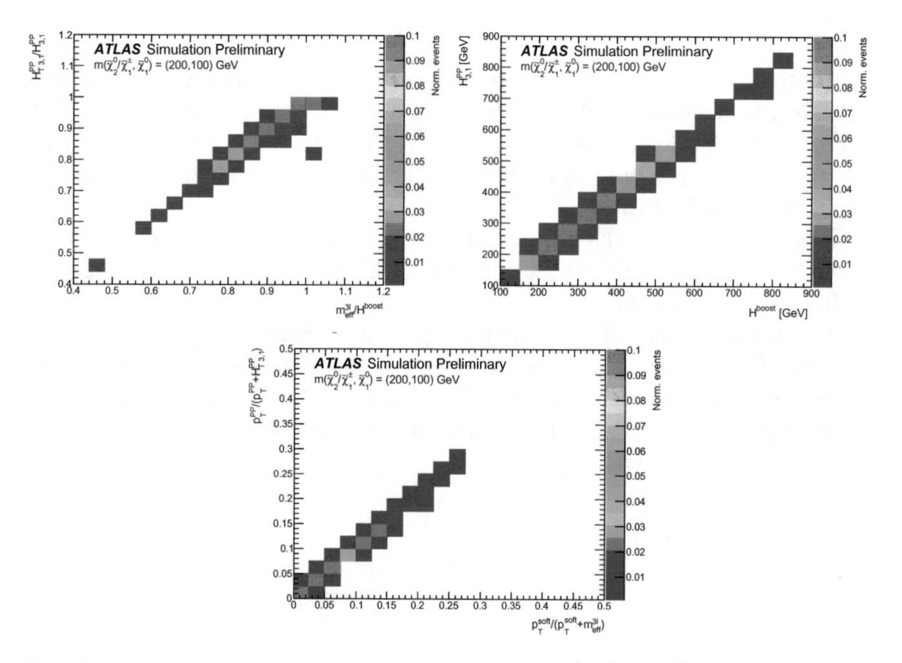

Fig. 8.14 Correlations of RJR and eRJR variables for the $m(\tilde{\chi}_1^{\pm}/\tilde{\chi}_2^0)$, $m(\tilde{\chi}_1^0) = (200, 100)$ signal point in low-mass preselection events with a jet veto applied

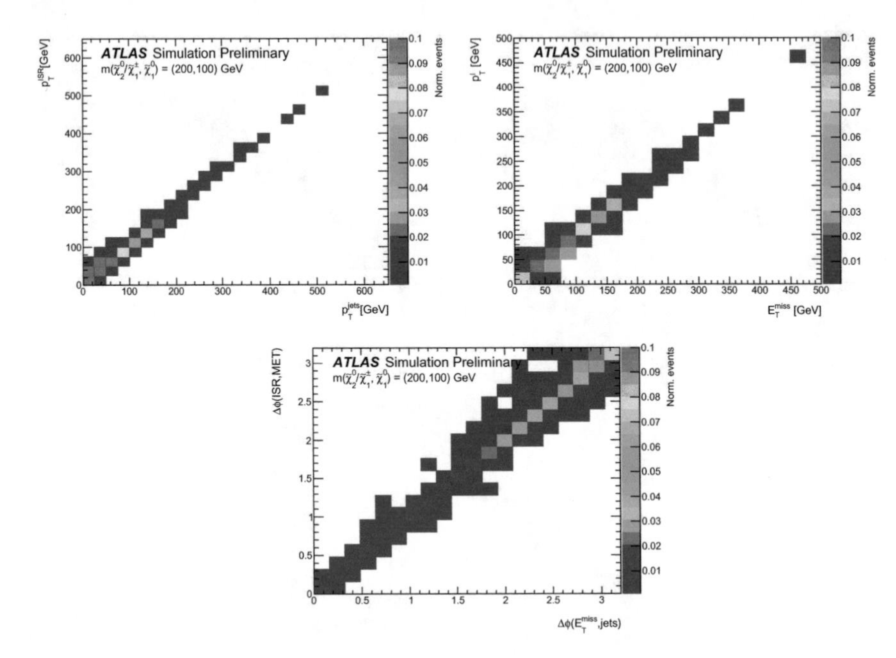

Fig. 8.15 Correlations of RJR and eRJR variables for the $m(\tilde{\chi}_1^\pm/\tilde{\chi}_2^0), m(\tilde{\chi}_1^0) = (200, 100)$ GeV signal point in ISR preselection events

$m_{\text{eff}}^{3\ell}$ is calculated in the laboratory frame while H^{boost} has a boost applied. In the RJR technique, both variables, $H_{3,1}^{\text{PP}}$ and $H_{\text{T }3,1}^{\text{PP}}$ are calculated in the same frame and as result, the ratio is always less than or equal to 1 because the transverse component is always smaller than or equal to the momentum calculated with the transverse and longitudinal components.

Correlations for the signal point of $m(\tilde{\chi}_1^\pm/\tilde{\chi}_2^0), m(\tilde{\chi}_1^0) = (200, 100)$ GeV are shown in the ISR preselection regions in Fig. 8.15. There is good correlation between the RJR and eRJR variables.

Distributions in the ISR preselections split by the number of jets are found in Fig. 8.16. The translation improves for $N_{\text{jets}} = 1$; however, there are discrepancies between the eRJR and RJR variables for $N_{\text{jets}} > 1$. The RJR algorithm selects a subset of jets as part of the ISR hemisphere; however, the eRJR selects all signal jets as part of the ISR system.

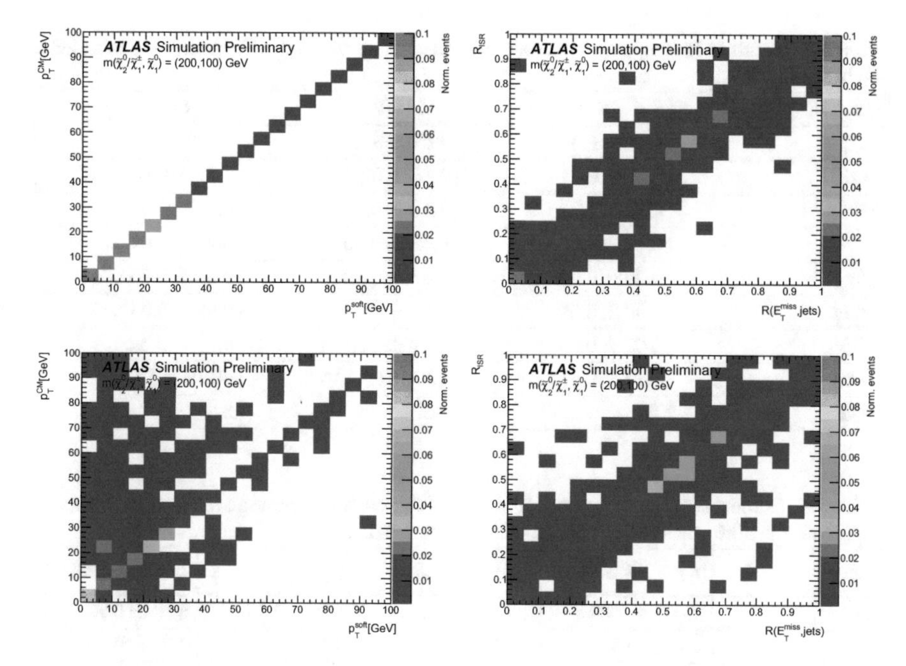

Fig. 8.16 Correlations of RJR and eRJR variables for the $m(\tilde{\chi}_1^\pm/\tilde{\chi}_2^0), m(\tilde{\chi}_1^0) = (200, 100)$ GeV signal point in ISR preselection events for (top) $N_{\text{jets}} = 1$ and (bottom) $N_{\text{jets}} > 1$

8.4 eRJR Yields and Study for the RJR Phase

8.4.1 Yields in SR-low and SR-ISR Using eRJR Variables

Yields after applying the SR-low and SR-ISR cuts on the eRJR variables are shown in Tables 8.7, 8.8 and along with the RJR yields in Table 8.9. The WZ normalization factors using eRJR variables is 1.04 ± 0.06 for CR-low and 0.92 ± 0.09 in CR-ISR, as compared with RJR WZ normalization factors of 1.09 ± 0.10 in CR-low and 1.13 ± 0.13 in CR-ISR. The normalization factors are similar in both.

The difference in the number of events in CR-low and in VR-low is due to the fact that p_T^{soft} was defined to include jets when p_T^{soft} should be defined without jets; however, in SR-low, since a jet veto is applied, p_T^{soft} is defined using leptons and E_T^{miss} only. This is corrected for the correlation plots and does not impact the normalization factor derived for the low-mass regions. The p_T^{soft} variable is defined with only leptons and E_T^{miss} in the work presented in Chap. 9.

In SR-low, the exact same data events were selected using the emulated variables as the RJR analysis with similar background expectation. In SR-ISR, because all signal jets are considered part of the ISR system in the eRJR method, additional data events were selected alongside a proportional increase in the number of expected background events, with the significance of the excess in agreement with the RJR

Table 8.7 Yields in ISR regions. Fakes are estimated using MC and 30% uncertainty is added to the backgrounds

Samples	SR-ISR	CR-ISR	VR-ISR
WZ	4.68 ± 0.39	120.09 ± 1.52	36.33 ± 0.76
ZZ	0.17 ± 0.12	7.45 ± 0.54	2.52 ± 0.44
ttV	0.10 ± 0.02	1.18 ± 0.07	0.76 ± 0.05
VVV	0.17 ± 0.04	0.42 ± 0.06	0.24 ± 0.04
Higgs	0.00 ± 0.00	0.33 ± 0.30	0.28 ± 0.26
other	0.00 ± 0.00	0.77 ± 0.28	0.08 ± 0.08
Z+jet	0.00 ± 0.00	0.89 ± 0.53	0.00 ± 0.00
Total	5.12 ± 0.41	130.10 ± 2.16	40.18 ± 1.08
Data	16	121	32

Table 8.8 Cutflow showing the impact of eRJR variables in the low-mass SR. Fakes are estimated using MC and 30% uncertainty is added to the backgrounds

Samples	m_T >100GeV	$\frac{p_T^{soft}}{p_T^{soft}+H_T} < 0.05$	$\frac{H_T}{H^{boost}}$ >0.9	H^{boost} >250GeV
WZ	66.61 ± 2.08	28.74 ± 1.32	12.45 ± 0.85	8.92 ± 0.72
ZZ	6.15 ± 0.36	2.02 ± 0.20	1.19 ± 0.18	0.91 ± 0.16
ttV	0.24 ± 0.03	0.04 ± 0.01	0.02 ± 0.01	0.02 ± 0.01
VVV	0.93 ± 0.09	0.58 ± 0.08	0.38 ± 0.07	0.31 ± 0.06
Higgs	0.25 ± 0.25	0.00 ± 0.00	0.00 ± 0.00	0.00 ± 0.00
Z+jet	-0.84 ± 1.47	-1.42 ± 1.43	-0.58 ± 0.99	-0.58 ± 0.99
singletop	0.00 ± 0.00	0.00 ± 0.00	0.00 ± 0.00	0.00 ± 0.00
$V\gamma$	1.70 ± 0.26	0.69 ± 0.12	0.36 ± 0.09	0.31 ± 0.07
$t\bar{t}$	0.54 ± 0.38	0.27 ± 0.27	0.27 ± 0.27	0.27 ± 0.27
Total	75.58 ± 2.62	30.91 ± 1.98	14.09 ± 1.35	10.16 ± 1.27
(200,100)	18.02 ± 1.22	10.99 ± 0.93	7.90 ± 0.77	5.97 ± 0.67
(200,100) Zn	0.50	0.74	1.05	1.00
Observed events	88	39	23	20

search. The signal significance in both the low-mass and ISR regions is comparable for both techniques.

Since the eRJR can reproduce the RJR SR phase space, this can be probed further to study the excesses seen in SR-low and SR-ISR.

Table 8.9 Yields in the RJR and eRJR CR, VR, and SRs. The WZ NF is applied in all regions for the RJR yields and only in the SR and VR for the eRJR yields

	SR-low	CR-low	VR-low	SR-ISR	CR-ISR	VR-ISR
RJR yields						
Expected	10 ± 2	331 ± 18	159 ± 37	3.9 ± 1.0	98 ± 10	109 ± 24
Observed	20	331	160	12	98	83
eRJR yields						
Expected	10.16 ± 1.27	465.98 ± 8.62	236.24 ± 4.06	5.01 ± 0.26	130.10 ± 2.16	40.18 ± 1.08
Observed	20	479	248	16	121	32

8.4.2 SR-low Excess Studies

SR-low distributions using eRJR variables are shown in Figs. 8.17, 8.18. The excess occurs at low $E_{\mathrm{T}}^{\mathrm{miss}}$ and at the kinematic edge of m_{T} ($m_{\mathrm{T}} \sim 100$ GeV). The $E_{\mathrm{T}}^{\mathrm{miss}}$ is further split by W lepton flavor in Fig. 8.18. The excess occurs at low $E_{\mathrm{T}}^{\mathrm{miss}}$, $E_{\mathrm{T}}^{\mathrm{miss}} < 25$ GeV, in events where the W lepton is a muon. These events are not seen in the conventional 3ℓ search due to requirement of $E_{\mathrm{T}}^{\mathrm{miss}} > 50$ GeV.

Further studies showing the impact on $E_{\mathrm{T}}^{\mathrm{miss}}$ and m_{T} of different combinations of eRJR variables can be found in Appendix A.2. There is no single eRJR variable that is responsible for the excess seen in SR-low.

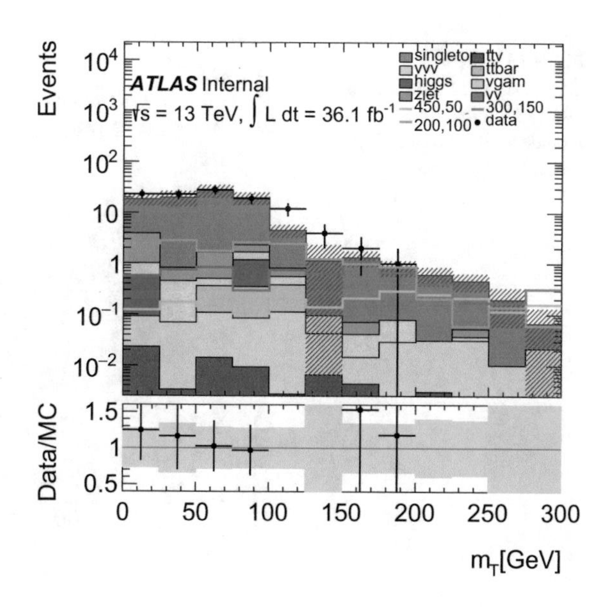

Fig. 8.17 The m_{T} distribution in SR-low using eRJR variables. Errors include a 30% uncertainty for all backgrounds

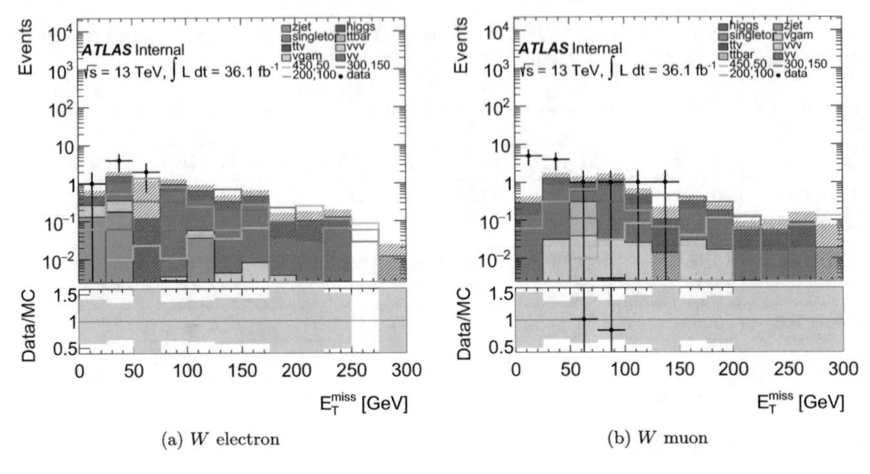

(a) W electron (b) W muon

Fig. 8.18 $E_{\mathrm{T}}^{\mathrm{miss}}$ distribution split by W lepton flavor

8.4.3 SR-ISR Excess Studies

Just as in SR-low, the excess in SR-ISR appears to be at the edge of the selection on the m_{T} distribution, $100 < m_{\mathrm{T}} < 120$ GeV, as shown in Fig. 8.19. The excess, which occurs where the m_{T} distribution starts to drop off, could be accounted for by the uncertainty in the $E_{\mathrm{T}}^{\mathrm{miss}}$ resolution.

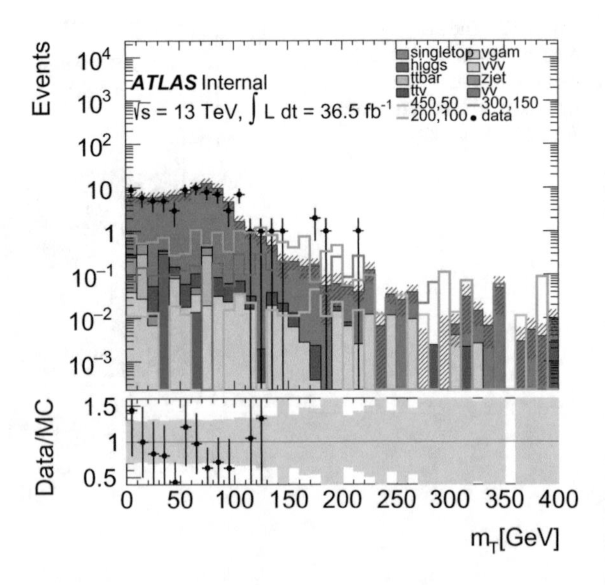

Fig. 8.19 The m_{T} distribution in SR-ISR using eRJR variables. Errors include a 30% uncertainty for all backgrounds

8.4.3.1 $E_{\mathrm{T}}^{\mathrm{miss}}$ Resolution Using Z+Jets Events

To determine if $E_{\mathrm{T}}^{\mathrm{miss}}$ resolution could account for the mismodeling of m_{T} in SR-ISR, a selection using Z+jets is used. Z+jets events are chosen because these events do not produce any real $E_{\mathrm{T}}^{\mathrm{miss}}$ so mismodeling in this variable are due to resolution.

The events are chosen to be as close to SR-ISR selection as possible. Events are required to have two signal leptons with $p_{\mathrm{T}} > 25$ GeVwhich form a same-flavor, opposite charge pair with invariant mass consistent with the Z-boson mass. Since the ISR topology is selected, at least one signal jet is required. An upper cut on the number of jets is applied to match the selection of SR-ISR and $p_{\mathrm{T}}^{\mathrm{ISR}}$ is also chosen to be greater than 100 GeV. A b-jet veto is required to minimize the $t\bar{t}$ contribution.

Two variables are defined to study the behavior of $E_{\mathrm{T}}^{\mathrm{miss}}$: the parallel and perpendicular components of $E_{\mathrm{T}}^{\mathrm{miss}}$ with respect to the dilepton system, and m_{T}. The parallel component of $E_{\mathrm{T}}^{\mathrm{miss}}$ is defined as,

$$E_{\mathrm{T}\parallel}^{\mathrm{miss}} = \frac{\overrightarrow{p_{\mathrm{T}}^{\mathrm{miss}}} \cdot \overrightarrow{p_{\mathrm{T}Z}}}{|\overrightarrow{p_{\mathrm{T}Z}}|}. \tag{8.4}$$

The perpendicular component of $E_{\mathrm{T}}^{\mathrm{miss}}$ is defined as,

$$E_{\mathrm{T}\perp}^{\mathrm{miss}} = \sqrt{(E_{\mathrm{T}}^{\mathrm{miss}})^2 - (E_{\mathrm{T}\parallel}^{\mathrm{miss}})^2}. \tag{8.5}$$

The transverse mass, m_{T} is calculated with the positive lepton and the negative lepton is added to the $E_{\mathrm{T}}^{\mathrm{miss}}$ vector. Since in SR-ISR, there is a cut on $E_{\mathrm{T}}^{\mathrm{miss}}$, a $E_{\mathrm{T}}^{\mathrm{miss}}$ proxy, $p_{\mathrm{T}}^{\ell\ell}$ is defined. Requiring a large $p_{\mathrm{T}}^{\ell\ell}$ value selects the ISR topology.

The perpendicular component of $E_{\mathrm{T}}^{\mathrm{miss}}$ and m_{T} after applying different requirements on $p_{\mathrm{T}}^{\ell\ell}$ are shown in Fig. 8.20. There is no mismodeling in the bulk in the $E_{\mathrm{T}}^{\mathrm{miss}}$ component distributions. Mismodeling appears at m_{T} edge and becomes more prominent as the requirement on $p_{\mathrm{T}}^{\ell\ell}$ increases. Even though there is mismodeling, it would not account fully for the discrepancy seen in SR-ISR.

8.4.3.2 Z-Boson Mass Side-Band Study to Check Muon Resolution

SR-ISR has an excess of three muon events over the background prediction, which could point to a muon resolution issue. This would manifest itself as a wider Z-mass peak than expected. In order to study this, a region is made in the side-band of Z-mass. The window is chosen to be of similar size as the Z-mass window of 30 GeV. The cuts applied are identical to the SR-ISR cuts except the invariant mass cut is chosen to be the Z-mass side-band: $m_{\ell\ell} \in (50, 75)$ or $(105, 130)$.

In the Z-side-band region, shown in Table 8.10, no excess is present; therefore, there is no obvious issue with muon resolution.

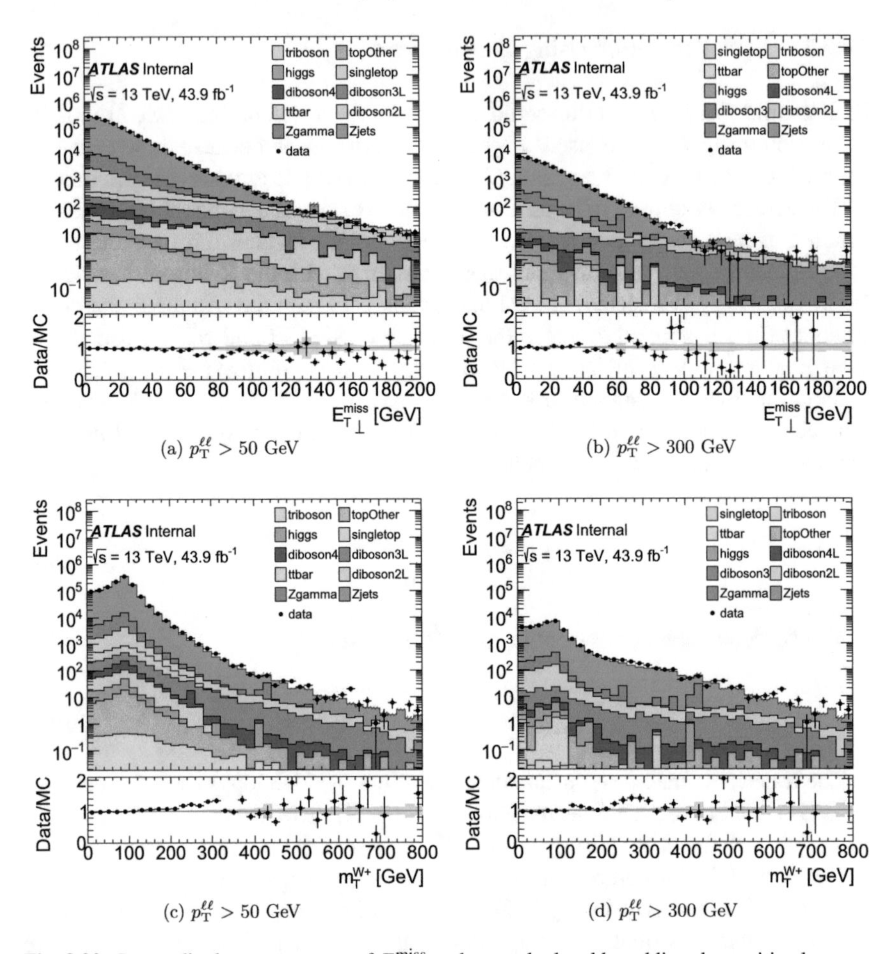

(a) $p_{\mathrm{T}}^{\ell\ell} > 50$ GeV

(b) $p_{\mathrm{T}}^{\ell\ell} > 300$ GeV

(c) $p_{\mathrm{T}}^{\ell\ell} > 50$ GeV

(d) $p_{\mathrm{T}}^{\ell\ell} > 300$ GeV

Fig. 8.20 Perpendicular components of $E_{\mathrm{T}}^{\mathrm{miss}}$ and m_{T} calculated by adding the positive lepton to the $E_{\mathrm{T}}^{\mathrm{miss}}$ for different $p_{\mathrm{T}}^{\ell\ell}$ selections

Table 8.10 Yields in ISR side-band region. WZ NF applied. Errors are statistical only

Sample	ISR side-band region
WZ	0.96 ± 0.13
ZZ	0.03 ± 0.01
Fakes	0.00 ± 0.61
Top-like	0.81 ± 0.40
Other	0.30 ± 0.05
Total	2.09 ± 0.74
Observed events	2

8.5 Next Steps

Two searches have targeted the same signal model and SUSY mass splittings near the electroweak scale. The conventional 3ℓ search[2] does not have any excess while the RJR search [1] has excesses in two orthogonal bins: SR-low and SR-ISR. Different requirements on the lepton assignment, m_{T}, and $E_{\mathrm{T}}^{\mathrm{miss}}$ result in most of the events observed in the RJR search to not be present in the $m_{\mathrm{T}}^{\mathrm{min}}$ search SRs.

To study the excess seen in the RJR search, a new technique, eRJR, is developed to emulate the RJR variables using simple, laboratory frame variables. Using this technique, the excesses seen by the RJR search can be reproduced. In SR-low, the excess is present at low $E_{\mathrm{T}}^{\mathrm{miss}}$ where the W-lepton is a muon and the excess in SR-ISR is seen at the m_{T} edge. Studies of both the $E_{\mathrm{T}}^{\mathrm{miss}}$ and muon resolutions did not show any mismodeling.

The eRJR search is expanded to add more data to determine if the excesses persist or if they are a statistical fluctuation. The next chapter discusses this work.

References

1. Collaboration ATLAS (2018) Search for chargino-neutralino production using recursive jigsaw reconstruction in final states with two or three charged leptons in proton-proton collisions at $\sqrt{s} = 13$ TeV with the ATLAS detector. Phys Rev D 98(9):092012. https://doi.org/10.1103/PhysRevD.98.092012. arXiv:1806.02293 [hep-ex]
2. Collaboration ATLAS (2018) Search for electroweak production of supersymmetric particles in final states with two or three leptons at $\sqrt{s} = 13$ TeV with the ATLAS detector. Eur Phys J C78(12):995. https://doi.org/10.1140/epjc/s10052-018-6423-7. arXiv:1803.02762 [hep-ex]
3. Jackson P, Rogan C, Santoni M (2017) Sparticles in motion: analyzing compressed SUSY scenarios with a new method of event reconstruction. Phys Rev D 95:035031. https://doi.org/10.1103/PhysRevD.95.035031. arXiv:1607.08307 [hep-ex]
4. Jackson P, Rogan C (2017) Recursive jigsaw reconstruction: HEP event analysis in the presence of kinematic and combinatoric ambiguities. Phys Rev D 96:112007. https://doi.org/10.1103/PhysRevD.96.112007. arXiv:1705.10733 [hep-ph]
5. ATLAS Collaboration (2010) The ATLAS simulation infrastructure. Eur Phys J C 70, 823. https://doi.org/10.1140/epjc/s10052-010-1429-9. arXiv:1005.4568 [physics.ins-det]
6. GEANT4 Collaboration, Agostinelli S, et al (2003) GEANT4–a simulation toolkit. Nucl Instrum Meth A 506:250. https://doi.org/10.1016/S0168-9002(03)01368-8
7. ATLAS Collaboration (2010) The simulation principle and performance of the ATLAS fast calorimeter simulation FastCaloSim, ATL-PHYS-PUB-2010-013. https://cds.cern.ch/record/1300517
8. Collaboration ATLAS (2011) Measurement of the top quark-pair production cross section with ATLAS in pp collisions at $\sqrt{s} = 7$ TeV. Eur Phys J C 71:1577. https://doi.org/10.1140/epjc/s10052-011-1577-6. arXiv:1012.1792 [hep-ex]
9. Jet calibration and systematic uncertainties for jets reconstructed in the ATLAS detector at $\sqrt{s} = 13$ TeV, Technical report, ATL-PHYS-PUB-2015-015, CERN, Geneva, July, 2015. https://cds.cern.ch/record/2037613
10. ATLAS Collaboration (2017) Jet energy scale measurements and their systematic uncertainties in proton-proton collisions at $\sqrt{s} = 13$ TeV with the ATLAS detector. Phys Rev D96:7, 072002. https://doi.org/10.1103/PhysRevD.96.072002. arXiv:1703.09665 [hep-ex]. 9.4

11. Expected performance of missing transverse momentum reconstruction for the ATLAS detector at $\sqrt{s} = 13$ TeV, Technical report, ATL-PHYS-PUB-2015-023, CERN, Geneva, July, 2015. https://cds.cern.ch/record/2037700
12. ATLAS Collaboration (2018) HepData record for Search for chargino-neutralino production using recursive jigsaw reconstruction in final states with two or three charged leptons in proton-proton collisions at $\sqrt{s} = 13$ TeV with the ATLAS detector. https://doi.org/10.17182/hepdata.83419

Chapter 9
Search for Wino-Bino Production Using the Emulated Recursive Jigsaw Reconstruction Technique with Run 2 Data

The previous chapter, Chap. 8, introduces the Emulated Recursive Jigsaw Reconstruction (eRJR) technique, developed to emulate the RJR technique using simplified, laboratory frame variables. This technique, defined and validated in Sect. 8.3, reproduces the three-lepton excesses in the low-mass region and ISR regions in the laboratory frame using 36.1 fb^{-1} of pp collision data collected between 2015 and 2016 by the ATLAS detector at the LHC.

This chapter presents the search for the chargino-neutralino ($\tilde{\chi}_1^{\pm}\tilde{\chi}_2^0$) pair-production with mass splitting near the electroweak scale. The targeted decay chain is shown in Fig. 8.2, with the chargino and neutralino decaying to the invisible LSP $\tilde{\chi}_1^0$ and either a W or Z gauge boson, respectively. Just like in the RJR and conventional 3ℓ searches, the $\tilde{\chi}_1^{\pm}$ and $\tilde{\chi}_2^0$ are assumed to be purely wino and mass degenerate, and decay with 100% branching ratio to W and Z bosons. The $\tilde{\chi}_1^0$ LSP is assumed to be pure bino. Both the W and Z bosons decay leptonically via SM branching ratios, leading to a final state with three leptons and missing momentum from two $\tilde{\chi}_1^0$ and a neutrino. The presence of initial state radiation (ISR) may lead to jets in the final state and boost the $\tilde{\chi}_1^{\pm}\tilde{\chi}_2^0$ system, enhancing the signature of the missing momentum. The search targets a range of $\tilde{\chi}_1^{\pm}/\tilde{\chi}_2^0$ masses between 100 GeV and 450 GeV and mass splittings with respect to the $\tilde{\chi}_1^0$ LSP, $\Delta m = m(\tilde{\chi}_1^{\pm}/\tilde{\chi}_2^0) - m(\tilde{\chi}_1^0)$, larger than the Z boson mass.

The object and region definitions using these new emulated Recursive Jigsaw Reconstruction (eRJR) variables are kept as close as possible to those from Ref. [1]. The excess is followed up using the eRJR technique using a larger dataset corresponding to 139 fb^{-1} of pp collision data collected between 2015 and 2018 [2, 3].

© The Editor(s) (if applicable) and The Author(s), under exclusive license
to Springer Nature Switzerland AG 2020

E. Resseguie, *Electroweak Physics at the Large Hadron Collider with the ATLAS Detector*, Springer Theses,
https://doi.org/10.1007/978-3-030-57016-3_9

To study the excess, the background modeling is validated for each year 2015–2016 (corresponding to 36 fb^{-1} just like the RJR search), 2017, and 2018. When the background modeling is validated, each year is unblinded and studied further before adding additional data.

9.1 Data Set and MC Samples

The data used for this search were collected between the years 2015 and 2018 by the ATLAS experiment and correspond to an integrated luminosity of 139 fb^{-1}. The LHC collided protons in bunch-crossing intervals of 25 ns, with the average number of interactions per crossing measured in the dataset to be $\langle\mu\rangle = 34$.

Monte Carlo (MC) simulation is used to model the expected contributions of various SM processes as well as possible SUSY signals. The MC simulation is also used to optimize the event selection criteria and estimate systematic uncertainties on event yield measurement. A full description of MC simulation samples used is given below, and summarized in Table 9.1. For most SM backgrounds, the expected contributions are taken directly from MC simulation or from MC simulation normalized to data in dedicated control regions. For Z+jets processes a data-driven method is used to predict the expected yield with MC simulation used for method development and deriving uncertainty estimates.

Diboson, triboson, and Z+jets samples [4, 5] are simulated with the SHERPA 2.2 [6] generator. Diboson samples include fully leptonic and semileptonic decays as well as loop-induced and electroweak $VVjj$ production, where V refers to a Z or W vector boson. For all SHERPA samples the additional hard parton emissions [7]

Table 9.1 Monte Carlo simulation details by physics process. Listed are the generators used for matrix-element calculation and for parton showering, the underlying event parameter tunes, the PDF sets, and the order in $\alpha_S$of cross-section calculations used for the yield normalization. "Other top" includes tZ, tWZ, $t\bar{t}WZ$, $t\bar{t}WW$, 3-top, 4-top, and rare top decay $t\bar{t}(t \to Wb\ell\ell)$ processes

Process	Event generator	PS and hadronization	UE tune	Cross section
$\tilde{\chi}_1^{\pm}\tilde{\chi}_2^0$	MADGRAPH 2.6	PYTHIA 8	A14	NLO+NLL
Diboson	SHERPA 2.2	SHERPA 2.2	Default	NLO
Triboson	SHERPA 2.2	SHERPA 2.2	Default	LO
Z+jets	SHERPA 2.2	SHERPA 2.2	Default	NNLO
$t\bar{t}$	POWHEG- BOX v2	PYTHIA 8	A14	NNLO+NNLL
Single top	POWHEG- BOX v2	PYTHIA 8	A14	NLO+NNLL
$t\bar{t}H$	POWHEG- BOX v2	PYTHIA 8	A14	NLO
$t\bar{t}V$	MADGRAPH5_aMC@NLO 2	PYTHIA 8	A14	NLO
Other top	MADGRAPH5_aMC@NLO 2	PYTHIA 8	A14	LO

are matched to a parton shower based on Catani-Seymour dipole factorization [8]. The NNPDF3.0nnlo [9] set of parton distribution functions (PDFs) and a dedicated set of tuned parton-shower parameters developed by the SHERPA authors are used [8]. The ME+PS matching [10] is employed for different jet multiplicities and then merged into an inclusive sample using an improved CKKW matching procedure [11] which is extended to next-to-leading order (NLO) accuracy using the MEPS@NLO prescription [12]. These samples are NLO accurate for up to one additional parton and leading order (LO) accurate for up to three additional parton emissions. The virtual QCD correction for matrix elements at NLO accuracy are provided by the OPENLOOPS library [13]. The Z+jets samples are normalised to a next-to-next-to-leading order (NNLO) prediction of the cross section [14].

The production of $t\bar{t}$ events is modeled using the POWHEGBOX [15] v2 generator at NLO with the NNPDF3.0nnlo PDF set and the h_{damp} parameter[1] is set to a factor of 1.5 of the top mass [16]. The events are interfaced with PYTHIA.230 [17] using the A14 tune [18] and the NNPDF2.3lo PDF set [19]. The NLO $t\bar{t}$ inclusive production cross section is corrected to the theory prediction at NNLO in QCD including the resummation of next-to-next-to-leading logarithmic (NNLL) soft-gluon terms calculated using TOP++2.0 [20].

Single-top s-channel, t-channel, and tW associated production are also modeled using the POWHEGBOX v2 generator at NLO in QCD with the NNPDF3.0nlo PDF set. The events are interfaced with PYTHIA 8.230 using the A3 tune [21] and the NNPDF2.3lo PDF set. Other rare top processes include tt+X (where X may be a Z, W, or Higgs boson), tZ, 3-top, and 4-top production and are modeled using the MADGRAPH5_AMC@NLO v2.3.3 [22] generator at NLO with the NNPDF3.0nlo PDF set. The events are interfaced with PYTHIA 8.210 using the A14 tune and the NNPDF2.3lo PDF set.

Higgs-boson production processes are generated using POWHEGBOX v2 and interfaced with PYTHIA 8.212 using the AZNLO [23] tune and CTEQ6L1 [24] PDF set. Higgs-boson production in association with a W or Z boson is produced with PYTHIA 8.186 using the A14 tune and NNPDF2.3lo PDF set. Higgs samples are normalized with cross sections calculated at NLO [25].

The SUSY $\tilde{\chi}_1^{\pm}\tilde{\chi}_2^0$ signal samples are produced at LO using MADGRAPH5_AMC@NLO v2.6.1 with up to two additional partons with the NNPDF2.3lo PDF set, and interfaced with PYTHIA.230 using the A14 tune and NNPDF2.3lo PDF set. The scale parameter for jet-parton CKKW-L matching was set at a quarter of the $\tilde{\chi}_1^{\pm}/\tilde{\chi}_2^0$ mass. Signal cross sections are calculated at NLO in α_S, adding the resummation of soft gluon emission at next-to-leading-logarithmic accuracy (NLL) [26]. The cross-section for $\tilde{\chi}_1^{\pm}\tilde{\chi}_2^0$ production, when each has a mass of 200 GeV, is 1.8 ± 0.1 pb.

The modeling of c- and b-hadron decays in samples generated with POWHEGBOX or MADGRAPH5_aMC@NLO was performed with EVTGEN 1.2.0 [27]. Gen-

[1] The h_{damp} parameter controls the transverse momentum p_T of the first additional emission beyond the leading-order Feynman diagram in the parton shower and therefore regulates the high-p_T emission against which the $t\bar{t}$ system recoils.

erated events are propagated through a full simulation of the ATLAS detector [28] using Geant4 [29], which describes the interactions of particles with the detector. A parametrized simulation of the ATLAS calorimeter called Atlfast-II [28] was used for faster detector simulation of signal samples, and is found to agree well with the full simulation. The effect of multiple interactions in the same and neighboring bunch crossings (pileup) is modeled by overlaying simulated minimum-bias events generated with PYTHIA.186 using the A14 tune and NNPDF2.31o PDF set over the original hard-scattering event.

9.2 Object and Event Selection

The object and event selection is kept as close to the one used in the RJR search. The changes made are to keep up with supported objects after the ATLAS collaboration changed to a new reconstruction algorithm and to make regions orthogonal.

9.2.1 Object Selection

Analysis events are recorded during stable beam conditions and must pass detector and data quality requirements for inclusion in the analysis. Each event is required to have a primary vertex that is associated with a minimum of two tracks of transverse momentum $p_T > 500$ MeV, where the primary vertex is defined as the reconstructed vertex with the largest Σp_T^2 of associated tracks.

Two identification levels are defined for leptons and jets, referred to as "baseline" and "signal", with signal objects being a subset of baseline. The baseline leptons are required to pass looser identification and isolation criteria, providing a higher selection efficiency for leptons and jets for use in calculating missing transverse momentum ($\mathbf{p}_{\mathrm{T}}^{\mathrm{miss}}$), resolving ambiguities between overlapping physics objects, and calculating the data-driven estimate of the background arising from fake or non-prompt leptons. The muon and electron selections are summarized in Table 9.2.

Electron candidates are reconstructed using energy clusters in the electromagnetic calorimeter which are matched to an ID track, and are calibrated using $Z \to ee$ decays. Baseline electrons must have $p_T > 10$ GeV and fall within the ID acceptance, $|\eta| < 2.47$. The electrons must also satisfy the "loose likelihood" quality criteria. The trajectory of baseline electrons must be consistent with the primary vertex to suppress electrons originating from pileup. Therefore, the tracks associated with baseline electrons must have a longitudinal impact parameter with respect to the primary vertex (z_0) such that $|z_0 \sin\theta| < 0.5$ mm. Signal electrons are required to satisfy the tighter "medium" identification criteria and must be well isolated from additional activity, passing a "tight", p_T-dependent isolation requirement that imposes fixed requirements on the value of the isolation criteria. The isolation is measured within a cone of $\Delta R = 0.2$ of the electron, and the amount of both non-associated calorimeter

Table 9.2 Summary of the baseline and signal levels for electron and muon criteria. Each new level contains the selection of the previous level

Cut	Value/description	
	Baseline electron	Baseline muon
Acceptance	$p_T > 10\,\text{GeV}$, $\lvert\eta^{\text{cluster}}\rvert < 2.47$	$p_T > 10\,\text{GeV}$, $\lvert\eta\rvert < 2.4$
Identification	`LooseAndBLayerLLH`	`Medium`
	Signal electron	Signal muon
Identification	`MediumLH`	`Medium`
Isolation	`FCTight`	`FCTight_FixedRad`
Impact parameter	$\lvert z_0 \sin\theta\rvert < 0.5$ mm, $\lvert d_0/\sigma_{d_0}\rvert < 5$	$\lvert z_0 \sin\theta\rvert < 0.5$ mm, $\lvert d_0/\sigma_{d_0}\rvert < 3$

energy and track p_T must be below 6%. Tracks are only considered by the isolation criteria if they are consistent with the primary vertex. For track isolation, the cone size decreases linearly with p_T above 50 GeV as the electron's energy becomes more collimated. The tracks associated with signal electrons must also pass a requirement on the transverse-plane distance of closest approach to the beamline (d_0) such that $|d_0/\sigma_{d_0}| < 5$, where σ_{d_0} is the uncertainty on d_0.

Muon candidates are reconstructed from either ID tracks matched to track segments in the MS or from tracks formed from a combined fit in the ID and MS, and are calibrated using $Z \to \mu\mu$ and $J/\psi \to \mu\mu$ decays. Baseline muons must have $p_T > 10\,\text{GeV}$, have $|\eta| < 2.4$, and pass the impact parameter cut of $|z_0 \sin\theta| < 0.5\,\text{mm}$. Signal muons must fulfill the "medium" identification criteria and a "tight" isolation criteria, defined similarly as for electrons but rejecting non-associated calorimeter energy at the level of 15% and non-associated track p_T at the level of 4%. The size of the track-isolation cone is $\Delta R = 0.3$ for muons of $p_T = 33\,\text{GeV}$ or below and decreases linearly to $\Delta R = 0.2$ at $p_T = 50\,\text{GeV}$, improving the selection efficiency for higher-p_T muons. Signal muons must all pass the impact parameter requirement of $|d_0/\sigma_{d_0}| < 3$.

Jet candidates are reconstructed from three-dimensional topological energy clusters using the anti-k_t algorithm with distance parameter $R = 0.4$. The jet energy scale (JES) and resolution (JER) are first calibrated to particle level using MC simulation and then through Z+jet, γ+jet, and multijet measurements. Baseline jets are required to have a $p_T > 20\,\text{GeV}$ and fall within the full calorimeter acceptance of $|\eta| < 4.5$. To suppress jets originating from pileup, jets are required to pass the "medium" working point of the track-based Jet Vertex Tagger if the jet has $p_T < 120\,\text{GeV}$ and falls within the ID acceptance of $|\eta| < 2.5$. Signal jets are required to have $|\eta| < 2.4$ to ensure full application of the pileup suppression, and events are rejected if they contain a jet that fails a "loose" quality criteria, reducing contamination from noise bursts and non-collision backgrounds. The jet selection is summarized in Table 9.3.

The identification of jets containing b-hadrons, called b-jets, is performed using a multivariate discriminant built with information from track impact parameters, the presence of displaced secondary vertices, and the reconstructed flight paths of b- and

Table 9.3 Summary of the baseline and signal selection for jets and b-jets

Cut	Value/description
	Baseline jet
Collection	`AntiKtEMTopo`
Acceptance	$p_T > 20\,\text{GeV}$, $\lvert\eta\rvert < 4.5$
	Signal jet
JVT	$\lvert JVT\rvert > 0.59$ for jets with $p_T < 120$ GeV and $\lvert\eta\rvert < 2.4$
Acceptance	$p_T > 20\,\text{GeV}$, $\lvert\eta\rvert < 2.4$
	Signal b-jet
b-tagger algorithm	MV2c10, 77% efficiency
Acceptance	$p_T > 20\,\text{GeV}$, $\lvert\eta\rvert < 2.4$

c-hadrons inside the jet. The identification criteria is tuned to an average identification efficiency of 77% as obtained for b-jets in simulated $t\bar{t}$ events, corresponding to rejection factors of 113, 4, and 16 for jets originating from light-quarks and gluons, c-quarks, and τ-leptons, respectively.

To avoid reconstructing a single detector signature as multiple leptons or jets, an overlap removal procedure is applied to baseline leptons and jets. For overlap removal ΔR is calculated using rapidity, rather than η, to ensure the distance measurement is Lorentz invariant with respect to jets that may have non-negligible masses. First, any electron that shares a track with a muon in the ID is removed, as the track is seen to be consistent with segments in the MS. Then, jets are removed if they are within $\Delta R < 0.2$ of a lepton, as they have likely formed from an electron shower or muon bremsstrahlung. For the overlap with associated muons, the nearby jet is only discarded if it is associated to less than three tracks of $p_T \geq 500\,\text{MeV}$. Finally, electrons and muons with $p_T \leq 50\,\text{GeV}$ that are close to a remaining jet are discarded to reject non-prompt or fake leptons originating from hadron decays. They are discarded if they are within a distance of $\Delta R < 0.4$ for leptons of 25 GeVor below, with the ΔR decreasing linearly with increasing lepton p_T down to $\Delta R < 0.2$ for leptons of 50 GeV.

The missing transverse momentum, $\mathbf{p}_{\mathrm{T}}^{\mathrm{miss}}$, with magnitude $E_{\mathrm{T}}^{\mathrm{miss}}$, is calculated as the negative vector sum of the transverse momenta of the baseline leptons, jets, and the soft term, the latter given by the sum of transverse momenta of additional low-momentum objects in the event. The soft term is reconstructed from particle tracks in the ID that are associated with the primary vertex but not to any reconstructed analysis objects.

Data events were collected with triggers requiring either two electrons, two muons or an electron plus a muon. The triggers have lepton p_T thresholds in the range of 8–22 GeV which are looser than the p_T thresholds required offline to ensure that trigger efficiencies are constant in the relevant phase space. The triggers used in this search are summarized in Table 9.4. All MC simulation samples emulate the triggers and have MC-to-data corrections applied to account for small differences in lepton identification, reconstruction, isolation and triggering efficiencies, as well as in jet pileup and flavor identification efficiencies.

9.2.2 Event Selection

The full set of event selections is summarized in Table 9.5 and described in detail below using the eRJR variables defined in Sect. 8.3. The signal region selection is identical to the one used in the RJR search in Tables 8.1 and 8.2 to follow-up on the excess with a nearly identical selection.

To target leptonically-decaying W and Z bosons from the electroweakinos, events must have exactly three leptons which pass the baseline and signal requirements defined in Table 9.2. The leptons must have at least one same-flavor opposite-charge (SFOS) pair (e^+e^- or $\mu^+\mu^-$) with an invariant mass of the pair $m_{\ell\ell}$ between 75 and 105 GeV, consistent with a Z boson. If there is more than one SFOS pair, the pair chosen is the one that has an invariant mass closest to that of a Z boson.

The leading source of SM background is WZ production, which when decaying fully leptonically has three leptons and $E_{\mathrm{T}}^{\mathrm{miss}}$ from a neutrino in the final state. To reduce the WZ contribution, the transverse mass is calculated from the unpaired third lepton and the $E_{\mathrm{T}}^{\mathrm{miss}}$. It is defined as $m_{\mathrm{T}} = \sqrt{2 p_T E_{\mathrm{T}}^{\mathrm{miss}} (1 - \cos(\Delta\phi))}$, where $\Delta\phi$ is the angular separation between the lepton and $\mathbf{p}_{\mathrm{T}}^{\mathrm{miss}}$, and will typically be at or below the W boson mass in SM events where the $E_{\mathrm{T}}^{\mathrm{miss}}$ is predominantly from the neutrino of the W decay. The m_{T} calculated in $\tilde{\chi}_1^{\pm}\tilde{\chi}_2^0$ events does not have such a constraint, and the SRs therefore require $m_{\mathrm{T}} \geq 100$ GeV to reduce the SM WZ background. Additionally, signal events usually have larger values of $E_{\mathrm{T}}^{\mathrm{miss}}$ due to the massive but undetected LSPs. The backgrounds where one or more leptons are fake or non-prompt are reduced by targeting the source of the additional leptons. Events containing b-tagged jets are rejected to minimize contributions from the top backgrounds $t\bar{t}$ and Wt. In the Z+jets background, a third signal lepton can arise from photon conversion, where the photon comes from the bremsstrahlung of a lepton originating from the Z boson. In this situation all three signal leptons originated from the Z boson, and this background can be reduced by requiring that the invariant mass of the three lepton system $m_{\ell\ell\ell}$ be larger than 105 GeV.

The signal regions are split into two different topologies: SR-low, the low-mass region that requires a jet veto, and SR-ISR, the ISR region that requires at least one central jet. Both SRs were optimized for signals with small mass splittings, which can lead to events with lower p_T leptons or smaller $E_{\mathrm{T}}^{\mathrm{miss}}$ in the final state. The inclusion of recoiling ISR boosts the invisible decay products in the same direction, enhancing the measured $E_{\mathrm{T}}^{\mathrm{miss}}$ and improving the discrimination against the lower $E_{\mathrm{T}}^{\mathrm{miss}}$ WZ background.

Table 9.4 Summary of trigger strategy

Lepton p_T	Trigger		
	Data15	Data16	Data17+Data18
Di-electron channel			
$p_T(e_{1(2)}) > 25$ GeV	HLT_2e12_lhloose_L12EM10VH	HLT_2e17_lhvloose_nod0	HLT_2e17_lhvloose_nod0_L12EM15VHI or HLT_2e24_lhvloose_nod0
Di-muon channel			
$p_T(\mu_{1(2)}) > 25$ GeV	HLT_mu18_mu8noL1	HLT_mu22_mu8noL1	HLT_mu22_mu8noL1
Electron-muon channel			
$p_T(e) > 25$ GeV and $p_T(\mu) > 25$ GeV	HLT_e17_lhloose_mu14	HLT_e17_lhloose_nod0_mu14	HLT_e17_lhloose_mu14

Table 9.5 Selection criteria for the low-mass and ISR regions. The variables are defined in the text. In addition, events are required to have three signal leptons, and a b-jet veto is applied. The invariant mass between the two leptons identified as coming from the Z boson decay is between 75 GeV and 105 GeV and the invariant mass of the three leptons is greater than 105 GeV

	Selection criteria									
Low-mass region	$p_T^{\ell_1}$ (GeV)	$p_T^{\ell_2}$ (GeV)	$p_T^{\ell_3}$ (GeV)	N_{jets}	m_T(GeV)	E_T^{miss}(GeV)	H^{boost} (GeV)	$\frac{m_{\text{eff}}^{3\ell}}{H^{\text{boost}}}$	$\frac{p_T^{\text{soft}}}{p_T^{\text{soft}}+m_{\text{eff}}^{3\ell}}$	
CR-low	>60	>40	>30	=0	∈ (0, 70)	>40	>250	>0.75	<0.2	
VR-low	>60	>40	>30	=0	∈ (70, 100)	–	>250	>0.75	<0.2	
SR-low	>60	>40	>30	=0	>100	–	>250	>0.9	<0.05	
ISR region	$p_T^{\ell_1}$ (GeV)	$p_T^{\ell_2}$ (GeV)	$p_T^{\ell_3}$ (GeV)	N_{jets}	m_T(GeV)	E_T^{miss}(GeV)	$\lvert\Delta\phi\left(E_T^{\text{miss}}, \text{jets}\right)\rvert$	$R\left(E_T^{\text{miss}}, \text{jets}\right)$	p_T^{jets} (GeV)	p_T^{soft} (GeV)
CR-ISR	>25	>25	>20	≥1	<100	>60	>2.0	∈ (0.55, 1.0)	>80	<25
VR-ISR	>25	>25	>20	≥1	>60	>60	>2.0	∈ (0.55, 1.0)	>80	>25
VR-ISR-small p_T^{soft}	>25	>25	>20	≥1	>60	>60	>2.0	∈ (0.55, 1.0)	<80	<25
VR-ISR-small $R\left(E_T^{\text{miss}}, \text{jets}\right)$	>25	>25	>20	≥1	>60	>60	>2.0	∈ (0.30, 0.55)	>80	<25
SR-ISR	>25	>25	>20	∈ [1, 3]	>100	>80	>2.0	∈ (0.55, 1.0)	>100	<25

The low-mass signal region requires the p_T of the first, second, and third leptons (ordered in p_T) to be greater than 60 GeV, 40 GeV, and 30 GeV, respectively, to minimize contributions from backgrounds with fake/non-prompt leptons. Tight selection thresholds on the eRJR variables H^{boost}, $\frac{p_{\text{T}}^{\text{soft}}}{p_{\text{T}}^{\text{soft}}+m_{\text{eff}}^{3\ell}}$, and $\frac{m_{\text{eff}}^{3\ell}}{H^{\text{boost}}}$ further reduce the WZ contribution in the signal region. The ISR region has a requirement of $E_{\text{T}}^{\text{miss}} \geq 80$ GeV to reduce the Z+jets background which does not have a source of real $E_{\text{T}}^{\text{miss}}$. The p_T requirement on the three leptons can then be relaxed to be greater than 25, 25, and 20 GeV, ensuring the dilepton triggers are fully efficient. To select the ISR topology in which the system of leptons and $E_{\text{T}}^{\text{miss}}$ is recoiling against the ISR jets, the angular separation between the signal jets and $\mathbf{p}_{\text{T}}^{\text{miss}}$, $\Delta\phi(E_{\text{T}}^{\text{miss}}, \text{jets})$, is required to be greater than 2.0. The ratio between the $\mathbf{p}_{\text{T}}^{\text{miss}}$ and the total transverse momenta of the jets is required to be $0.55 \leq R(E_{\text{T}}^{\text{miss}}, \text{jets}) \leq 1.0$ to ensure the majority of transverse momentum along the jet axis is carried by the invisible particles and not by the high-p_T leptons from the WZ background. A requirement of p_T^{soft} less than 25 GeV further reduces background contamination.

9.2.3 Impact of New Reconstruction Algorithm

The reconstruction algorithm has changed since the published RJR search result. The eRJR 2015–16 yields in the CRs and VRs have only 60–70% overlap between the two reconstruction algorithms. To understand where this discrepancy arises, a cutflow is done where each selection requirement is added sequentially, as shown in Table 9.6. There is a larger disagreement for events with three electron than for events with three muons, which is expected since the reconstructed of electrons was changed. Except for this discrepancy, there remains about 80% overlap between R20 and R21 until the $E_{\text{T}}^{\text{miss}}$ cut where the overlap drops to 73%.

About 30% in CR-ISR events have different jet multiplicities when moving to the new reconstruction algorithm while only 10–15% have different jet multiplicity

Table 9.6 Overlap between reconstruction algorithms in a cutflow. Each cut is applied sequentially except for the requirement on $\mu\mu\mu$ and eee events

Cut	Overlap of reconstruction algorithms (%)
3 leptons ($p_T > 25, 25, 20$ GeV) + SFOS	79.7
For $\mu\mu\mu$ only	86.0
For eee only	73.8
b-jet veto	78.0
Dilepton trigger	78.3
$m_{\ell\ell} \in (75, 105)$ GeV	79.6
$m_{\text{T}} < 100$ GeV	78.1
$E_{\text{T}}^{\text{miss}} > 50$ GeV	72.9

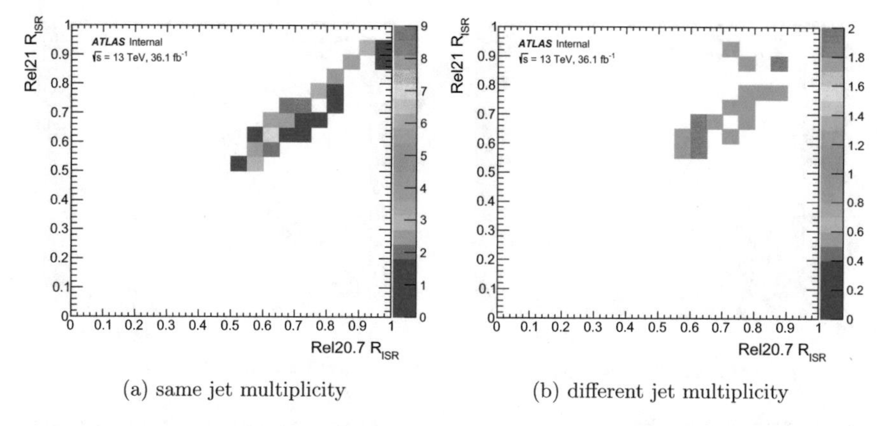

(a) same jet multiplicity (b) different jet multiplicity

Fig. 9.1 Correlation of $R\left(E_{\mathrm{T}}^{\mathrm{miss}}, \mathrm{jets}\right)$ between the reconstruction algorithms in CR-ISR for events with the same and different jet multiplicities

in CR-low and VR-low. The correlation of kinematic variables between the two reconstruction algorithm in CR-ISR are split for events with the same and different jet multiplicities, as shown in Fig. 9.1. The correlation of variables improves in events with same N_{jets} while the different N_{jets} events contain the tail events. Thus, the reconstruction change not only affects the number of jets but since ISR eRJR use jet kinematic in their variable definition, the values of the variables change in the new reconstruction.

In the low-mass regions, CR-low and VR-low, the correlation of variables improve when adding a requirement on $E_{\mathrm{T}}^{\mathrm{miss}}$, as shown in Fig. 9.2. Since VR-low and SR-low do not have a requirement on $E_{\mathrm{T}}^{\mathrm{miss}}$, there will be less overlap in those regions between the two reconstructions.

9.3 Background Estimation

The backgrounds in this analysis can be classified into two groups: irreducible backgrounds with at least three prompt leptons in the final state, and reducible backgrounds containing at least one fake or non-prompt lepton. The background strategy is summarized in Table 9.7.

The dominant irreducible background is WZ production which is estimated from MC simulation whose yields are normalized to data in CRs. The methodology on how to calculate normalization factors can be found in Sect. 5.6.2.

Other irreducible backgrounds include ZZ, VVV, $t\bar{t}V$, and Higgs processes, and are estimated directly from MC simulation due to their small contribution. The reducible backgrounds can be categorized into the top-quark like $t\bar{t}$, Wt, and WW processes, which mostly consists of non-prompt leptons from heavy-flavor hadron decays, and the Z+jets process, which also accounts for the Z+γ process, with fake

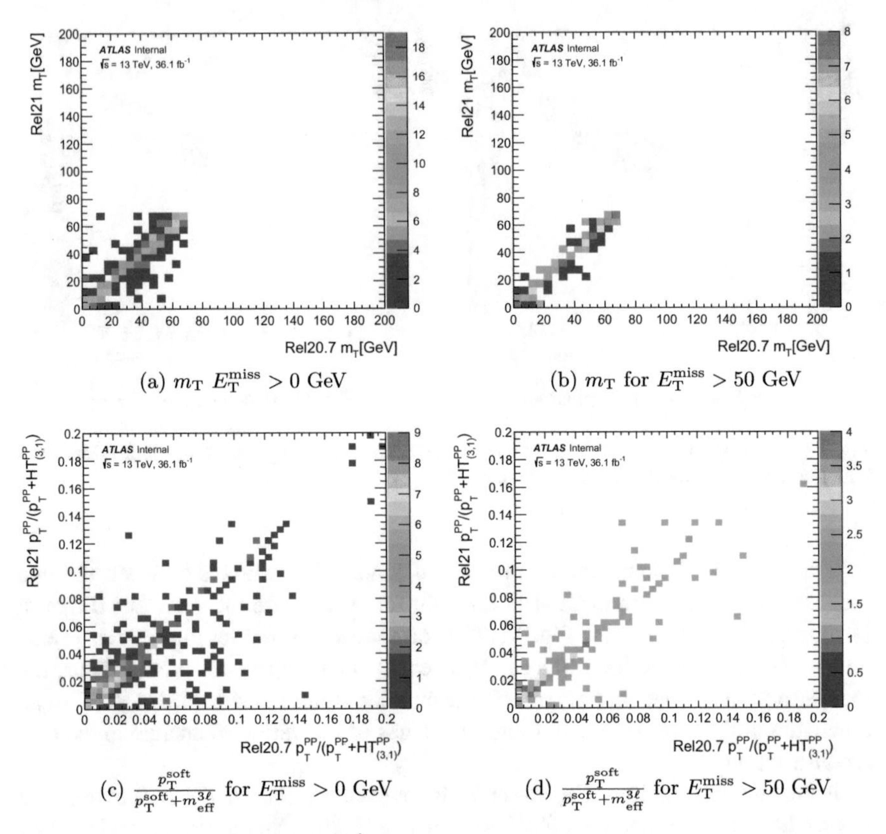

(a) m_{T} $E_{\mathrm{T}}^{\mathrm{miss}} > 0$ GeV

(b) m_{T} for $E_{\mathrm{T}}^{\mathrm{miss}} > 50$ GeV

(c) $\frac{p_{\mathrm{T}}^{\mathrm{soft}}}{p_{\mathrm{T}}^{\mathrm{soft}}+m_{\mathrm{eff}}^{3\ell}}$ for $E_{\mathrm{T}}^{\mathrm{miss}} > 0$ GeV

(d) $\frac{p_{\mathrm{T}}^{\mathrm{soft}}}{p_{\mathrm{T}}^{\mathrm{soft}}+m_{\mathrm{eff}}^{3\ell}}$ for $E_{\mathrm{T}}^{\mathrm{miss}} > 50$ GeV

Fig. 9.2 Correlation of m_{T} and $\frac{p_{\mathrm{T}}^{\mathrm{soft}}}{p_{\mathrm{T}}^{\mathrm{soft}}+m_{\mathrm{eff}}^{3\ell}}$ between the reconstruction algorithms in CR-low for events with different $E_{\mathrm{T}}^{\mathrm{miss}}$ requirements

Table 9.7 Summary of the estimation methods for each background process

Background	Estimation
Z+jets/Z+γ	FF (under "Fake/non-prompt")
$t\bar{t} + Wt + WW$	NF (under "Top-like")
WZ	NF (under "WZ")
ZZ	MC
Higgs/ VVV/ ttV	MC (under "others")

and non-prompt leptons coming primarily from photon conversions or misidentified jets. The reducible backgrounds are estimated separately in regions enriched in fake and non-prompt leptons, one targeting top-quark like processes, described in Sect. 5.6.2, and another targeting other fake/non-prompt sources, usually from Z+jets processes using a Fake Factor method [30, 31], described in Sect. 5.6.1.

Table 9.8 Anti-ID electron and muon definitions used in the fake factor method

Electrons	Muons
Pass *Loose+BL* identification $\|\Delta z_0 \sin\theta\| < 0.5$	Pass *Medium* identification $\|\Delta z_0 \sin\theta\| < 0.5$
(!*Medium* identification || $\|d_0$significance$\| > 5$ || !*FCTight* isolation)	($\|d_0$significance$\| > 3$ || !*FCTight_FixedRad* isolation)

9.3.1 *Z+jet/ $Z + \gamma$ Background*

The data-driven fake factor method is used to estimate the fake/non-prompt lepton background associated with the Z+jets process. The fake factor method uses two levels of lepton identification criteria. The regular identification criteria, "ID", corresponds to the signal lepton criteria used in the analysis. A reversed identification criteria, "anti-ID", has one or more of the identification, isolation, or impact parameter criteria inverted relative to signal leptons to obtain a selection enriched in fake leptons. The anti-ID criteria is described in Table 9.8.

A fake factor is then defined as the ratio of the yield of ID leptons to anti-ID leptons in a given region of phase space. The fake factors are measured in a region dominated by Z+jets events, requiring $E_{\mathrm{T}}^{\mathrm{miss}} < 40\,\mathrm{GeV}$, $m_{\mathrm{T}} < 30\,\mathrm{GeV}$, $|m_{\ell\ell} - m_Z| < 15\,\mathrm{GeV}$, and a b-jet veto. Events are required to pass the dilepton triggers, summarized in Table 9.4, with an offline Z lepton p_{T} of 25 GeV. The leptons paired with the Z-boson are required to pass the signal lepton criteria while the remaining lepton, fake-lepton candidate, must satisfy either the signal or the anti-ID lepton criteria. Just as in the SR, a requirement $m_{\ell\ell\ell} > 105\,\mathrm{GeV}$ is imposed to remove photon conversion from Z+γ events. To allow for more statistics of fake muons, no overlap removal is applied for the unpaired muons considered in the Z+jets/Z+γ fake factor measurement region. In both the derivation and application of the fake factors, the prompt lepton and top-quark like backgrounds that have one or more anti-ID leptons are subtracted to avoid double counting.

To primarily look at Z+jets events with exactly one jet (which fakes a lepton), a jet veto is also applied. To allow for a jet veto in spite of the relaxed muon overlap removal that is used, events are required to either have no reconstructed jets with $p_T > 30\,\mathrm{GeV}$ or no more than one $p_T > 30\,\mathrm{GeV}$ jet for events with a $p_T > 30\,\mathrm{GeV}$ muon.

The selection criteria used to select the Z+jets/Z+γ-dominated fake factor measurement region are summarized in Table 9.9.

Electron and muon fake factors are then measured separately and as a function of lepton p_T and are shown in Fig. 9.3. Fake Factors as a function η is shown in Fig. 9.4. The dependence of the fake factor as a function of η is accounted for as a parametrization systematic.

The fake factors are validated in a statistically independent region with a similar selection but requiring $E_{\mathrm{T}}^{\mathrm{miss}} < 40\,\mathrm{GeV}$ and $30 < m_{\mathrm{T}} < 50\,\mathrm{GeV}$, defined to be closer

Table 9.9 Selection criteria used to define the Z+jets/Z+γ-dominated fake factor measurement region

Three baseline leptons
Same-flavor, opposite-sign pair of signal leptons with $\|m_{\ell\ell} - m_Z\| < 15\,\text{GeV}$
$m_T < 30\,\text{GeV}$ (computed with the remaining "unpaired" lepton, which must satisfy either the signal or anti-ID criteria)
$E_T^{\text{miss}} < 40\,\text{GeV}$
Dilepton triggers
Z leptons > 25 GeV (to be on plateau)
$m_{\ell\ell\ell} > 105\,\text{GeV}$
$p_T^{3\ell} > 10$ GeV (baseline lepton selection)
No overlap removal applied for the "unpaired" muons
$N_{\text{jets}}^{20\,\text{GeV}} \leq 1$ for events with a $p_T > 30\,\text{GeV}$ muon, and $N_{\text{jets}}^{20\,\text{GeV}} == 0$ for all other events
$N_{\text{b-jets}}^{20\,\text{GeV}} == 0$

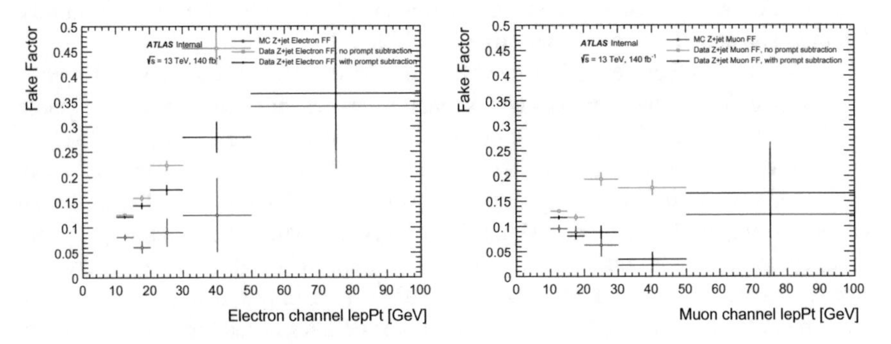

Fig. 9.3 Z+jets/Z+γ fake factors for electrons (left) and muons (right) as a function of p_T. Note that the uncertainties shown are only statistical

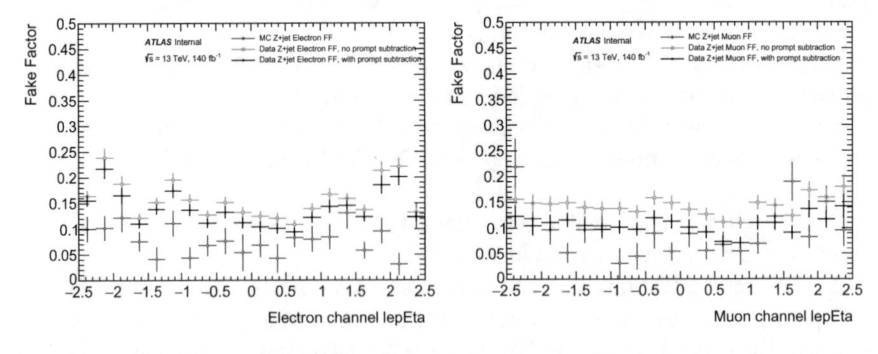

Fig. 9.4 Z+jets/Z+γ fake factors for electrons (left) and muons (right) as a function of η. Note that the uncertainties shown are only statistical

Table 9.10 Selection criteria used to define the Z+jets/$Z+\gamma$ fake factor validation region

Selection criteria
Three baseline leptons
Same-flavor, opposite-sign pair of signal leptons with $\lvert m_{\ell\ell} - m_Z \rvert < 15\,\text{GeV}$
$30\,\text{GeV} < m_{\text{T}} < 50\,\text{GeV}$ (computed with the remaining "unpaired" lepton, which must satisfy either the signal or anti-ID criteria)
$E_{\text{T}}^{\text{miss}} < 40\,\text{GeV}$
$m_{\ell\ell\ell} > 105\,\text{GeV}$
Dilepton triggers
$p_T^{\ell 1} > 25$ GeV, $p_T^{\ell 2} > 25$ GeV, $p_T^{\ell 3} > 10$ GeV
$N_{\text{b-jets}}^{20\text{ GeV}} == 0$

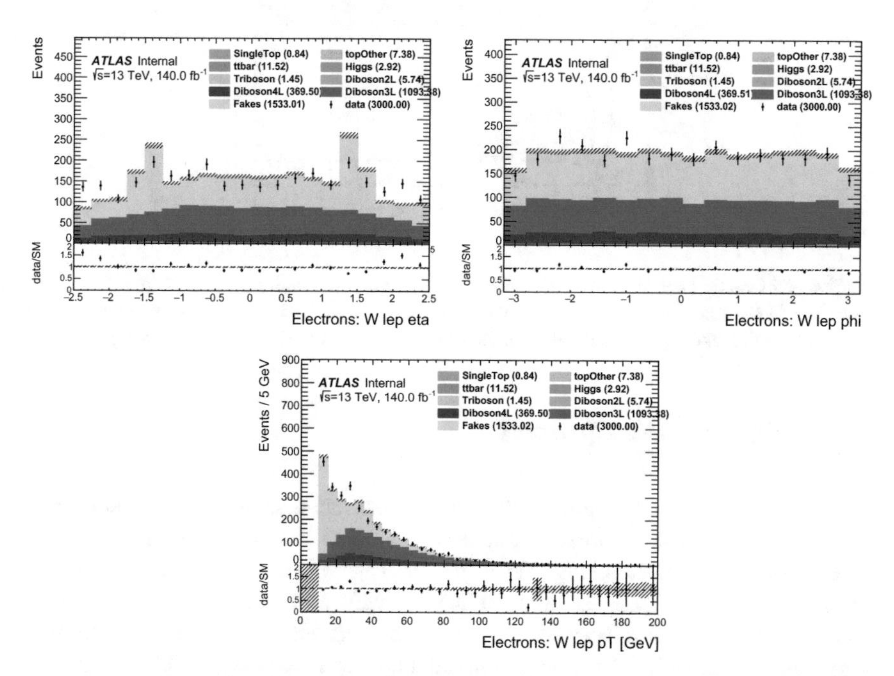

Fig. 9.5 Kinematic distributions in the Z+jets/$Z+\gamma$ validation region for events with a fake electron candidate. Good agreement is seen between data and the background estimate. Note that the lepton referred to is the fake lepton candidate, i.e. the lepton which is not used in the Z pairing

to the signal region. The selection criteria for the fake factor validation region is summarized in Table 9.10.

Good agreement between data and the fake lepton estimate is observed, as seen in Figs. 9.5 and 9.6.

The derived fake factors are applied to events in the CRs, VRs, and SRs (defined in Table 9.5), but for which at least one of the signal leptons is replaced by an anti-ID lepton.

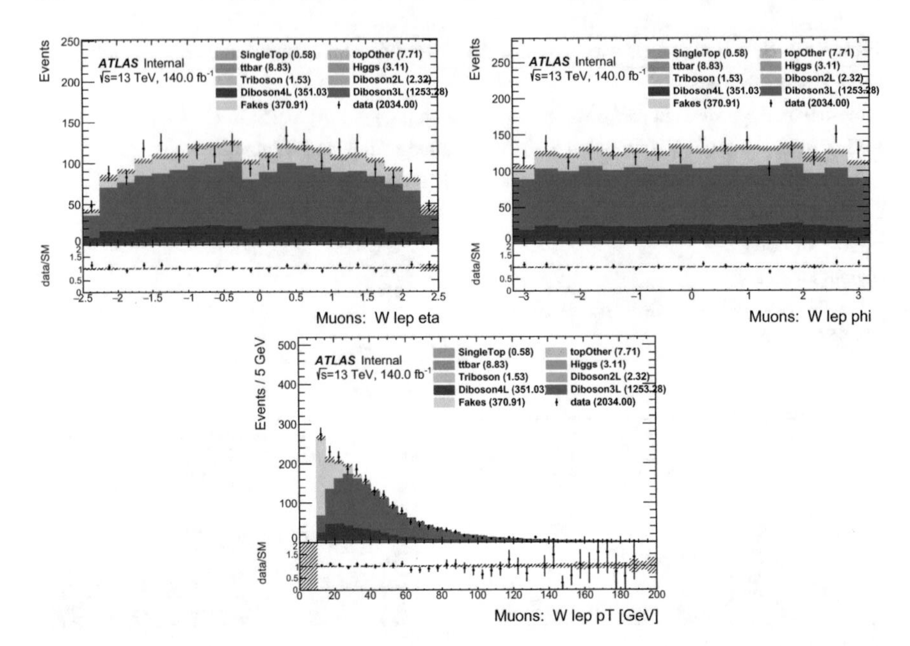

Fig. 9.6 Kinematic distributions in the Z+jets/Z+γ validation region for events with a fake muon candidate. Good agreement is seen between data and the background estimate. Note that the lepton referred to is the fake lepton candidate, i.e. the lepton which is not used in the Z pairing

9.3.2 Top-Like Backgrounds

The top-quark like background contribution is estimated using MC simulation normalized to data in a top-quark dominated CR. The region is constructed using different-flavor, opposite-charge ($e^{\pm}e^{\pm}\mu^{\mp}$ or $\mu^{\pm}\mu^{\pm}e^{\mp}$) trilepton events with lepton p_T thresholds of 25, 25, and 20 GeV as well as a b-jet veto. Due to electron charge flip, the $m_{\ell\ell}$ between the two same-sign electrons is required to have $|m_{\ell\ell} - m_Z| > 15$ GeV. Moreover, events are required to have $m_{\ell\ell} > 1$ GeV between the same-sign leptons, to avoid events where leptons are incorrectly misreconstructed as the same object twice. The selection criteria used to define this top-like CR is summarized in Table 9.11.

The normalization factors are applied to the same-flavor opposite-charge events in the top-quark like MC simulation. The normalization factors for events with three signal leptons are found to be 1.32 ± 0.35 for events with fake electrons and 1.22 ± 0.19 for events with fake muons. Distributions are shown in Fig. 9.7. Note, the contribution of WW is small in this control region as well as in the other regions in the analysis.

As the MC-based top-like fakes must also be subtracted from the Z+jets/Z+γ fake lepton estimate, this procedure must also be repeated for events containing two ID and one anti-ID lepton. Using the same control region but allowing for the presence of an anti-ID lepton results in region which is even more pure in top-like events than

Table 9.11 Selection criteria used to define the control region for determining normalization factors for the top-like fakes

Events with either three signal leptons or two signal leptons and one anti-ID lepton
Only $e^{\pm}e^{\pm}\mu^{\mp}$ and $\mu^{\pm}\mu^{\pm}e^{\mp}$ events
When measuring the normalization factors for events with an anti-ID lepton, the anti-ID lepton must be one of the same-flavor, same-sign leptons
$\lvert m_{\ell\ell}^{\text{same-sign}} - m_Z \rvert > 15$ GeV for events with two same-sign electrons
$m_{\ell\ell}^{\text{same-sign}} > 1$ GeV
Dilepton triggers
$p_T^{\ell 1} > 25$ GeV, $p_T^{\ell 2} > 25$ GeV, $p_T^{\ell 3} > 20$ GeV
$N_{\text{b-jets}}^{20\text{ GeV}} == 0$

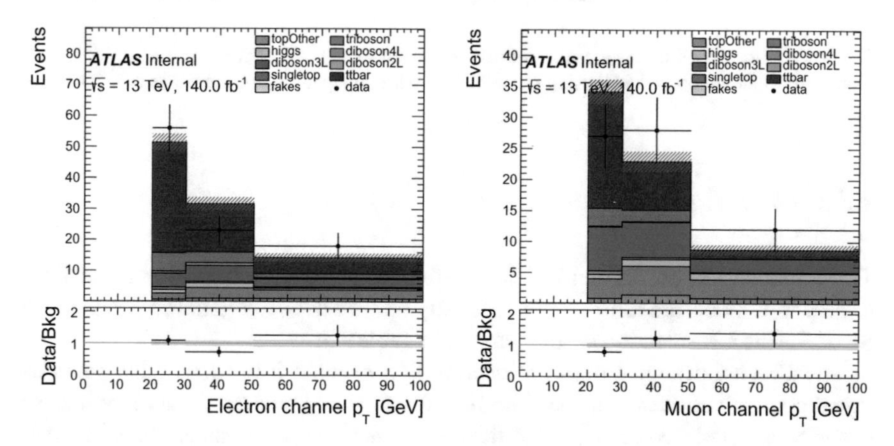

Fig. 9.7 Distributions in the top-like control region for events with fake electrons (left) and for events with fake muons (right), using only events with three signal leptons

the one previously shown. The distributions for this region are shown in Fig. 9.8, and the normalization factors are 0.82 ± 0.03 and 0.82 ± 0.04 for fake electron and fake muon events, respectively.

A top-like validation region was attempted, but any selection which would be reasonably close to the signal region and enriched in top-like events (without being dominated by other top processes such as ttV) was found to be poor in statistics. As the normalization factors are near 1 and the distributions appear to be well-modeled, it was decided that such a validation region was not necessary, and a large uncertainty is used instead.

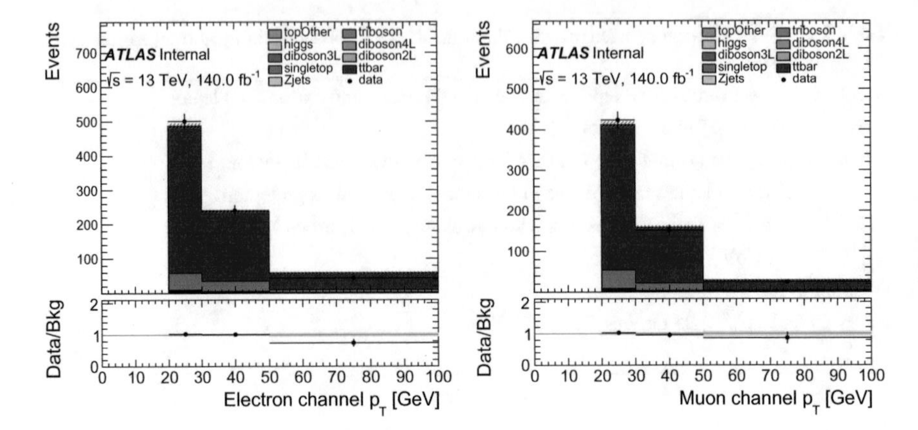

Fig. 9.8 Distributions in the top-like control region for events with fake electrons (left) and for events with fake muons (right), using only events with two signal leptons and one anti-ID lepton leptons

9.3.3 *WZ* Background

The dominant SM background, WZ, is estimated using MC simulation normalized to data in CRs designed to be kinematically similar but orthogonal to SR-low and SR-ISR. The CR selection is summarized in Table 9.5.

The CRs are designed to be enriched in WZ events while keeping the potential signal contamination small, less than 10% for all signal models. To achieve this, an upper bound is placed on the m_{T} of the CRs, targeting events that are likely to have a leptonically decaying W boson and no other sources of $E_{\mathrm{T}}^{\mathrm{miss}}$. Therefore the low-mass CR (CR-low) requires $m_{\mathrm{T}} < 70\,\mathrm{GeV}$while the ISR CR (CR-ISR) has a slightly looser requirement of $m_{\mathrm{T}} < 100\,\mathrm{GeV}$, benefiting from the boost of the $E_{\mathrm{T}}^{\mathrm{miss}}$system by the ISR. The other kinematic selections are similar to the corresponding SRs, with some loosened to enhance statistics and reduce signal contamination.

The CR-ISR definition is identical to the one used in RJR search. In the RJR search, CR-low and VR-low are inclusive in jets, requiring at most two additional jets in the events. In order to be orthogonal to CR-ISR, which has at least one jet, and to be kinematically closer to SR-low, which has a jet veto, a jet veto is added to CR-low and VR-low. Correlation between the RJR and eRJR variables also improves in the jet veto region, as shown in Fig. 8.14. Additionally, a requirement on $E_{\mathrm{T}}^{\mathrm{miss}}$ to be greater than 40 GeV is added to make CR-low orthogonal to the region where the fake factors are derived. Neither the jet veto nor the additional requirement on $E_{\mathrm{T}}^{\mathrm{miss}}$have an impact on the normalization factor derived in CR-low.

The validation regions are designed to check the agreement of the background estimation with data in regions kinematically closer to the SRs, typically targeting the extrapolation from CR to SR of a specific variable. The full VR definitions

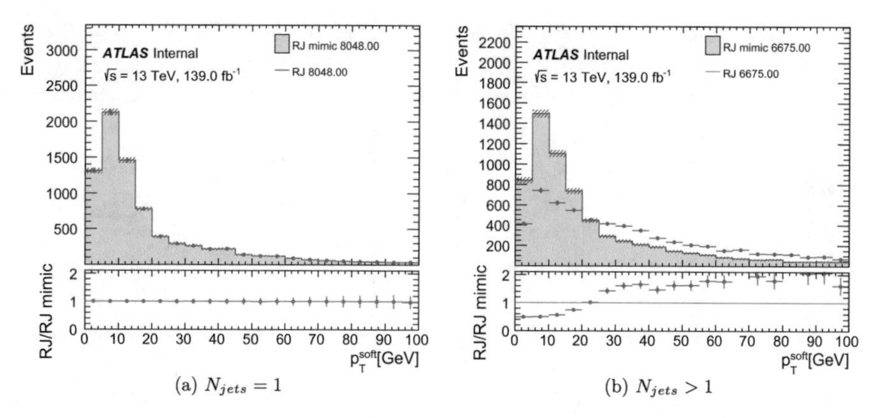

Fig. 9.9 Comparing RJR $p_{\mathrm{T}}^{\mathrm{CM}}$ and eRJR $p_{\mathrm{T}}^{\mathrm{soft}}$ in WZ ISR preselection events split by the number of jets

are summarized in Table 9.5. The VR definitions are also chosen to keep signal contamination below 10%.

Two VRs are from the RJR search: VR-low and VR-ISR. A low-mass VR-low, with the addition of a jet veto, is designed to test the extrapolation over m_{T} between CR-low and SR-low, requiring $70 < m_{\mathrm{T}} < 100$ GeV.

The ISR VR, VR-ISR, inverts the $p_{\mathrm{T}}^{\mathrm{soft}}$ requirement: VR-ISR is defined at $p_{\mathrm{T}}^{\mathrm{soft}} >$ 25 GeV, while CR-ISR and SR-ISR are defined at $p_{\mathrm{T}}^{\mathrm{soft}} < 25$ GeV. When comparing the shapes of the RJR $p_{\mathrm{T}}^{\mathrm{CM}}$ and the eRJR $p_{\mathrm{T}}^{\mathrm{soft}}$ in Fig. 9.9, in events with greater than one jet, the ratio of $p_{\mathrm{T}}^{\mathrm{CM}}$ to $p_{\mathrm{T}}^{\mathrm{soft}}$ changes at $p_{\mathrm{T}}^{\mathrm{soft}} = 25$ GeV. For $p_{\mathrm{T}}^{\mathrm{soft}} < 25$ GeV, where SR-ISR and CR-ISR are defined, the ratio of RJR to eRJR variables is less than 1 but at for $p_{\mathrm{T}}^{\mathrm{soft}} > 25$ GeV, where VR-ISR is defined, the ratio of RJR to eRJR variables is greater than 1.

The impact of the $p_{\mathrm{T}}^{\mathrm{soft}}$ as a function of pileup is also checked because the data taking conditions are different for 2015–2016 and 2017. The average pileup in 2015–16 is 22.9, while in 2017 $\langle\mu\rangle = 37.8$. The efficiency is defined as the ratio of ISR preselection events with a requirement on $p_{\mathrm{T}}^{\mathrm{soft}}$ over ISR preselection events. As seen in Fig. 9.10, there is a different dependence with pileup for events at $p_{\mathrm{T}}^{\mathrm{soft}} < 25$ GeV as opposed to $p_{\mathrm{T}}^{\mathrm{soft}} > 25$ GeV for both 2015+16 and 2017.

As a result, two additional ISR VRs are defined at $p_{\mathrm{T}}^{\mathrm{soft}} < 25$ to be closer kinematically to CR-ISR and CR-ISR: VR-ISR-small $p_{\mathrm{T}}^{\mathrm{soft}}$ and VR-ISR-small $R\left(E_{\mathrm{T}}^{\mathrm{miss}}, \mathrm{jets}\right)$. VR-ISR-small $p_{\mathrm{T}}^{\mathrm{soft}}$ is defined by inverting the requirement on $p_{\mathrm{T}}^{\mathrm{jets}}$. VR-ISR-small $R\left(E_{\mathrm{T}}^{\mathrm{miss}}, \mathrm{jets}\right)$ is defined in the side-band of the $R\left(E_{\mathrm{T}}^{\mathrm{miss}}, \mathrm{jets}\right)$ cuts. Thus, the three ISR validation regions VR-ISR, VR-ISR-small $p_{\mathrm{T}}^{\mathrm{soft}}$, and VR-ISR-small $R\left(E_{\mathrm{T}}^{\mathrm{miss}}, \mathrm{jets}\right)$ invert different selections to validate the modeling in a varied phase space.

The normalization factors derived from the CRs per year and for the published RJR result are summarized in Table 9.12. The normalization factors per year are

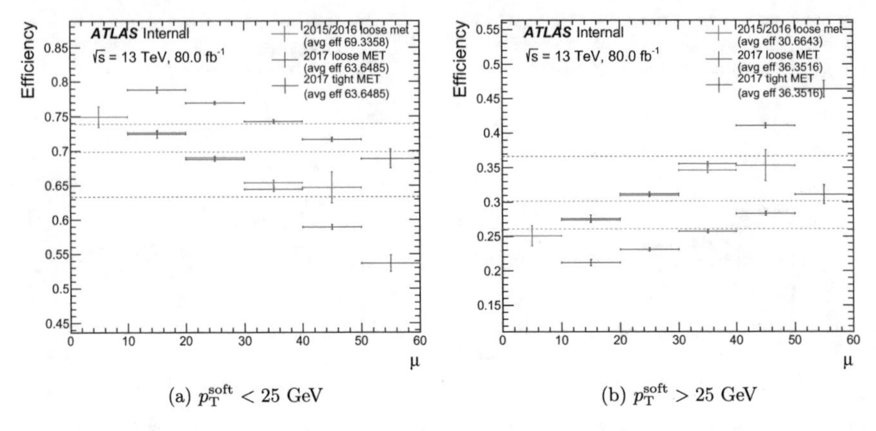

(a) $p_{\mathrm{T}}^{\mathrm{soft}} < 25$ GeV (b) $p_{\mathrm{T}}^{\mathrm{soft}} > 25$ GeV

Fig. 9.10 Efficiency of $p_{\mathrm{T}}^{\mathrm{soft}}$ as a function of μ for WZ events

Table 9.12 Background only CR-only fit normalization factors with statistical and systematic uncertainties included

Year	μ_{WZlow}	μ_{WZISR}
RJR	1.09 ± 0.10	1.13 ± 0.13
eRJR previous reconstruction (2015–16)	1.04 ± 0.06	0.92 ± 0.09
2015–16	0.89 ± 0.11	0.89 ± 0.09
2017	0.85 ± 0.11	0.98 ± 0.09
2018	0.81 ± 0.09	0.95 ± 0.08
2015–18	0.84 ± 0.07	0.94 ± 0.05

within one sigma of each other so the modeling is consistent across the years. To unblind each year, the background modelling is checked per year.

9.3.4 Background Modeling in 2015–16

The total yields in the CRs and VRs are shown for the 2015–16 dataset in Table 9.13 for the low-mass regions and Table 9.14 for the ISR regions. Figure 9.11 shows the kinematic distributions in the CR-low and CR-ISR regions, with good agreement seen between data and the background prediction. Figure 9.12 shows distributions in VR-low, VR-ISR, and the two additional ISR VRs, VR-ISR-small $p_{\mathrm{T}}^{\mathrm{soft}}$, and VR-ISR-small $R\left(E_{\mathrm{T}}^{\mathrm{miss}}, \mathrm{jets}\right)$ for the full background prediction. There is good agreement seen between the expected background prediction and the observed data in all four VRs.

Table 9.13 The observed and expected yields after the background-only fit in the low-mass CR and VR for 36.1 fb^{-1}. The normalization factors of the WZ sample for the low-mass and ISR regions are different and are treated separately in the combined fit. The Others category includes triboson, Higgs boson, and rare top-quark processes. Combined statistical and systematic uncertainties are presented. The individual uncertainties can be correlated and do not necessarily add in quadrature to the total background uncertainty

	CR-low	VR-low
Observed events	118	97
Fitted SM events	118 ± 11	84 ± 9
WZ	99 ± 12	75 ± 10
ZZ	5.2 ± 0.5	4.7 ± 0.5
Others	1.1 ± 0.9	$0.5^{+0.6}_{-0.5}$
Top-quark like	0.12 ± 0.10	0.0 ± 0.0
Fake/non-prompt leptons	13 ± 5	4.3 ± 2.6

Table 9.14 The observed and expected yields after the background-only fit in the ISR CR and VRs for 36.1 fb^{-1}. The normalization factors of the WZ sample for the low-mass and ISR regions are different and are treated separately in the combined fit. The Others category includes triboson, Higgs boson, and rare top-quark processes. Combined statistical and systematic uncertainties are presented. The individual uncertainties can be correlated and do not necessarily add in quadrature to the total background uncertainty

	CR-ISR	VR-ISR	VR-ISR-small $p_{\mathrm{T}}^{\mathrm{soft}}$	VR-ISR-small $R\left(E_{\mathrm{T}}^{\mathrm{miss}}, \mathrm{jets}\right)$
Observed events	119	23	24	72
Fitted SM events	119 ± 11	25 ± 4	26.0 ± 3.2	66 ± 7
WZ	110 ± 11	23 ± 4	22.7 ± 2.8	62 ± 7
ZZ	2.52 ± 0.21	0.50 ± 0.13	0.66 ± 0.11	0.63 ± 0.20
Others	3.3 ± 1.7	1.4 ± 0.7	0.8 ± 0.7	1.5 ± 0.8
Top-quark like	1.1 ± 0.5	0.7 ± 0.4	0.46 ± 0.27	0.41 ± 0.21
Fake/non-prompt leptons	$2.1^{+2.5}_{-2.1}$	$0.01^{+0.05}_{-0.01}$	$1.3^{+1.4}_{-1.3}$	$1.1^{+1.6}_{-1.1}$

9.3.5 Background Modeling in 2017

The total yields in the CRs and VRs for the 2017 dataset are shown in Table 9.15 for the low-mass regions and Table 9.16 for the ISR regions. Figure 9.13 shows the background composition in the CR-low and CR-ISR regions, with good agreement seen between data and the background prediction. Figure 9.14 shows distributions in VR-low, VR-ISR, VR-ISR-small $p_{\mathrm{T}}^{\mathrm{soft}}$, and VR-ISR-small $R\left(E_{\mathrm{T}}^{\mathrm{miss}}, \mathrm{jets}\right)$ for the full background prediction.

There is generally good agreement seen between the expected background prediction and the observed data. The agreement between data and the prediction seen in VR-low and VR-ISR-small $p_{\mathrm{T}}^{\mathrm{soft}}$ is within 2σ, and good agreement is seen in the shape of relevant kinematic distributions.

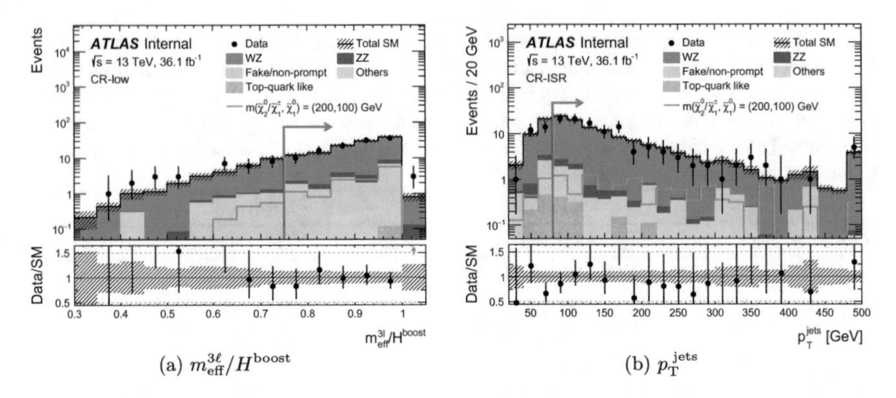

Fig. 9.11 Kinematic distributions after the background-only fit showing the data and the post-fit background in a CR-low for $m_{\mathrm{eff}}^{3\ell}/H^{\mathrm{boost}}$ and b CR-ISR for $p_{\mathrm{T}}^{\mathrm{jets}}$ for 36.1 fb^{-1}. The corresponding CR event selections are applied for each distribution except for the variable shown where the selection is indicated by a red arrow

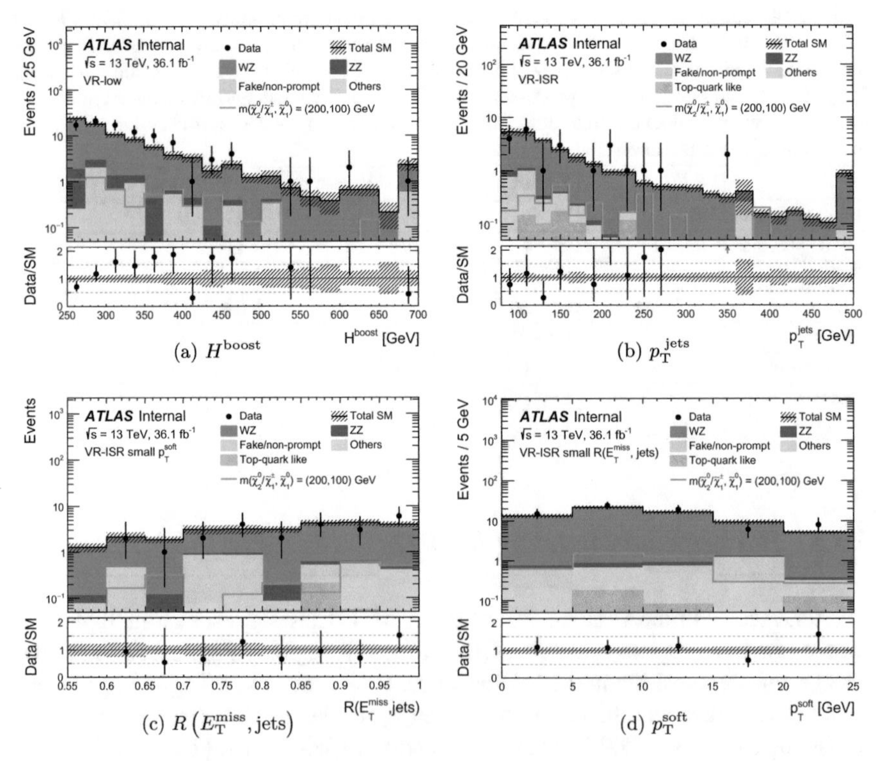

Fig. 9.12 Kinematic distributions showing the data and post-fit background in a VR-low for H^{boost}, b VR-ISR for $p_{\mathrm{T}}^{\mathrm{jets}}$, c VR-ISR-small $p_{\mathrm{T}}^{\mathrm{soft}}$ for $R\left(E_{\mathrm{T}}^{\mathrm{miss}}, \mathrm{jets}\right)$, and d VR-ISR-small $R\left(E_{\mathrm{T}}^{\mathrm{miss}}, \mathrm{jets}\right)$ for $p_{\mathrm{T}}^{\mathrm{soft}}$ for 36.1 fb^{-1}

Table 9.15 The observed and expected yields after the background-only fit in the low-mass CR and VR for 43.9 fb^{-1}. The normalization factors of the WZ sample for the low-mass and ISR regions are different and are treated separately in the combined fit. The Others category includes triboson, Higgs boson, and rare top-quark processes. Combined statistical and systematic uncertainties are presented. The individual uncertainties can be correlated and do not necessarily add in quadrature to the total background uncertainty

	CR-low	VR-low
Observed events	131	108
Fitted SM events	131 ± 11	90 ± 11
WZ	109 ± 13	83 ± 11
ZZ	5.4 ± 0.8	5.9 ± 1.0
Others	1.1 ± 0.8	0.39 ± 0.21
Top-quark like	$0.10^{+0.12}_{-0.10}$	$0.03^{+0.12}_{-0.03}$
Fake/non-prompt leptons	16 ± 6	$0.6^{+1.0}_{-0.6}$

Table 9.16 The observed and expected yields after the background-only fit in the ISR CR and VRs for 43.9 fb^{-1}. The normalization factors of the WZ sample for the low-mass and ISR regions are different and are treated separately in the combined fit. The Others category includes triboson, Higgs boson, and rare top-quark processes. Combined statistical and systematic uncertainties are presented. The individual uncertainties can be correlated and do not necessarily add in quadrature to the total background uncertainty

	CR-ISR	VR-ISR	VR-ISR-small $p_{\mathrm{T}}^{\mathrm{soft}}$	VR-ISR-small $R\left(E_{\mathrm{T}}^{\mathrm{miss}}, \mathrm{jets}\right)$
Observed events	142	32	18	70
Fitted SM events	142 ± 12	36 ± 10	32 ± 4	82 ± 8
WZ	129 ± 12	32 ± 7	28.8 ± 3.5	78 ± 8
ZZ	2.83 ± 0.24	0.62 ± 0.24	0.83 ± 0.16	0.85 ± 0.16
Others	3.0 ± 1.5	1.8 ± 1.0	0.48 ± 0.29	1.8 ± 1.0
Top-quark like	1.0 ± 0.5	0.7 ± 0.4	0.8 ± 0.6	0.5 ± 0.4
Fake/non-prompt leptons	5.7 ± 3.2	1^{+7}_{-1}	$1.1^{+1.3}_{-1.1}$	$0.6^{+1.2}_{-0.6}$

VR-ISR-small $p_{\mathrm{T}}^{\mathrm{soft}}$ has modelling issues due to the fact that it is defined for a low statistics region. It is defined at large values of $E_{\mathrm{T}}^{\mathrm{miss}}$, which would suggest that $p_{\mathrm{T}}^{\mathrm{jets}}$ should be large to create the ISR topology; however, VR-ISR-small $p_{\mathrm{T}}^{\mathrm{soft}}$ is created at low $p_{\mathrm{T}}^{\mathrm{jets}}$ so this is a more difficult topology to achieve, resulting in a small number of events.

9.3.6 Background Modeling in 2018

The total yields in the CRs and VRs for the 2018 dataset are shown in Table 9.17 for the low-mass regions and Table 9.18 for the ISR regions. Figure 9.15 shows the

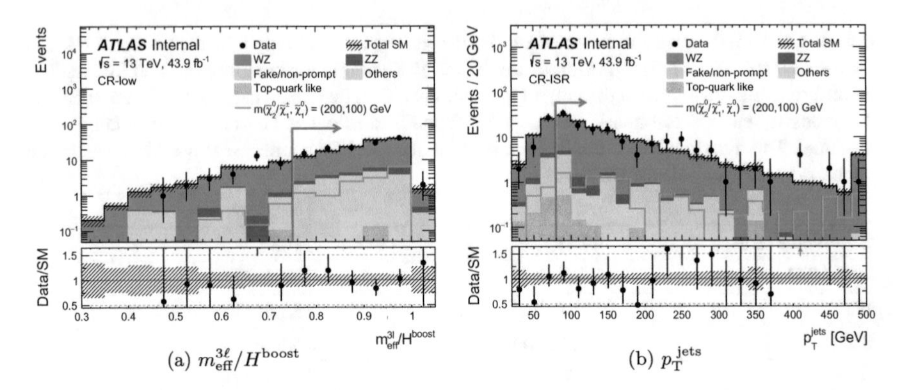

(a) $m_{\text{eff}}^{3\ell}/H^{\text{boost}}$ (b) $p_{\text{T}}^{\text{jets}}$

Fig. 9.13 Kinematic distributions after the background-only fit showing the data and the post-fit background in a CR-low for $m_{\text{eff}}^{3\ell}/H^{\text{boost}}$ and b CR-ISR for $p_{\text{T}}^{\text{jets}}$ for 43.9 fb^{-1}. The corresponding CR event selections are applied for each distribution except for the variable shown where the selection is indicated by a red arrow

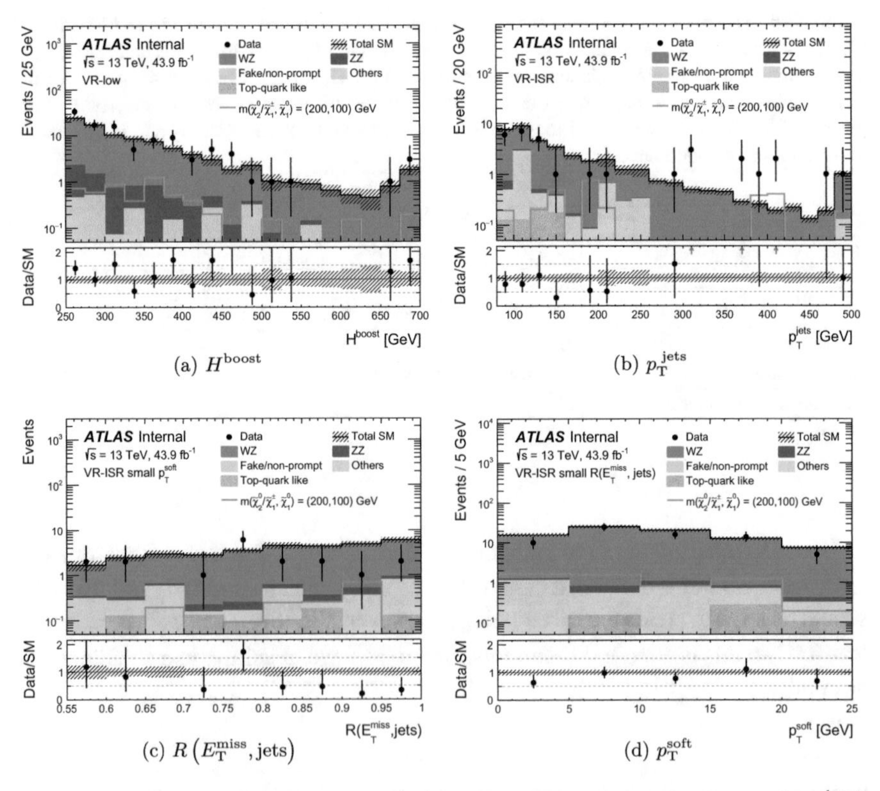

(a) H^{boost} (b) $p_{\text{T}}^{\text{jets}}$

(c) $R\left(E_{\text{T}}^{\text{miss}}, \text{jets}\right)$ (d) $p_{\text{T}}^{\text{soft}}$

Fig. 9.14 Kinematic distributions showing the data and post-fit background in a VR-low for H^{boost}, b VR-ISR for $p_{\text{T}}^{\text{jets}}$, c VR-ISR-small $p_{\text{T}}^{\text{soft}}$ for $R\left(E_{\text{T}}^{\text{miss}}, \text{jets}\right)$, and d VR-ISR-small $R\left(E_{\text{T}}^{\text{miss}}, \text{jets}\right)$ for $p_{\text{T}}^{\text{soft}}$ for 43.9 fb^{-1}

Table 9.17 The observed and expected yields after the background-only fit in the low-mass CR and VR for 59 fb^{-1}. The normalization factors of the WZ sample for the low-mass and ISR regions are different and are treated separately in the combined fit. The Others category includes triboson, Higgs boson, and rare top-quark processes. Combined statistical and systematic uncertainties are presented. The individual uncertainties can be correlated and do not necessarily add in quadrature to the total background uncertainty

	CR-low	VR-low
Observed events	163	133
Fitted SM events	163 ± 13	117 ± 12
WZ	136 ± 15	105 ± 12
ZZ	8.6 ± 1.2	7.6 ± 0.9
Others	0.9 ± 0.5	0.45 ± 0.24
Top-quark like	$0.24^{+0.35}_{-0.24}$	0.0 ± 0.0
Fake/non-prompt leptons	18 ± 7	4.3 ± 3.0

Table 9.18 The observed and expected yields after the background-only fit in the ISR CR and VRs for 59 fb^{-1}. The normalization factors of the WZ sample for the low-mass and ISR regions are different and are treated separately in the combined fit. The Others category includes triboson, Higgs boson, and rare top-quark processes. Combined statistical and systematic uncertainties are presented. The individual uncertainties can be correlated and do not necessarily add in quadrature to the total background uncertainty

	CR-ISR	VR-ISR	VR-ISR-small p_T^{soft}	VR-ISR-small $R\left(E_T^{miss}, \text{jets}\right)$
Observed events	181	46	30	110
Fitted SM events	181 ± 15	48 ± 9	39 ± 5	109 ± 9
WZ	172 ± 14	43 ± 8	37 ± 4	103 ± 9
ZZ	3.8 ± 0.4	1.01 ± 0.29	1.10 ± 0.35	1.24 ± 0.17
Others	2.7 ± 1.4	1.6 ± 0.8	0.49 ± 0.33	1.7 ± 0.9
Top-quark like	2.6 ± 1.3	1.2 ± 0.9	$0.3^{+0.9}_{-0.3}$	1.1 ± 0.9
Fake/non-prompt leptons	0^{+6}_{-0}	$0.5^{+2.8}_{-0.5}$	$0.0^{+0.0}_{-0.0}$	$1.9^{+2.2}_{-1.9}$

background composition in the CR-low and CR-ISR regions, with good agreement seen between data and the background prediction.

Figure 9.16 shows distributions in VR-low, VR-ISR, VR-ISR-small p_T^{soft}, and VR-ISR-small $R\left(E_T^{miss}, \text{jets}\right)$ for the full background prediction. There is good agreement seen between the expected background prediction and the observed data in all four VRs. The is an improvement in agreement between data and the prediction seen in VR-low and VR-ISR-small p_T^{soft}, within 1σ, and good agreement is seen in the shape of relevant kinematic distributions.

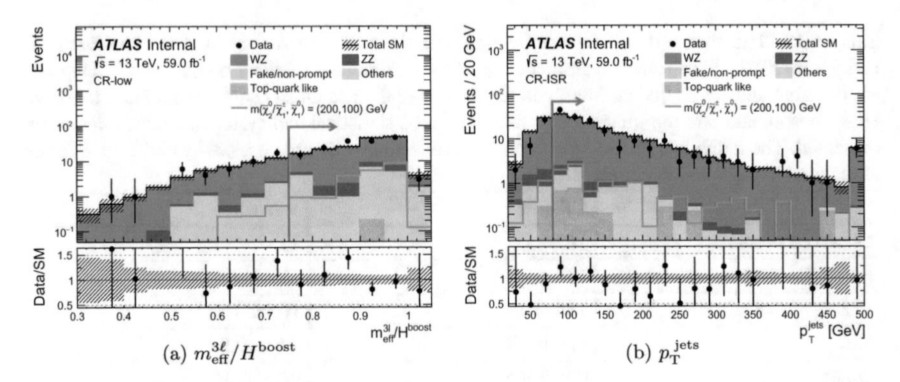

Fig. 9.15 Kinematic distributions after the background-only fit showing the data and the post-fit background in a CR-low for $m_{\text{eff}}^{3\ell}/H^{\text{boost}}$ and b CR-ISR for $p_{\text{T}}^{\text{jets}}$ for 59 fb^{-1}. The corresponding CR event selections are applied for each distribution except for the variable shown where the selection is indicated by a red arrow

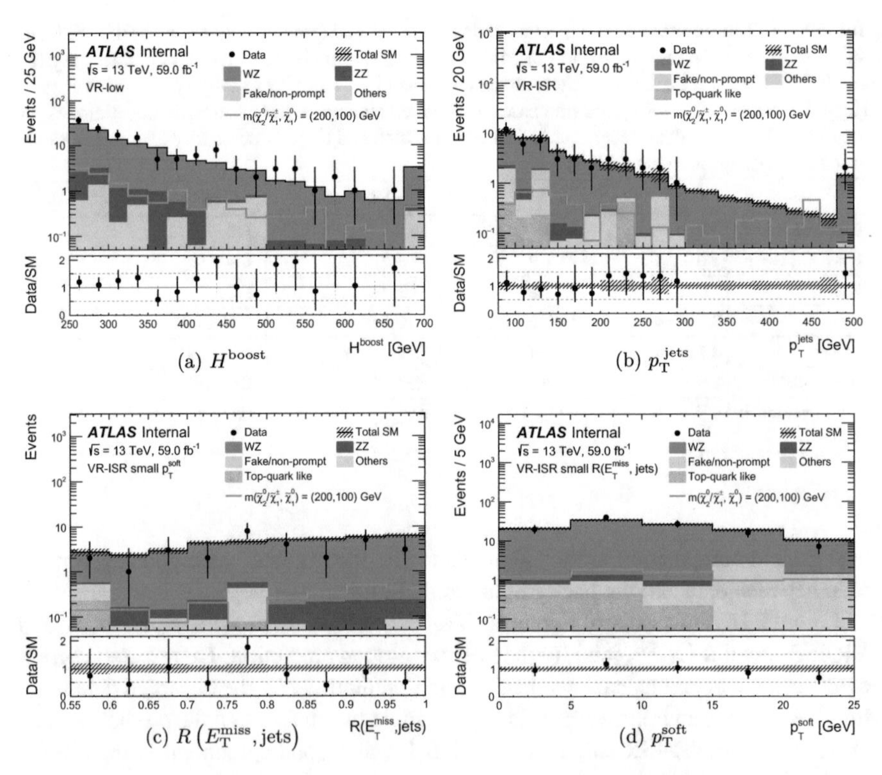

Fig. 9.16 Kinematic distributions showing the data and post-fit background in a VR-low for H^{boost}, b VR-ISR for $p_{\text{T}}^{\text{jets}}$, c VR-ISR-small $p_{\text{T}}^{\text{soft}}$ for $R\left(E_{\text{T}}^{\text{miss}}, \text{jets}\right)$, and d VR-ISR-small $R\left(E_{\text{T}}^{\text{miss}}, \text{jets}\right)$ for $p_{\text{T}}^{\text{soft}}$ for 59 fb^{-1}

Table 9.19 The observed and expected yields after the background-only fit in the low-mass CR and VR for 139 fb^{-1}. The normalization factors of the WZ sample for the low-mass and ISR regions are different and are treated separately in the combined fit. The Others category includes triboson, Higgs boson, and rare top-quark processes. Combined statistical and systematic uncertainties are presented. The individual uncertainties can be correlated and do not necessarily add in quadrature to the total background uncertainty

	CR-low	VR-low
Observed events	412	338
Fitted SM events	412 ± 20	291 ± 20
WZ	343 ± 27	262 ± 22
ZZ	19.2 ± 1.6	18.2 ± 1.7
Others	3.0 ± 1.5	1.6 ± 0.8
Top-quark like	0.5 ± 0.4	$0.02\ ^{+0.25}_{-0.02}$
Fake/nonprompt leptons	46 ± 17	10 ± 5

9.3.7 Background Modeling in 2015–18

The background modelling is also checked for the full 2015–18 dataset. The total yields in the CRs and VRs are shown in Table 9.19 for the low-mass regions and Table 9.20 for the ISR regions. Figure 9.17 shows the background composition in the CR-low and CR-ISR regions, with good agreement seen between data and the background prediction.

Figure 9.18 shows distributions in VR-low, VR-ISR, VR-ISR-small $p_{\mathrm{T}}^{\mathrm{soft}}$, and VR-ISR-small $R\left(E_{\mathrm{T}}^{\mathrm{miss}}, \mathrm{jets}\right)$ for the full background prediction. There is generally good agreement seen between the expected background prediction and the

Table 9.20 The observed and expected yields after the background-only fit in the ISR CR and VRs for 139 fb^{-1}. The normalization factors of the WZ sample for the low-mass and ISR regions are different and are treated separately in the combined fit. The Others category includes triboson, Higgs boson, and rare top-quark processes. Combined statistical and systematic uncertainties are presented. The individual uncertainties can be correlated and do not necessarily add in quadrature to the total background uncertainty

	CR-ISR	VR-ISR	VR-ISR-small $p_{\mathrm{T}}^{\mathrm{soft}}$	VR-ISR-small $R\left(E_{\mathrm{T}}^{\mathrm{miss}}, \mathrm{jets}\right)$
Observed events	442	101	72	253
Fitted SM events	442 ± 21	111 ± 19	96 ± 7	256 ± 13
WZ	415 ± 22	98 ± 17	89 ± 7	245 ± 13
ZZ	9.1 ± 0.8	2.1 ± 0.5	2.6 ± 0.4	2.7 ± 0.4
Others	12 ± 6	6.9 ± 3.5	1.7 ± 0.9	6.2 ± 3.2
Top-quark like	4.7 ± 1.6	2.7 ± 1.1	1.5 ± 1.2	2.0 ± 1.0
Fake/nonprompt leptons	$1.5^{+2.3}_{-1.5}$	$0.9^{+1.6}_{-0.9}$	$1.3^{+1.6}_{-1.3}$	$0.01^{+0.05}_{-0.01}$

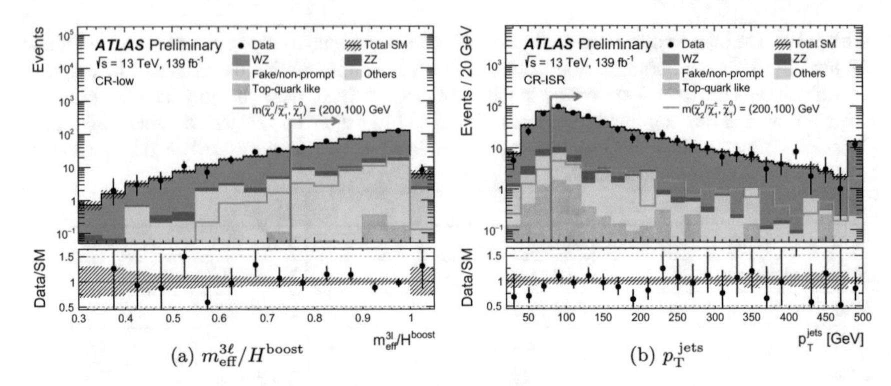

Fig. 9.17 Kinematic distributions after the background-only fit showing the data and the post-fit background in a CR-low for $m_{\rm eff}^{3\ell}/H^{\rm boost}$ and b CR-ISR for $p_{\rm T}^{\rm jets}$ for 139 fb^{-1}. The corresponding CR event selections are applied for each distribution except for the variable shown where the selection is indicated by a red arrow

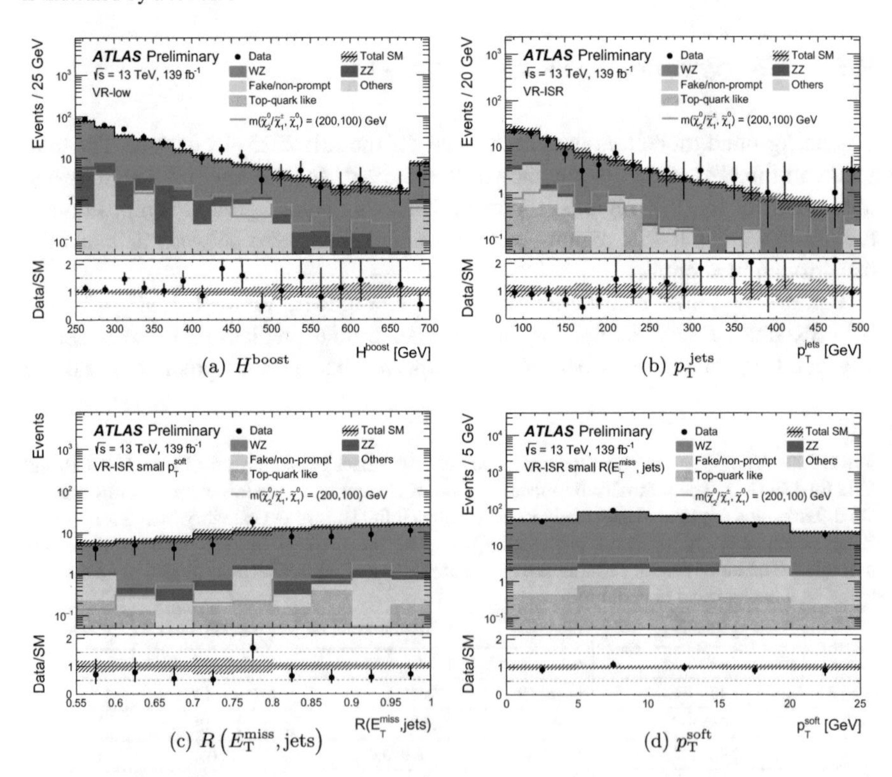

Fig. 9.18 Kinematic distributions showing the data and post-fit background in a VR-low for $H^{\rm boost}$, b VR-ISR for $p_{\rm T}^{\rm jets}$, c VR-ISR-small $p_{\rm T}^{\rm soft}$ for $R\left(E_{\rm T}^{\rm miss}, {\rm jets}\right)$, and d VR-ISR-small $R\left(E_{\rm T}^{\rm miss}, {\rm jets}\right)$ for $p_{\rm T}^{\rm soft}$ for 139 fb^{-1}

observed data. The agreement between data and the prediction seen in VR-low and VR-ISR-small $p_{\mathrm{T}}^{\mathrm{soft}}$ is within 2σ, and good agreement is seen in the shape of relevant kinematic distributions.

9.4 Uncertainties

Systematic uncertainties are derived for the signal and background predictions and account for experimental sources related to detector measurements as well as theoretical sources on the expected yields and MC simulation modeling.

Experimental uncertainties reflect the precision of the experimental measurements of jets, electrons, muons, and $E_{\mathrm{T}}^{\mathrm{miss}}$. The jet energy scale (JES) and resolution (JER) uncertainties [32, 33] are derived as a function of jet p_T and η, and account for dependencies on the pileup conditions and on the flavor composition of jets. The JES reflects the uncertainty on the average jet p_T measurement, varying from about 4% for 20 GeV jets down to 1% above 300 GeV, while the JER reflects the uncertainty on the precision of the jet p_T measurement, varying from about 2% to 0.4% across the same p_T range. Varying the JES and JER can alter the jet multiplicity of an event, affecting its inclusion into SR-low or SR-ISR regions, as well as affecting the eRJR variables that are dependent on jet and $E_{\mathrm{T}}^{\mathrm{miss}}$ kinematics. Similar uncertainties account for the energy scale and resolution of electrons [34] and muons [35], with the muon uncertainties having a negligible impact on the analysis. Variations on the per-object uncertainties are propagated through the $E_{\mathrm{T}}^{\mathrm{miss}}$ calculation, with additional uncertainties for the scale and resolution of the soft-term [36, 37].

Additional experimental uncertainties account for differences between data and MC simulation in the efficiency of the identification, reconstruction, isolation requirements, and triggering of electrons [38] and muons [35], as well as on the identification of pileup jets by the Jet Vertex Tagger [39, 40]. These uncertainties are found to have a negligible effect in both signal regions. An uncertainty is applied on the measured $\langle\mu\rangle$ distribution, which is shifted by $+14\%/-6\%$ in data before reweighting the $\langle\mu\rangle$ in MC simulation to match that of data, and is found to have an effect below 1% in both signal regions. The uncertainty in the combined 2015–2018 integrated luminosity is 1.7%, but has a greatly reduced impact on the background estimation due to the CR normalization procedure. It is derived from calibration of the luminosity scale using x-y beam-separation scans, following a methodology similar to that detailed in Ref. [41], and using the LUCID-2 detector for the baseline luminosity measurements [42].

9.4.1 Fake Factor Uncertainties

For the Z+jets/Z+γ estimation using the fake factor method, the uncertainties include:

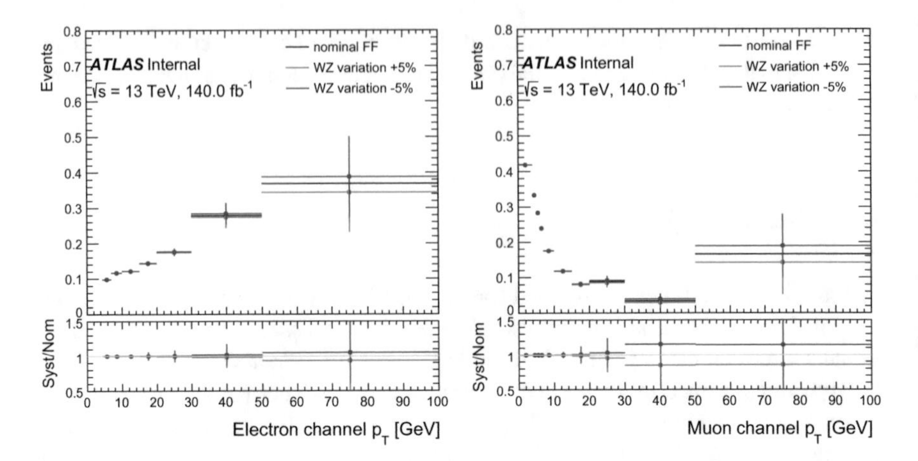

Fig. 9.19 Evaluation of the Z+jets/$Z + \gamma$ diboson subtraction systematic for electrons (left) and muons (right). The difference between the nominal Fake Factor and the scaled diboson yield is then taken as the uncertainty on the diboson subtraction

- **Propagation of FF statistical uncertainties to the "anti-ID" events**. These uncertainties arise from the limited sample sizes available used to measure the fake-factors, and are visible as the statistical uncertainty bars in Fig. 9.3 for electrons and for muons. Good agreement is generally seen given the poorer statistics of fake events in MC simulation, and uncertainties of 12% and 18% are assigned for electrons and muons, respectively.
- **Effect on the FF value of varying prompt subtraction up/down, based on the** WZ **cross-section uncertainty**. The WZ MC is scaled up and down by 5% [43], and the Fake Factors are recalculated. The largest difference with respect to the nominal fake factor is then used as the fake factor's uncertainty on the diboson subtraction (and this is assigned as a symmetric uncertainty). This is demonstrated in Fig. 9.19. For electrons, the systematic uncertainty is 6.7%, and for muons, it is 15.3%. The largest impact of the prompt subtraction is on the lepton p_T distribution.
- **Parametrization of the FFs.** A systematic uncertainty related to the choice of parametrizing the fake factors in p_T is assessed by binning in another kinematic variable. In this case, the only kinematic variable found to have a notable residual dependence is η. A two-dimensional fake parametrization in p_T and η was not found to be feasible due to a lack of statistics, so the overall dependence is taken as an additional uncertainty. Shown in Fig. 9.20 are the fake factors binned in η, as well as the average fake factors. All error bars are statistical only. Flat parametrization systematics of 25% and 21.3% are assigned for electrons and muons, respectively.
- **Closure of the FF method.** Fake factors are derived from Z+jets/$Z + \gamma$ MC in the FF measurement region and applied to Z+jets/$Z + \gamma$ MC "anti-ID" events in the FF validation region. This MC fake estimate is then compared with the Z+jets/$Z + \gamma$ MC yields for events with three signal leptons. The p_T distributions

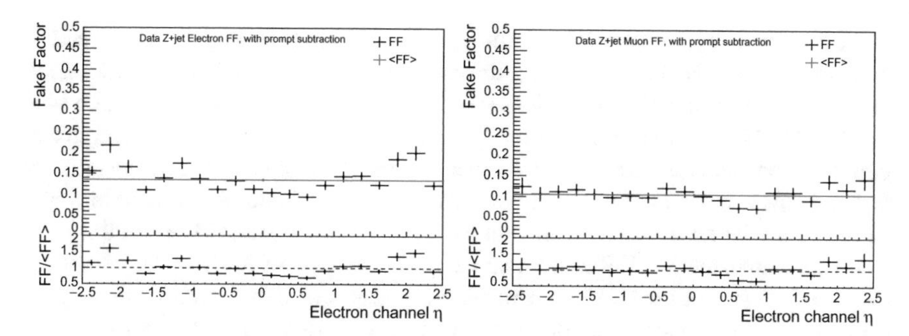

Fig. 9.20 Evaluation of the FF parametrization systematic for electrons (left) and muons (right). Shown are the η-binned fake factors, as well as the average fake factors. Flat systematic uncertainties are assigned for electrons and muons separately

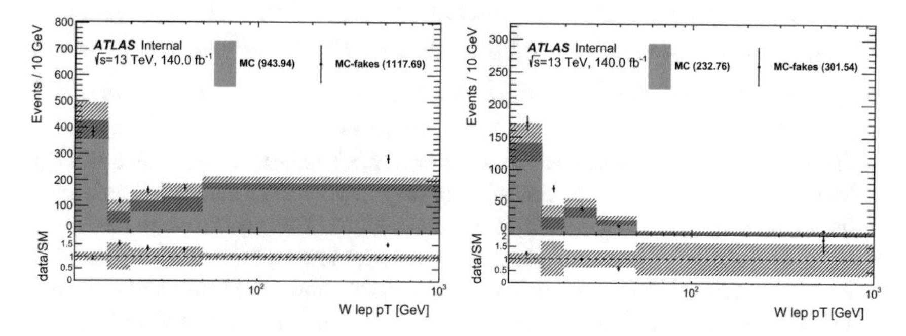

Fig. 9.21 Evaluation of the Fake Factor closure systematic for electrons (left) and muons (right). The fake lepton candidate p_T distributions are shown for both Z+jets/$Z+\gamma$ MC and the MC fake estimate. Flat systematic uncertainties are assigned for electrons and muons separately

of the fake lepton candidates are shown in Fig. 9.21, and good agreement is seen between the MC fake estimate and the out-of-the-box MC given the MC statistics. Flat closure systematics of 12% and 18% are assigned for electrons and muons, respectively.

For the top-like fakes, the only uncertainty is from the propagation the statistical uncertainty on the top-like fakes NF to the background estimate.

9.4.2 *WZ* Theory Uncertainties

The theoretical uncertainties account for any mismodeling of the MC simulation, particularly for the WZ process. They include QCD scale uncertainties on the WZ cross-section, PDF uncertainties, and varying α_S within its uncertainty. QCD scale uncertainties are evaluated using seven-point variations of the factorization and renormalization scales in the matrix elements. The scales are varied independently upwards and downwards by a factor 2, but without shifting them both in the same direction simultaneously. PDF uncertainties for the nominal PDF set are evaluated by taking the envelope of the 100 variation replicas and the central values of the CT14nnlo [44] and MMHT2014 NNLO [45] PDF sets. The impact of ± 0.001 shifts of α_S, the strong coupling constant, on the acceptance is also considered. The QCD scale uncertainty is dominant and affects the prediction of the amount of additional radiation, and therefore the jet multiplicity, within an event. It has an impact of 17% on the yield in the jet-populated SR-ISR region, and a smaller impact of 3.4% in the jet-vetoed SR-low. The size of the QCD scale uncertainty grows with the number of jets in an event, but the total uncertainty on the transfer factor is reduced by similarities in the jet multiplicity distribution in the control and signal regions.

Figures 9.22 and 9.23, show the impact on the WZ background of the choice of QCD scales, α_s and PDFs on the normalization and shapes of the m_{T} and $E_{\mathrm{T}}^{\mathrm{miss}}$ distributions in low-mass and ISR preselections, respectively. The low-mass preselection region removes the selections on the m_{T}, H^{boost}, $\frac{p_{\mathrm{T}}^{\mathrm{soft}}}{p_{\mathrm{T}}^{\mathrm{soft}}+m_{\mathrm{eff}}^{3\ell}}$, and $\frac{m_{\mathrm{eff}}^{3\ell}}{H^{\mathrm{boost}}}$ variables, and the ISR preselection region removes the selections on the $E_{\mathrm{T}}^{\mathrm{miss}}$, m_{T}, $p_{\mathrm{T}}^{\mathrm{soft}}$, $|\Delta\phi\left(E_{\mathrm{T}}^{\mathrm{miss}}, \mathrm{jets}\right)|$, $R\left(E_{\mathrm{T}}^{\mathrm{miss}}, \mathrm{jets}\right)$, and $p_{\mathrm{T}}^{\mathrm{jets}}$ variables.

The impact of α_s and PDFs is small, generally below $\sim$3%. The choice of QCD scales is by far the dominant effect, with an impact on the normalization as high as $\sim$30% in the ISR regions, and $\sim$5% in the low regions. This difference in the impact of the choice of QCD scale is due to the fact that the low region has a jet veto while the ISR region requires at least one jet. Figure 9.24 shows the impact of choice of QCD scales, α_s and PDFs as a function of N_{jets} in both loose low and ISR regions with the requirement on the number of jets removed. The QCD scale uncertainty is much larger when requiring at least one jet than with a jet veto.

Table 9.21 summarizes the background modeling uncertainties derived in the different regions of the search. Because the normalization of the WZ process is extracted from control regions, the uncertainties on the estimated yields of this background in the signal (or validation) regions will be smaller than reported on this table. This is the desired effect of using control regions, where the uncertainties impacting the estimated yields in the SRs(VRs) enter through a SR(VR)/CR ratio (transfer factor), and therefore largely cancel out.

As a final cross-check, the m_{T} and $E_{\mathrm{T}}^{\mathrm{miss}}$ shape predictions between POWHEG and SHERPA were compared, in the low-mass and ISR signal regions, to ensure the assigned modelling uncertainties cover any possible modelling differences. This comparison is displayed in Fig. 9.25. The differences seen in the modelling of the different generators are not statistically significant, and therefore no additional uncertainties are considered. N_{jets} is the only distribution that shows discrepancies between POWHEG and SHERPA; however, when taking the ratio of SR/CR, the ratio of SHERPA to POWHEG is flat, as shown in Fig. 9.26. As a result, no additional uncertainty is added.

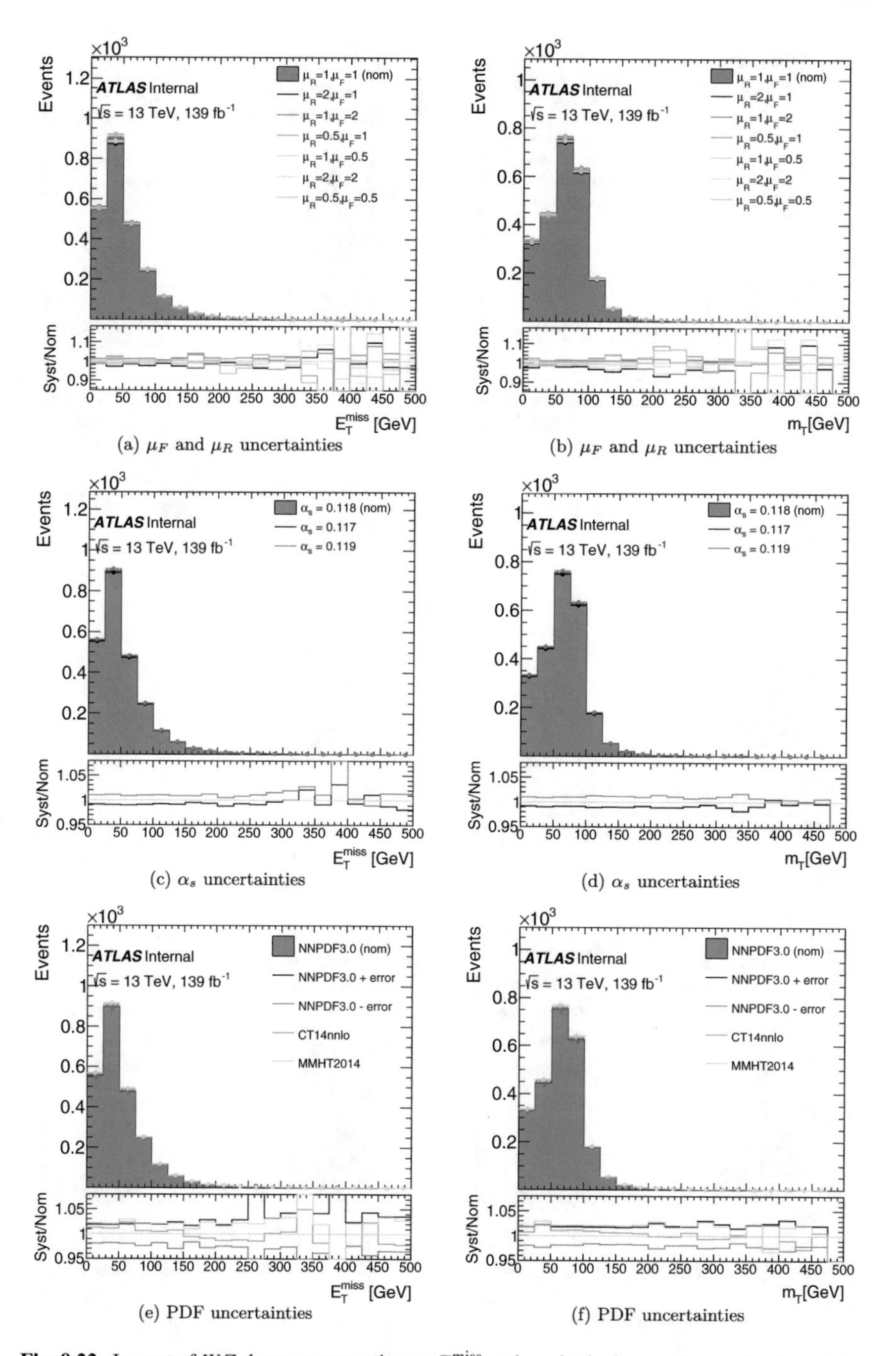

(a) μ_F and μ_R uncertainties

(b) μ_F and μ_R uncertainties

(c) α_s uncertainties

(d) α_s uncertainties

(e) PDF uncertainties

(f) PDF uncertainties

Fig. 9.22 Impact of WZ theory systematics on E_T^{miss} and m_T in the low-mass preselection region

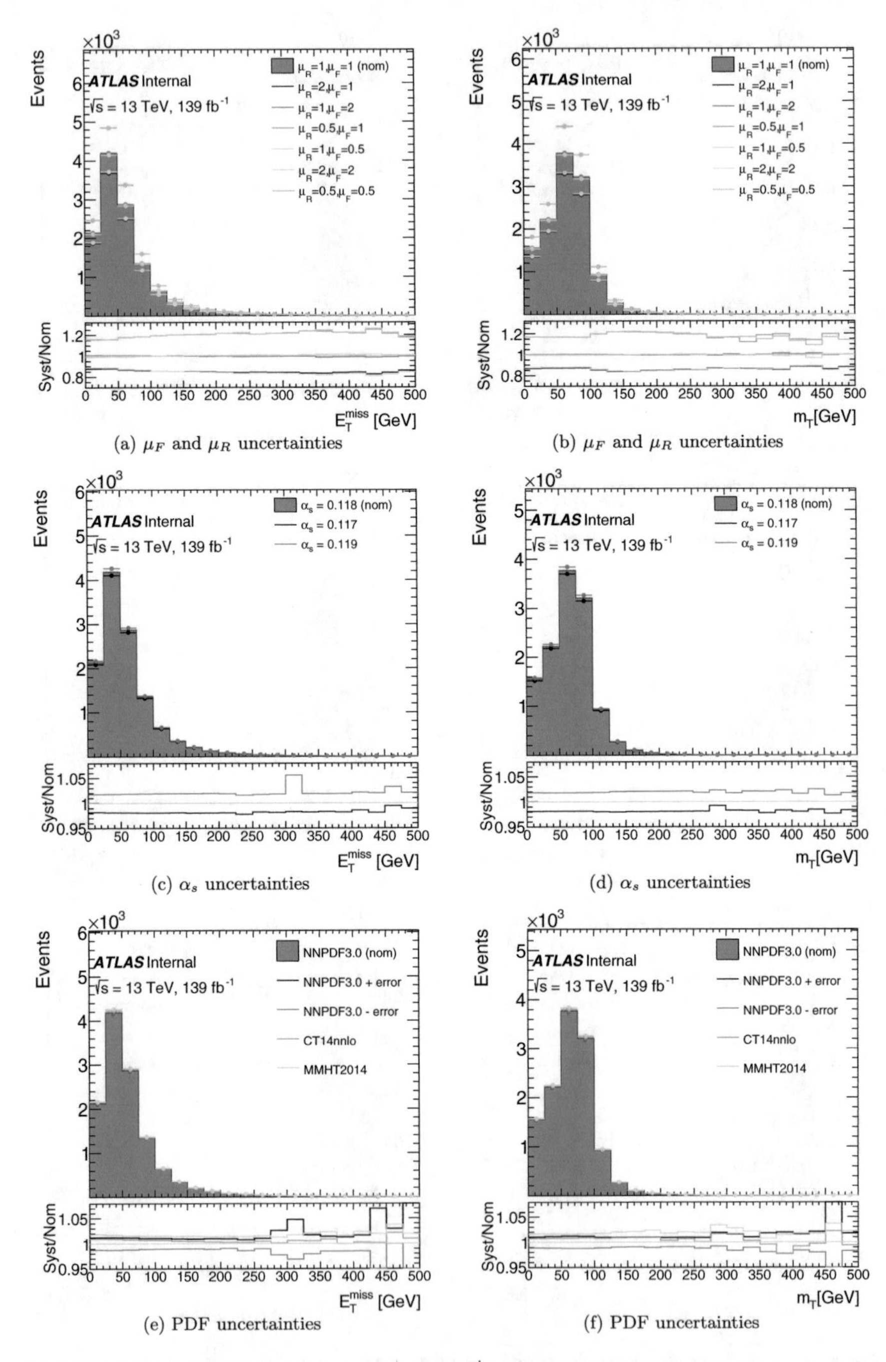

(a) μ_F and μ_R uncertainties

(b) μ_F and μ_R uncertainties

(c) α_s uncertainties

(d) α_s uncertainties

(e) PDF uncertainties

(f) PDF uncertainties

Fig. 9.23 Impact of WZ theory systematics on $E_{\mathrm{T}}^{\mathrm{miss}}$ and m_{T} in the ISR preselection region

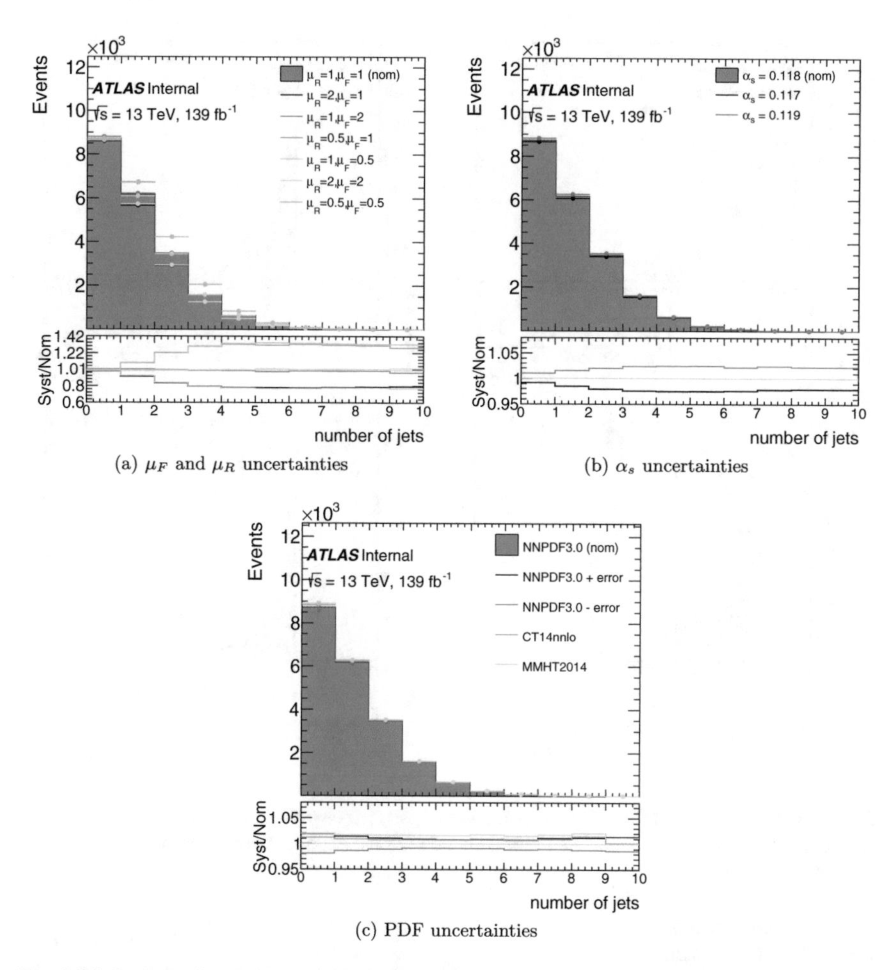

(a) μ_F and μ_R uncertainties

(b) α_s uncertainties

(c) PDF uncertainties

Fig. 9.24 Impact of WZ theory systematics on N_{jets} distribution in the ISR preselection region without the requirement on jets applied

Table 9.21 Relative uncertainties (in %) on the modelling applied to the normalization of the WZ in all regions

Region	QCD	α_s	PDF
SR-low	+4.24, −4.14	±0.98	+1.82, −1.83
SR-ISR	+23.78, −16.92	±1.98	+1.97, −1.97
CR-low	+1.49, −2.27	±0.86	+2.04, −1.92
VR-low	+1.30, −2.01	±0.83	+1.22, −1.10
CR-ISR	+21.60, −15.49,	±1.93	+1.20, −1.00
VR-ISR	+27.62, −19.35	±2.22	+1.48, −0.90
VR-ISR-small $p_{\mathrm{T}}^{\mathrm{soft}}$	+15.88, −12.46	±1.85	+1.38, −1.38
VR-ISR-small $R\left(E_{\mathrm{T}}^{\mathrm{miss}}, \mathrm{jets}\right)$	+23.19, −16.36	±1.97	+1.27, −0.96

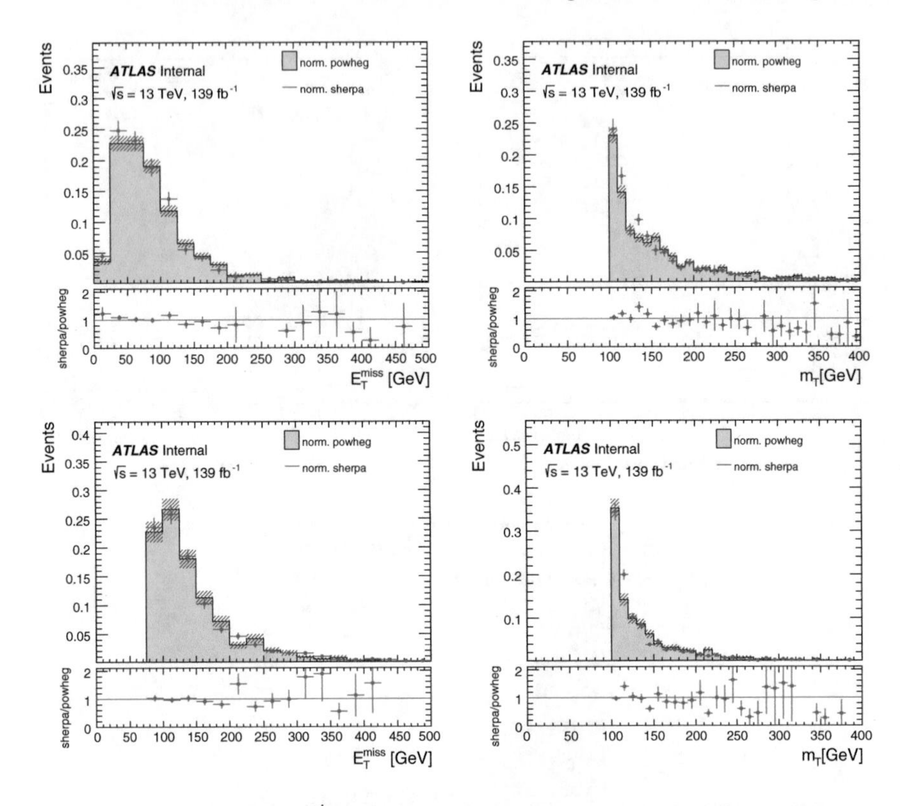

Fig. 9.25 Comparison of the $E_{\mathrm{T}}^{\mathrm{miss}}$ (left) and m_{T} (right) shapes predicted by different WZ generators, in the low-mass (top) and ISR (bottom) signal regions. All distributions are normalized to the same number of entries

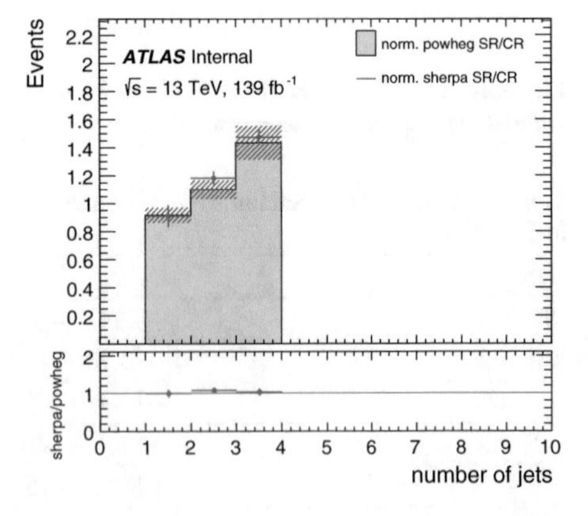

Fig. 9.26 Extrapolation of CR to SR of the POWHEG and SHERPA ratio

Table 9.22 Summary of the dominant experimental and theoretical uncertainties on the SM background prediction in the low-mass and ISR signal regions. The individual uncertainties can be correlated, and do not necessarily add in quadrature to the total post-fit background uncertainty

Uncertainty in signal regions	SR-low (%)	SR-ISR (%)
Jet energy scale and resolution	7.1	6.1
WZ normalization procedure	6.6	4.5
$E_{\mathrm{T}}^{\mathrm{miss}}$	3.3	2.1
MC statistics	2.9	3.9
Anti-ID CR statistics	2.7	0.21
WZ theory	1.9	1.3
30% uncertainty in minor backgrounds ("Others")	1.7	3.1
Fake-factor estimation	1.2	<0.01
Muon momentum scale and resolution	0.30	0.04
Electron energy scale and resolution	0.22	0.34
Pileup	0.20	0.94
Top-quark-like background estimation	<0.01	1.4
Flavor tagging	<0.01	0.47

9.4.3 *Summary of Systematics*

The dominant uncertainties are summarized in Table 9.22 for both the SR-low and SR-ISR regions. The largest experimental uncertainties reflect the unknowns of the energy and p_T calibration of jets and the measurement of the soft term of the $E_{\mathrm{T}}^{\mathrm{miss}}$. The largest theoretical source is the uncertainty on the QCD factorization and renormalization scales on the WZ cross-section. The analysis also accounts for the statistical uncertainty of the MC simulation samples.

9.5 Results

Two unconstrained normalization factors μ_{WZlow}, μ_{WZISR} are determined by measurements to data in the corresponding control regions. The μ_{WZlow} is used to normalize WZ Monte Carlo samples in the jet-vetoed low regions while the μ_{WZISR} is used to normalize WZ Monte Carlo samples in the jet-required ISR regions. The low-mass and ISR regions are fit simultaneously in these results.

A likelihood is constructed as the product of Poisson distributions with mean taken as the nominal MC yield in each of the control regions. This likelihood is extremized using the HistFitter package [46], which constrains the values and uncertainties on μ_{WZlow}, μ_{WZISR}. These are used to extrapolate the background prediction into validation regions, where modelling is verified, and the signal regions. Systematic uncertainties are treated as nuisance parameters in the likelihood fit. 1% pruning is applied to systematics.

9.5.1 2015–16 Results

After moving to the new reconstruction algorithm and validating the background modelling, the observed event yields in the low-mass and ISR regions are compared to the fitted background estimation derived from the log-likelihood fits in Table 9.23. There is no longer a significant excess in SR-low but there still is an excess in SR-ISR.

The overlap of data events between the eRJR result in the old reconstruction, summarized in Table 8.9, and the yields in Table 9.23 is checked. Only 70% of SR-low events overlap between the two reconstructions. Some of the events present in the old reconstruction are no longer present in SR-low in the new reconstruction because they were now reconstructed with one jet since the number of jets changed between reconstruction algorithms while SR-low requires a jet veto. Some events do not pass signal or baseline requirement due to the change in electron reconstruction. Finally, some of the events have slight changes in the values of the kinematic variables because the correlations seen in Fig. 9.2 have some residuals. The overlap in SR-ISR is 86%. One event does not pass the signal lepton requirement and two events have changes in the values of kinematic variables.

Kinematic distributions for these SRs are shown in Figs. 9.27 and 9.28. The excess in SR-ISR appears at low p_T^{soft} but does not appear right at the edge of m_T unlike in Fig. 8.19.

Table 9.23 The observed and expected yields after the background-only fit in the SRs for 36.1 fb^{-1}. The normalization factors of the WZ sample for the low-mass and ISR regions are different and are treated separately in the combined fit. The Other category includes triboson, Higgs boson, and rare top-quark processes. Combined statistical and systematic uncertainties are presented. The individual uncertainties can be correlated and do not necessarily add in quadrature to the total background uncertainty

	SR-low	SR-ISR
Observed events	18	10
Fitted SM events	14.2 ± 2.0	5.7 ± 0.8
WZ	11.9 ± 1.9	4.7 ± 0.8
ZZ	1.43 ± 0.23	0.11 ± 0.04
Others	0.37 ± 0.18	0.48 ± 0.25
Top-quark like	0.0 ± 0.0	0.47 ± 0.18
Fake/non prompt leptons	$0.5^{+0.6}_{-0.5}$	$0.01^{+0.05}_{-0.01}$

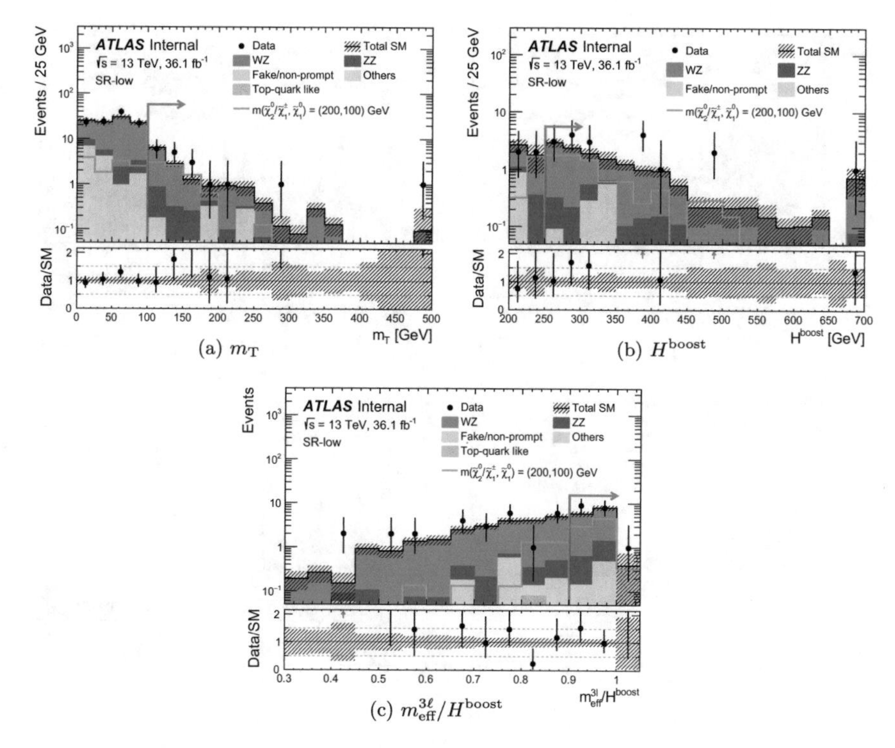

Fig. 9.27 Distributions in SR-low of the data and post-fit background prediction for 36.1 fb^{-1}for a m_{T}, b H^{boost}, and c $m_{\mathrm{eff}}^{3\ell}/H^{\mathrm{boost}}$. The SR-low event selections are applied for each distribution except for the variable shown where the selection is indicated by a red arrow. The normalization factor for the WZ background is derived from the background-only estimation. The expected distribution for a benchmark signal model is included for comparison

9.5.2 2017 Results

The observed event yields in the low-mass and ISR regions are compared to the fitted background estimation derived from the log-likelihood fits in Table 9.24 for the 2017 dataset. There still is no longer a significant excess in SR-low but the SR-ISR excess is still present.

Using Z+jet events, the $E_{\mathrm{T}}^{\mathrm{miss}}$ components were examined to determine if $E_{\mathrm{T}}^{\mathrm{miss}}$ resolution could be the cause of the excess seen in SR-ISR. Even though there is some excess present, this could not account for the entire SR-ISR excess. The Z-mass side-band is also examined to determine if the excess present is due to muon resolution and there is no issue found there. These studies are discussed in Sect. 8.4.3.

Kinematic distributions for these SRs are shown in Figs. 9.29 and 9.30. There is an excess of data events over the background predictions in different values of the kinematic variables in 2017 than in 2015–16. The only consistent excess is at low $p_{\mathrm{T}}^{\mathrm{soft}}$ values, $p_{\mathrm{T}}^{\mathrm{soft}} < 10$ GeV.

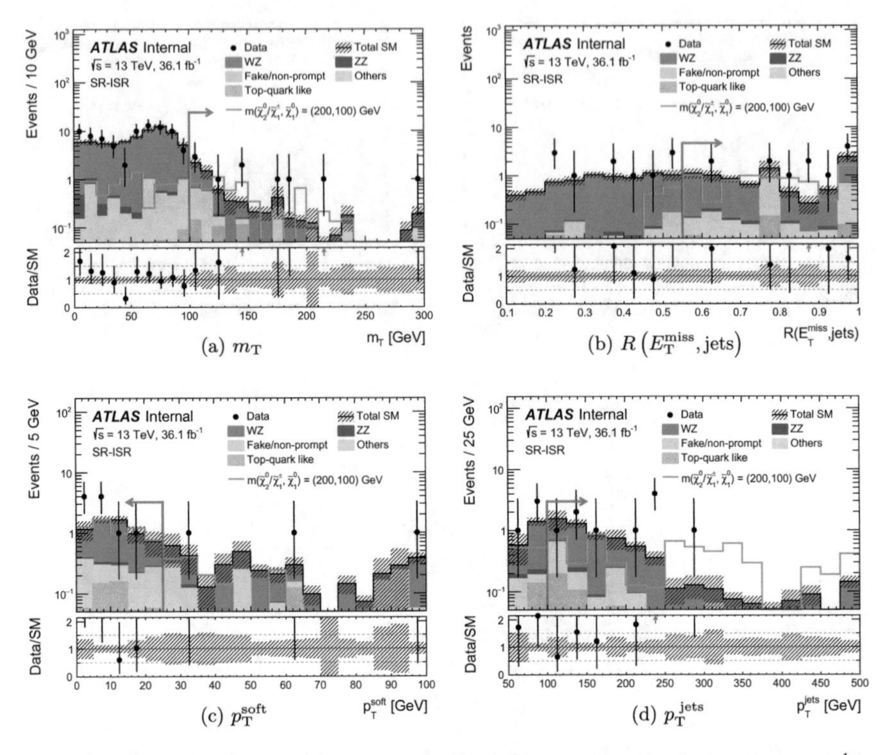

Fig. 9.28 Distributions in SR-ISR of the data and post-fit background prediction for 36.1 fb^{-1}for a m_T, b $R\left(E_T^{miss}, \text{jets}\right)$, c p_T^{soft}, and d p_T^{jets}. The SR-ISR event selections are applied for each distribution except for the variable shown where the selection is indicated by a red arrow. The normalization factor for the WZ background is derived from the background-only estimation. The expected distribution for a benchmark signal model is included for comparison

Table 9.24 The observed and expected yields after the background-only fit in the SRs for 43.9 fb^{-1}. The normalization factors of the WZ sample for the low-mass and ISR regions are different and are treated separately in the combined fit. The Other category includes triboson, Higgs boson, and rare top-quark processes. Combined statistical and systematic uncertainties are presented. The individual uncertainties can be correlated and do not necessarily add in quadrature to the total background uncertainty

	SR-low	SR-ISR
Observed events	11	16
Fitted SM events	13.2 ± 1.8	7.6 ± 1.0
WZ	11.1 ± 1.7	6.6 ± 0.8
ZZ	1.46 ± 0.24	0.090 ± 0.031
Others	0.45 ± 0.23	0.43 ± 0.23
Top-quark like	0.0 ± 0.0	$0.5^{+0.5}_{-0.5}$
Fake/non-prompt leptons	$0.2^{+0.4}_{-0.2}$	$0.01^{+0.05}_{-0.01}$

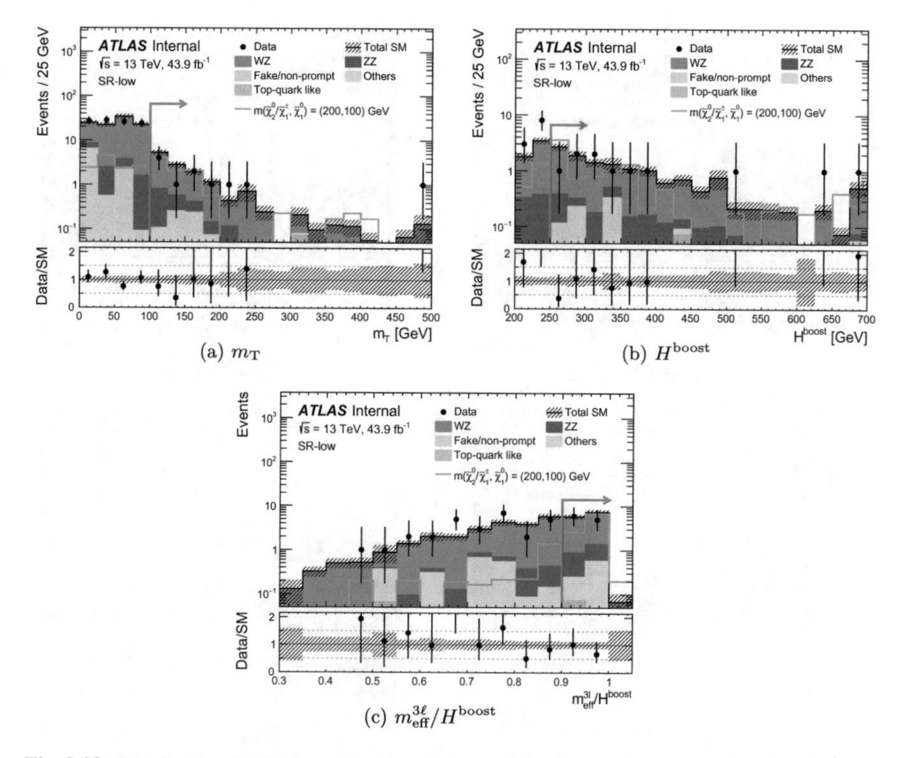

(a) m_{T} (b) H^{boost} (c) $m_{\mathrm{eff}}^{3\ell}/H^{\mathrm{boost}}$

Fig. 9.29 Distributions in SR-low of the data and post-fit background prediction for 43.9 fb^{-1}for a m_{T}, b H^{boost}, and c $m_{\mathrm{eff}}^{3\ell}/H^{\mathrm{boost}}$. The SR-low event selections are applied for each distribution except for the variable shown where the selection is indicated by a red arrow. The normalization factor for the WZ background is derived from the background-only estimation. The expected distribution for a benchmark signal model is included for comparison

9.5.3 2018 Results

The observed event yields in the low-mass and ISR regions are compared to the fitted background estimation derived from the log-likelihood fits in Table 9.25 for the 2018 dataset. There still is no longer a significant excess in SR-low and there is a deficit in SR-ISR.

Kinematic distributions for these SRs are shown in Figs. 9.31 and 9.32. In SR-low, there are some fluctuations in the H^{boost} distributions and a bit of excess in the tail of m_{T} distribution. Neither of these look like the signal model. In SR-ISR, the events are still concentrated at low $p_{\mathrm{T}}^{\mathrm{soft}}$.

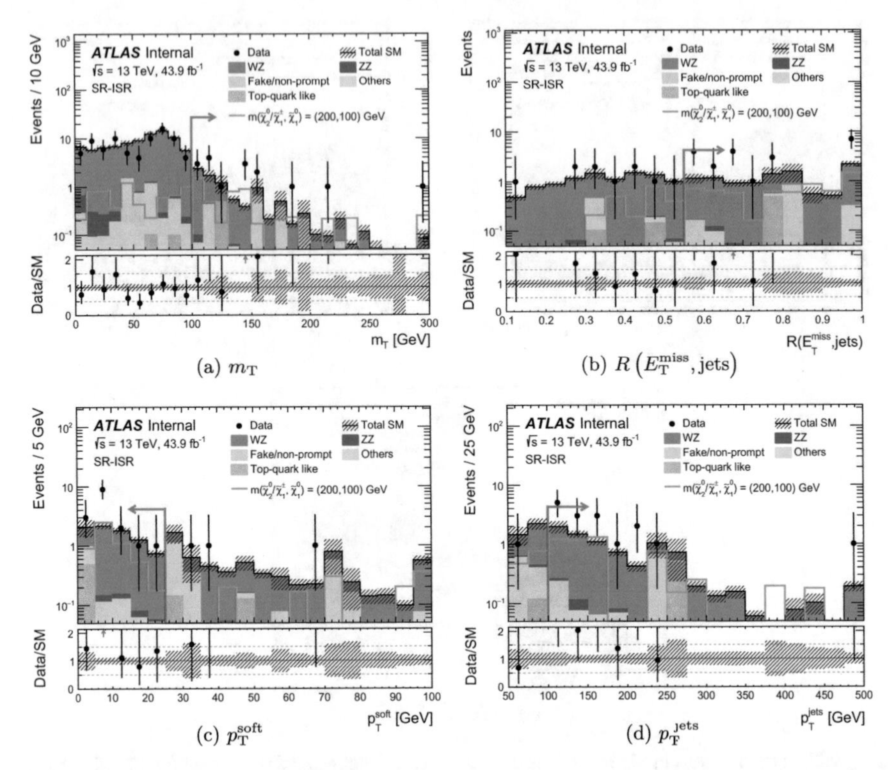

Fig. 9.30 Distributions in SR-ISR of the data and post-fit background prediction for 43.9 fb^{-1}for a m_{T}, b $R\left(E_{\mathrm{T}}^{\mathrm{miss}}, \mathrm{jets}\right)$, c $p_{\mathrm{T}}^{\mathrm{soft}}$, and d $p_{\mathrm{T}}^{\mathrm{jets}}$. The SR-ISR event selections are applied for each distribution except for the variable shown where the selection is indicated by a red arrow. The normalization factor for the WZ background is derived from the background-only estimation. The expected distribution for a benchmark signal model is included for comparison

Table 9.25 The observed and expected yields after the background-only fit in the SRs for 59 fb^{-1}. The normalization factors of the WZ sample for the low-mass and ISR regions are different and are treated separately in the combined fit. The Other category includes triboson, Higgs boson, and rare top-quark processes. Combined statistical and systematic uncertainties are presented. The individual uncertainties can be correlated and do not necessarily add in quadrature to the total background uncertainty

	SR-low	SR-ISR
Observed events	22	4
Fitted SM events	18.4 ± 2.9	9.8 ± 1.4
WZ	14.9 ± 2.6	8.3 ± 1.3
ZZ	2.03 ± 0.30	0.18 ± 0.04
Others	0.50 ± 0.26	0.33 ± 0.26
Top-quark like	$0.04^{+0.17}_{-0.04}$	1.0 ± 0.5
Fake/non-prompt leptons	$0.9^{+0.9}_{-0.9}$	$0.01^{+0.06}_{-0.01}$

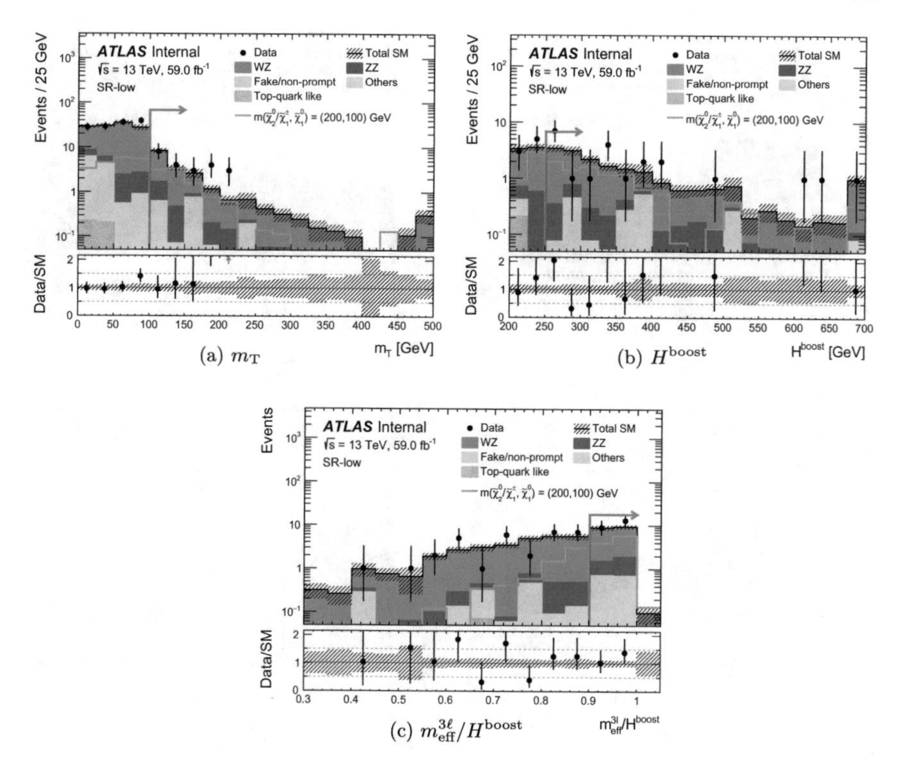

Fig. 9.31 Distributions in SR-low of the data and post-fit background prediction for 59 fb^{-1}for a m_T, b H^boost, and c $m_\mathrm{eff}^{3\ell}/H^\mathrm{boost}$. The SR-low event selections are applied for each distribution except for the variable shown where the selection is indicated by a red arrow. The normalization factor for the WZ background is derived from the background-only estimation. The expected distribution for a benchmark signal model is included for comparison

9.5.4 2015–18 Results

9.5.4.1 Background-Only Fit

To determine the background prediction, the control regions are used to constrain the fit parameters assuming no signal events in the CR, referred to as a background-only fit. Normalization factors on the WZ MC simulation are derived from a simultaneous background-only fit of the two orthogonal CRs with all other background processes held constant. The normalization factors are found to be 0.84 ± 0.07 for CR-low and 0.94 ± 0.05 for CR-ISR. The two normalization factors are compatible within their uncertainties, with small differences expected due to the difficulties in accurately modeling higher-order radiation in the electroweak WZ process.

The observed event yields in the low-mass and ISR regions are compared to the fitted background estimation derived from the log-likelihood fits in Table 9.26 and visualized alongside the validation regions in Fig. 9.33. The data agrees well with

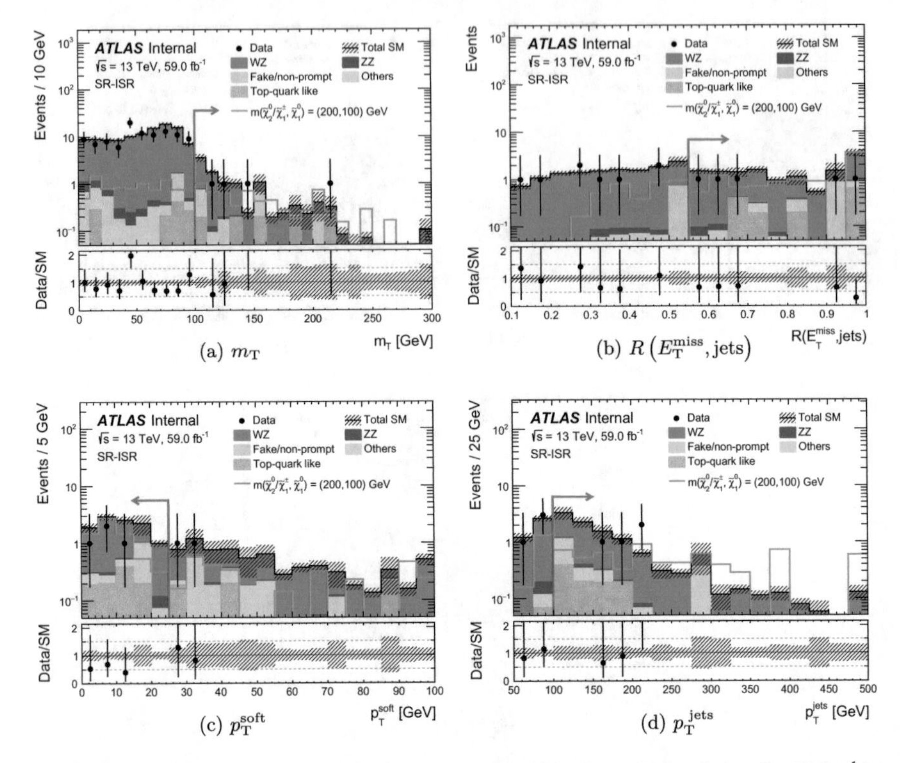

(a) m_{T} (b) $R\left(E_{\mathrm{T}}^{\mathrm{miss}}, \mathrm{jets}\right)$ (c) $p_{\mathrm{T}}^{\mathrm{soft}}$ (d) $p_{\mathrm{T}}^{\mathrm{jets}}$

Fig. 9.32 Distributions in SR-ISR of the data and post-fit background prediction for 59 fb^{-1}for a m_{T}, b $R\left(E_{\mathrm{T}}^{\mathrm{miss}}, \mathrm{jets}\right)$, c $p_{\mathrm{T}}^{\mathrm{soft}}$, and d $p_{\mathrm{T}}^{\mathrm{jets}}$. The SR-ISR event selections are applied for each distribution except for the variable shown where the selection is indicated by a red arrow. The normalization factor for the WZ background is derived from the background-only estimation. The expected distribution for a benchmark signal model is included for comparison

Table 9.26 The observed and expected yields after the background-only fit in the SRs for 139 fb^{-1}. The normalization factors of the WZ sample for the low-mass and ISR regions are different and are treated separately in the combined fit. The Other category includes triboson, Higgs boson, and rare top-quark processes. Combined statistical and systematic uncertainties are presented. The individual uncertainties can be correlated and do not necessarily add in quadrature to the total background uncertainty

	SR-low	SR-ISR
Observed events	51	30
Fitted SM events	46 ± 5	23.4 ± 2.1
WZ	38 ± 5	19.7 ± 2.0
ZZ	4.9 ± 0.6	0.38 ± 0.08
Others	1.6 ± 0.8	1.5 ± 0.8
Top-quark like	$0.03^{+0.18}_{-0.03}$	1.9 ± 0.8
Fake/non-prompt leptons	1.6 ± 1.3	$0.01^{+0.05}_{-0.01}$

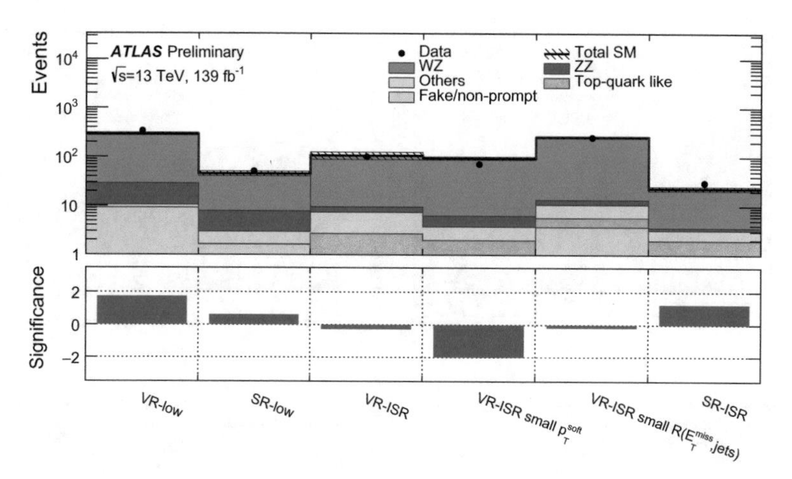

Fig. 9.33 The observed data and expected SM background yields in the VRs and SRs. The SM background prediction is derived with the background-only fit configuration, and the hatched band includes the experimental, theoretical, and statistical uncertainties. The bottom panel shows the significance [47] of the differences between the observed and expected yields

the background estimation in both signal regions, with SR-ISR showing only a small $1.27\,\sigma$ excess of data with respect to the predictions. Kinematic distributions for these SRs are shown in Figs. 9.34 and 9.35, demonstrating good agreement between data and the background estimation in the SRs and across the boundaries of the SR selections.

9.5.4.2 Discovery Fit

As no significant excess is observed, model-independent limits are derived at 95% confidence level (CL) using the CL_sprescription [48]. An upper limit on the visible cross section of beyond-the-SM processes is derived for each SR. A log-likelihood fit is performed to the number of observed events in the target SR and the associated CR, and a generic BSM process is assumed to contribute to the SR only. No theoretical or systematic uncertainties are considered for the signal model except the luminosity uncertainty. The observed (S^{95}_{obs}) and expected (S^{95}_{exp}) limits on the number of BSM events are shown in Table 9.27. Also shown are the observed limits on the visible cross section σ_{vis}, defined as S^{95}_{obs} normalized to the integrated luminosity, which represents the product of the production cross section, acceptance, and selection efficiency of a generic BSM signal. Limits on σ_{vis} are set at 0.16 fb in SR-low and 0.13 fb in SR-ISR. The p-value, representing the probability of the SM background alone fluctuating to the observed number of events, and the associated significance Z are also shown.

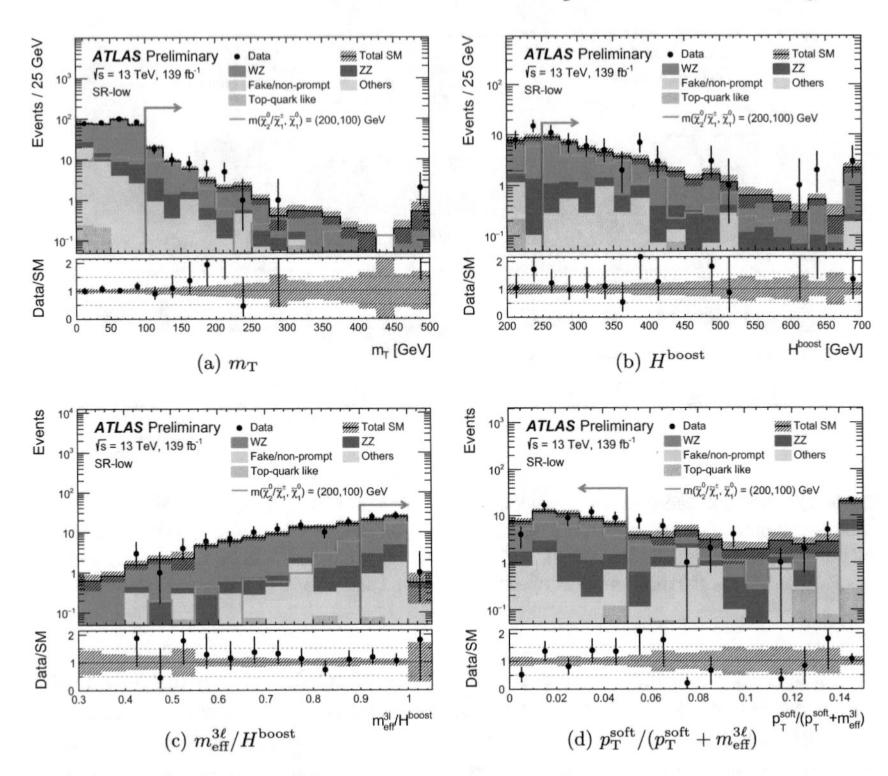

(a) m_{T} (b) H^{boost} (c) $m_{\mathrm{eff}}^{3\ell}/H^{\mathrm{boost}}$ (d) $p_{\mathrm{T}}^{\mathrm{soft}}/(p_{\mathrm{T}}^{\mathrm{soft}}+m_{\mathrm{eff}}^{3\ell})$

Fig. 9.34 Distributions in SR-low of the data and post-fit background prediction for a m_{T}, b H^{boost}, c $m_{\mathrm{eff}}^{3\ell}/H^{\mathrm{boost}}$, and d $p_{\mathrm{T}}^{\mathrm{soft}}/(p_{\mathrm{T}}^{\mathrm{soft}}+m_{\mathrm{eff}}^{3\ell})$. The SR-low event selections are applied for each distribution except for the variable shown where the selection is indicated by a red arrow. The normalization factor for the WZ background is derived from the background-only estimation. The expected distribution for a benchmark signal model is included for comparison

Table 9.27 Summary of the expected background and data yields in SR-low and SR-ISR. The second and third columns show the data and total expected background with systematic uncertainties. The fourth column gives the model-independent upper limits at 95% CL on the visible cross section (σ_{vis}). The fifth and sixth columns give the visible number of observed (S_{obs}^{95}) and expected (S_{exp}^{95}) events of a generic beyond-the-SM process, where uncertainties on S_{exp}^{95} reflect the $\pm 1\sigma$ uncertainties on the background estimation. The last column shows the discovery p-value and Gaussian significance Z assuming no signal

Signal channel	N_{obs}	N_{exp}	σ_{vis}[fb]	S_{obs}^{95}	S_{exp}^{95}	$p(s=0)$ (Z)
SR-low	51	46 ± 5	0.16	22.1	$19.9^{+7.8}_{-3.6}$	0.27 (0.61)
SR-ISR	30	23.4 ± 2.1	0.13	17.8	$12.0^{+5.3}_{-1.8}$	0.11 (1.21)

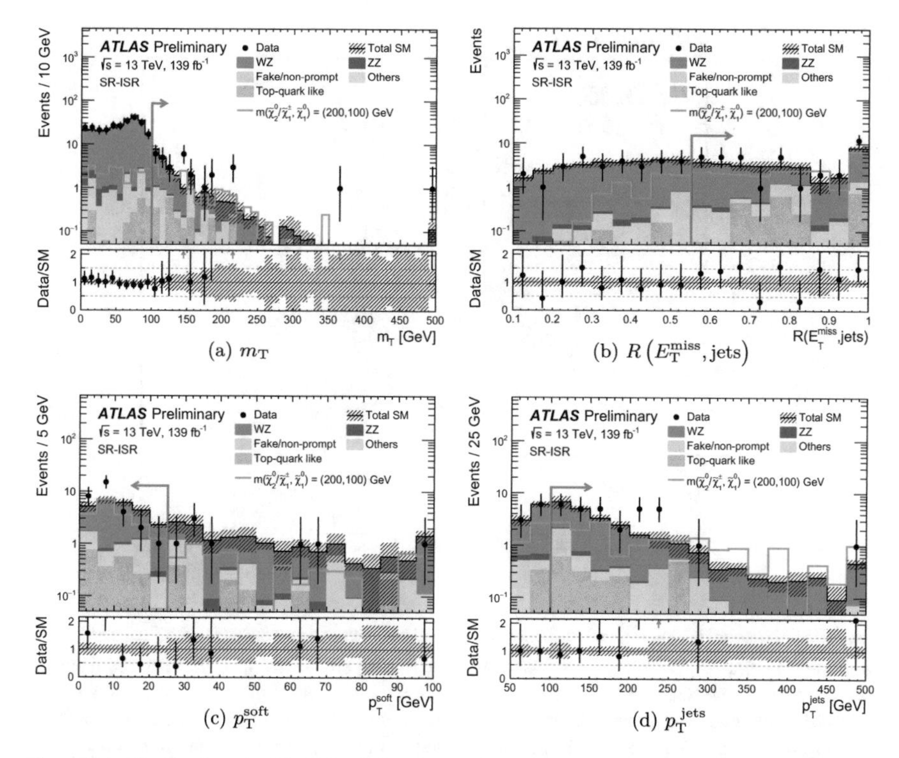

(a) m_{T} (b) $R\left(E_{\mathrm{T}}^{\mathrm{miss}}, \mathrm{jets}\right)$ (c) $p_{\mathrm{T}}^{\mathrm{soft}}$ (d) $p_{\mathrm{T}}^{\mathrm{jets}}$

Fig. 9.35 Distributions in SR-ISR of the data and post-fit background prediction for a m_{T}, b $R\left(E_{\mathrm{T}}^{\mathrm{miss}}, \mathrm{jets}\right)$, c $p_{\mathrm{T}}^{\mathrm{soft}}$, and d $p_{\mathrm{T}}^{\mathrm{jets}}$. The SR-ISR event selections are applied for each distribution except for the variable shown where the selection is indicated by a red arrow. The normalization factor for the WZ background is derived from the background-only estimation. The expected distribution for a benchmark signal model is included for comparison

9.5.4.3 Exclusion Fit

Exclusion limits are derived at 95% CL for the $\tilde{\chi}_1^{\pm}\tilde{\chi}_2^0$ models which decay exclusively into W and Z bosons. Limits are obtained through a profile log-likelihood ratio test using the CL$_s$prescription, following the simultaneous fit to the low-mass and ISR CRs and SRs [46]. The low-mass and ISR regions do not affect the nominal fit in the other region due to their orthogonality, but uncertainties that are correlated across regions may be constrained. Experimental uncertainties are treated as correlated between signal and background events and across low-mass and ISR regions. The theoretical uncertainty on the signal cross section is accounted for by repeating the limit-setting procedure with the varied signal cross sections and reporting the effect on the observed limit.

The expected and observed exclusion contours as a function of the signal $\tilde{\chi}_1^{\pm}/\tilde{\chi}_2^0$ and LSP $\tilde{\chi}_1^0$ masses are shown in Fig. 9.36. Masses can be excluded when the Z/W bosons of the decay are on mass-shell, such that the mass splittings Δm are close

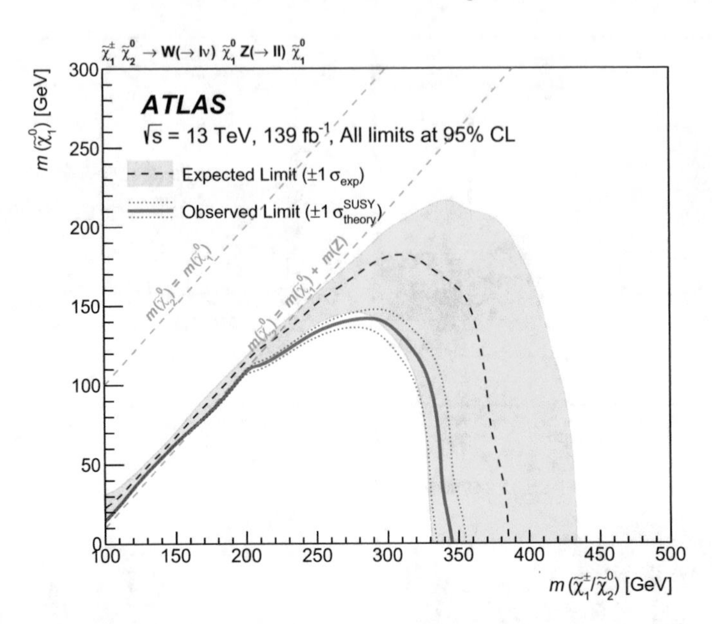

Fig. 9.36 Expected (dashed blue) and observed (solid red) exclusion contours on $\tilde{\chi}_1^{\pm}\tilde{\chi}_2^0$ production assuming on-shell W/Z decays as a function of the $\tilde{\chi}_1^{\pm}/\tilde{\chi}_2^0$ and $\tilde{\chi}_1^0$ masses, and derived from the combined fit of low-mass and ISR regions. The yellow band reflects the $\pm 1\sigma$ uncertainty on the expected limits due to uncertainties in the background prediction and experimental uncertainties affecting the signal. The dotted red lines correspond to the $\pm 1\sigma$ cross section uncertainty of the observed limit derived by varying the signal cross section within its uncertainty

to or larger than the Z boson mass. Signal $\tilde{\chi}_1^{\pm}/\tilde{\chi}_2^0$ are excluded for masses up to 350 GeV for small $\tilde{\chi}_1^0$ masses in which Δm is large.

These results extend the exclusion limits of the low-mass and ISR regions compared to those of the RJR analysis [1]. The excesses from the RJR analysis were validated in the 36 fb^{-1} of data from the 2015 and 2016 datasets, and found to be reduced with the inclusion of 103 fb^{-1} of data from the 2017 and 2018 datasets, corresponding to local significances of 0.6σ in SR-low and 1.3σ in SR-ISR.

9.6 Conclusion

The search targets electroweakino production for which current limits derived from the recursive jigsaw reconstruction technique and from conventional techniques in the laboratory frame are in disagreement. This new search uses 139 fb^{-1} of proton–proton collisions collected at $\sqrt{s} = 13$ TeV by the ATLAS detector between 2015 and 2018. The data are analyzed with a new emulated recursive jigsaw reconstruction method that uses conventional variables in the laboratory frame to target low-mass electroweakinos and those produced in the presence of initial-state radiation. A subset of the data collected between 2015 and 2016 is analyzed and excesses are seen in the laboratory frame for two signal regions of similar construction to those of the recursive jigsaw reconstruction search [1]. In the full dataset the observed event yields are found to be in agreement with Standard Model expectations, with no significant excess seen in either signal region. The results are interpreted with simplified models of electroweakino pair-production, excluding neutralinos and charginos with masses between 100 and 350 GeV at 95% confidence level when the W and Z bosons are on mass-shell.

References

1. ATLAS Collaboration (2018) Search for chargino-neutralino production using recursive jigsaw reconstruction in final states with two or three charged leptons in proton-proton collisions at $\sqrt{s} = 13$ TeV with the ATLAS detector. Phys Rev D98(9):092012. http://dx.doi.org/10.1103/PhysRevD.98.092012, arXiv:1806.02293 [hep-ex]
2. ATLAS Collaboration (2019) Search for chargino-neutralino production with mass splittings near the electroweak scale in three-lepton final states in $\sqrt{s} = 13$ TeV pp collisions with the ATLAS detector. Technical Report ATLAS-CONF-2019-020, CERN, Geneva, May. http://cds.cern.ch/record/2676597
3. ATLAS Collaboration, Aad G et al (2020) Search for chargino-neutralino production with mass splittings near the electroweak scale in three-lepton final states in $\sqrt{s} = 13$ TeV pp collisions with the ATLAS detector. Phys Rev D101(7):072001. http://dx.doi.org/10.1103/PhysRevD.101.072001, arXiv:1912.08479 [hep-ex]
4. ATLAS Collaboration (2017) Multi-Boson Simulation for *13*TeV ATLAS Analyses, ATL-PHYS-PUB-2017-005. https://cds.cern.ch/record/2261933
5. ATLAS Collaboration (2017) ATLAS simulation of boson plus jets processes in Run 2, ATL-PHYS-PUB-2017-006. https://cds.cern.ch/record/2261937
6. Bothmann E et al (2019) Event Generation with Sherpa 2.2. SciPost Phys 7(3):034. http://dx.doi.org/10.21468/SciPostPhys.7.3.034, arXiv:1905.09127 [hep-ph]
7. Gleisberg T, Höche S (2008) Comix, a new matrix element generator. JHEP 12:039. http://dx.doi.org/10.1088/1126-6708/2008/12/039, arXiv:0808.3674 [hep-ph]
8. Schumann S, Krauss F (2008) A Parton shower algorithm based on Catani-Seymour dipole factorisation. JHEP 03:038. http://dx.doi.org/10.1088/1126-6708/2008/03/038, arXiv:0709.1027 [hep-ph]
9. NNPDF Collaboration, Ball RD et al (2015) Parton distributions for the LHC Run II. JHEP 04:040. http://dx.doi.org/10.1007/JHEP04(2015)040, arXiv:1410.8849 [hep-ph]
10. Höche S, Krauss F, Schonherr M, Siegert F (2012) A critical appraisal of NLO+PS matching methods. JHEP 09:049. http://dx.doi.org/10.1007/JHEP09(2012)049, arXiv:1111.1220 [hep-ph]
11. Catani S, Krauss F, Kuhn R, Webber BR (2001) QCD matrix elements + parton showers. JHEP 11:063. http://dx.doi.org/10.1088/1126-6708/2001/11/063, arXiv:hep-ph/0109231
12. Höche S, Krauss F, Schönherr M, Siegert F (2013) QCD matrix elements + parton showers: the NLO case. JHEP 04:027. http://dx.doi.org/10.1007/JHEP04(2013)027, arXiv:1207.5030 [hep-ph]
13. Cascioli F, Maierhofer P, Pozzorini S (2012) Scattering amplitudes with open loops. Phys Rev Lett 108:111601. http://dx.doi.org/10.1103/PhysRevLett.108.111601, arXiv:1111.5206 [hep-ph]
14. Anastasiou C, Dixon LJ, Melnikov K, Petriello F (2004) High precision QCD at hadron colliders: electroweak gauge boson rapidity distributions at NNLO. Phys Rev D 69:094008. http://dx.doi.org/10.1103/PhysRevD.69.094008, arXiv:hep-ph/0312266

15. Alioli S, Nason P, Oleari C, Re E (2010) A general framework for implementing NLO calculations in shower Monte Carlo programs: the POWHEG BOX. JHEP 06:043. http://dx.doi.org/10.1007/JHEP06(2010)043, arXiv:1002.2581 [hep-ph]
16. ATLAS Collaboration (2016) Studies on top-quark Monte Carlo modelling for Top2016, ATL-PHYS-PUB-2016-020. https://cds.cern.ch/record/2216168
17. Sjöstrand T, Ask S, Christiansen JR, Corke R, Desai N, Ilten P, Mrenna S, Prestel S, Rasmussen CO, Skands PZ (2015) An introduction to PYTHIA 8.2. Comput Phys Commun 191:159. http://dx.doi.org/10.1016/j.cpc.2015.01.024, arXiv:1410.3012 [hep-ph]
18. ATLAS Collaboration (2014) ATLAS Pythia 8 tunes to 7TeV data, ATL-PHYS-PUB-2014-021. https://cds.cern.ch/record/1966419
19. Ball RD et al (2013) Parton distributions with LHC data. Nucl Phys B 867:244. http://dx.doi.org/10.1016/j.nuclphysb.2012.10.003, arXiv:1207.1303 [hep-ph]
20. Czakon M, Mitov A (2014) Top++: A program for the calculation of the top-pair cross-section at hadron colliders. Comput Phys Commun 185:2930. http://dx.doi.org/10.1016/j.cpc.2014.06.021, arXiv:1112.5675 [hep-ph]
21. ATLAS Collaboration (2016) The Pythia 8 A3 tune description of ATLAS minimum bias and inelastic measurements incorporating the Donnachie–Landshoff diffractive model, ATL-PHYS-PUB-2016-017. https://cds.cern.ch/record/2206965
22. Alwall J, Frederix R, Frixione S, Hirschi V, Maltoni F, Mattelaer O, Shao HS, Stelzer T, Torrielli P, Zaro M (2014) The automated computation of tree-level and next-to-leading order differential cross sections, and their matching to parton shower simulations. JHEP 07:079. http://dx.doi.org/10.1007/JHEP07(2014)079, arXiv:1405.0301 [hep-ph]
23. ATLAS Collaboration (2014) Measurement of the Z/γ^* boson transverse momentum distribution in pp collisions at $\sqrt{s} = 7$ TeV with the ATLAS detector. JHEP 09:145. http://dx.doi.org/10.1007/JHEP09(2014)145, arXiv:1406.3660 [hep-ex]
24. Pumplin J et al (2002) New generation of parton distributions with uncertainties from global QCD analysis. JHEP 07:012. http://dx.doi.org/10.1088/1126-6708/2002/07/012, arXiv:hep-ph/0201195
25. LHC Higgs Cross Section Working Group Collaboration, Handbook of LHC Higgs Cross Sections: 4. Deciphering the Nature of the Higgs Sector. arXiv:1610.07922 [hep-ph]
26. Fuks B, Klasen M, Lamprea DR, Rothering M (2013) Precision predictions for electroweak superpartner production at hadron colliders with Resummino. Eur Phys J C 73:2480. http://dx.doi.org/10.1140/epjc/s10052-013-2480-0, arXiv:1304.0790 [hep-ph]
27. Lange DJ (2001) The EvtGen particle decay simulation package. Nucl Instrum Meth A 462:152. http://dx.doi.org/10.1016/S0168-9002(01)00089-4
28. ATLAS Collaboration (2010) The ATLAS simulation infrastructure. Eur Phys J C 70:823. http://dx.doi.org/10.1140/epjc/s10052-010-1429-9, arXiv:1005.4568 [physics.ins-det]
29. GEANT4 Collaboration, Agostinelli S et al (2003) GEANT4—a simulation toolkit. Nucl Instrum Meth A 506:250 http://dx.doi.org/10.1016/S0168-9002(03)01368-8
30. ATLAS Collaboration (2012) Measurement of the WW cross section in $\sqrt{s} = 7$ TeV pp collisions with the ATLAS detector and limits on anomalous gauge couplings. Phys Lett B 712:289. http://dx.doi.org/10.1016/j.physletb.2012.05.003, arXiv:1203.6232 [hep-ex]
31. Prospects for Higgs Boson Searches using the $H \to WW^{(*)} \to \ell\nu\ell\nu$ Decay Mode with the ATLAS Detector for 10 TeV. Technical Report, ATL-PHYS-PUB-2010-005, CERN, Geneva, Jun, 2010. https://cds.cern.ch/record/1270568
32. ATLAS Collaboration (2017) Jet energy scale measurements and their systematic uncertainties in proton–proton collisions at $\sqrt{s} = 13$ TeV with the ATLAS detector. Phys Rev D 96:072002. http://dx.doi.org/10.1103/PhysRevD.96.072002, arXiv:1703.09665 [hep-ex]
33. ATLAS Collaboration (2015) Jet energy measurement and its systematic uncertainty in proton–proton collisions at $\sqrt{s} = 7$ TeV with the ATLAS detector. Eur Phys J C 75:17. http://dx.doi.org/10.1140/epjc/s10052-014-3190-y, arXiv:1406.0076 [hep-ex]
34. ATLAS Collaboration (2019) Electron and photon energy calibration with the ATLAS detector using 2015–2016 LHC proton–proton collision data. JINST 14:P03017. http://dx.doi.org/10.1088/1748-0221/14/03/P03017, arXiv:1812.03848 [hep-ex]

35. ATLAS Collaboration (2016) Muon reconstruction performance of the ATLAS detector in proton–proton collision data at $\sqrt{s} = 13$ TeV. Eur Phys J C 76:292. http://dx.doi.org/10.1140/epjc/s10052-016-4120-y, arXiv:1603.05598 [hep-ex]
36. ATLAS Collaboration (2018) Performance of missing transverse momentum reconstruction with the ATLAS detector using proton-proton collisions at $\sqrt{s} = 13$ TeV. Eur Phys J C78(11):903. http://dx.doi.org/10.1140/epjc/s10052-018-6288-9, arXiv:1802.08168 [hep-ex]
37. ATLAS Collaboration (2015) Performance of missing transverse momentum reconstruction with the ATLAS detector in the first proton–proton collisions at $\sqrt{s} = 13$TeV, ATL-PHYS-PUB-2015-027. https://cds.cern.ch/record/2037904
38. ATLAS Collaboration (2019) Electron reconstruction and identification in the ATLAS experiment using the 2015 and 2016 LHC proton-proton collision data at $\sqrt{s} = 13$ TeV. Submitted to: Eur Phys J. arXiv:1902.04655 [physics.ins-det]
39. ATLAS Collaboration (2016) Performance of pile-up mitigation techniques for jets in pp collisions at $\sqrt{s} = 8$ TeV using the ATLAS detector. Eur Phys J C 76:581. http://dx.doi.org/10.1140/epjc/s10052-016-4395-z, arXiv:1510.03823 [hep-ex]
40. ATLAS Collaboration (2014) Tagging and suppression of pileup jets with the ATLAS detector, ATLAS-CONF-2014-018. https://cds.cern.ch/record/1700870
41. ATLAS Collaboration (2016) Luminosity determination in pp collisions at $\sqrt{s} = 8$ TeV using the ATLAS detector at the LHC. Eur Phys J C 76:653. http://dx.doi.org/10.1140/epjc/s10052-016-4466-1, arXiv:1608.03953 [hep-ex]
42. Avoni G et al (2018) The new LUCID-2 detector for luminosity measurement and monitoring in ATLAS. JINST 13(07):P07017. http://dx.doi.org/10.1088/1748-0221/13/07/P07017
43. ATLAS Collaboration (2018) Measurement of $W^{\pm}Z$ production cross sections and gauge boson polarisation in pp collisions at $\sqrt{s} = 13$TeV with the ATLAS detector, ATLAS-CONF-2018-034. https://cds.cern.ch/record/2630187
44. Dulat S, Hou T-J, Gao J, Guzzi M, Huston J, Nadolsky P, Pumplin J, Schmidt C, Stump D, Yuan CP (2016) New parton distribution functions from a global analysis of quantum chromodynamics. Phys Rev D 93(3):033006. http://dx.doi.org/10.1103/PhysRevD.93.033006, arXiv:1506.07443 [hep-ph]
45. Harland-Lang LA, Martin AD, Motylinski P, Thorne RS (2015) Parton distributions in the LHC era: MMHT 2014 PDFs. Eur Phys J C 75(5):204. http://dx.doi.org/10.1140/epjc/s10052-015-3397-6, arXiv:1412.3989 [hep-ph]
46. Baak M, Besjes GJ, Cte D, Koutsman A, Lorenz J, Short D (2015) HistFitter software framework for statistical data analysis. Eur Phys J C 75:153. http://dx.doi.org/10.1140/epjc/s10052-015-3327-7, arXiv:1410.1280 [hep-ex]
47. Cousins RD, Linnemann JT, Tucker J (2008) Evaluation of three methods for calculating statistical significance when incorporating a systematic uncertainty into a test of the background-only hypothesis for a Poisson process. Nuclear Inst. Methods Phys Res Sect. A: Accel Spectrom Detect Assoc Equip 595(2):480–501. http://dx.doi.org/https://doi.org/10.1016/j.nima.2008.07.086, http://www.sciencedirect.com/science/article/pii/S0168900208010255
48. Read AL (2002) Presentation of search results: the CL(s) technique. J Phys G28:2693–2704. http://dx.doi.org/10.1088/0954-3899/28/10/313

Chapter 10
Search for the EWK Production of Compressed SUSY with Soft Leptons

The previous chapters discuss the production of $\tilde{\chi}_1^{\pm}\tilde{\chi}_2^0$ decaying via W and Z bosons to two or three leptons and missing energy. The assumption is that $\tilde{\chi}_1^{\pm}$ and $\tilde{\chi}_2^0$ were mass degenerate winos and the LSP, $\tilde{\chi}_1^0$, is a bino. This chapter will also present a search of $\tilde{\chi}_1^{\pm}\tilde{\chi}_2^0$ decaying via W and Z bosons to two leptons and missing energy. In this case, the assumption is that all SUSY particles are higgsinos, and the mass of the chargino, $m(\tilde{\chi}_1^{\pm})$, is in between the mass of the two neutralinos. The production of higgsino particles result a compressed spectrum, as shown in Sect. 6.3.

Compressed spectra can also arise for the wino production with bino LSP scenario, where the winos decay via sleptons to leptons. Supersymmetric explanations for the tension in the observed anomalous magnetic moment of the muon $(g-2)_{\mu}$ [1, 2] suggest that sleptons and neutralinos should have masses on the order of the weak-scale [3, 4].

Searching for compressed scenarios is limited by the smaller production cross-sections but also due to the difficulty in reconstructing the final states due to the small momenta of the decay products. The strongest limits from previous searches are from combinations of results from the LEP experiments [5–15]. These experiments set lower bounds on direct chargino production of $m(\widetilde{\chi}_1^{\pm}) > 103.5$ GeV for $\Delta m(\widetilde{\chi}_1^{\pm}, \widetilde{\chi}_1^0) > 3$ GeV and $m(\widetilde{\chi}_1^{\pm}) > 92.4$ GeV for smaller mass differences. For sleptons, lower limits on the mass of the scalar partner of the right-handed muon $(\widetilde{\mu}_R)$ are $m(\widetilde{\mu}_R) \gtrsim 94.6$ GeV for mass splittings down to $\Delta m(\widetilde{\mu}_R, \widetilde{\chi}_1^0) \gtrsim 2$ GeV. For the scalar partner of the right-handed electron $(\widetilde{e}_R)$ a lower bound of $m(\widetilde{e}_R) \gtrsim 73$ GeV independent of $\Delta m(\widetilde{e}_R, \widetilde{\chi}_1^0)$ exists.

Searches with the ATLAS detector at $\sqrt{s} = 8$ TeV [16–19] set limits on the production of winos decaying via W or Z bosons for mass splittings of $\Delta m(\widetilde{\chi}_1^{\pm}, \widetilde{\chi}_1^0) \gtrsim 35$ GeV, and $\Delta m(\tilde{\ell}, \widetilde{\chi}_1^0) \gtrsim 55$ GeV for slepton production. Searches with the CMS detector, at $\sqrt{s} = 8$ TeV [20, 21] and at $\sqrt{s} = 13$ TeV [22], set limit on winos decaying via W or Z bosons for mass splittings $\Delta m(\widetilde{\chi}_1^{\pm}, \widetilde{\chi}_1^0) \gtrsim 23$ GeV.

© The Editor(s) (if applicable) and The Author(s), under exclusive license to Springer Nature Switzerland AG 2020

E. Resseguie, *Electroweak Physics at the Large Hadron Collider with the ATLAS Detector*, Springer Theses,
https://doi.org/10.1007/978-3-030-57016-3_10

Phenomenological studies propose that natural SUSY can be found by probing compressed mass spectra in the electroweak SUSY sector by using leptons with small transverse momentum, p_{T}, referred to as soft leptons [23–30].

This chapter will cover two searches for the electroweak production of compressed SUSY decaying to two soft leptons and missing transverse energy [31]. Collectively these searches will be referred to as the compressed searches.

10.1 Signal Signature

Figure 10.1 shows the two compressed SUSY signal models considered in this search, interpreted using simplified models, introduced in Sect. 7.2. Compressed spectra result in soft leptons and events with small missing transverse momenta. The ISR topology, described in Sect. 7.5.3.1, is used for both models, resulting in the missing energy summing to a large value due to the recoil against the jet. For the small mass splittings considered in this analysis, the $E_{\mathrm{T}}^{\mathrm{miss}}$ sums up to a larger value than the lepton p_{T} from the ISR jet boost. This increase in $E_{\mathrm{T}}^{\mathrm{miss}}$ helps distinguish between the signal and the background.

The first diagram shows the production of $\tilde{\chi}_1^{\pm}\tilde{\chi}_2^0$ decaying via W and Z bosons decaying to leptons, jets, and missing energy. The SUSY particles are higgsinos which results in small mass splittings. As a result of the small mass splitting between the masses of $\tilde{\chi}_2^0$ and $\tilde{\chi}_1^0$, the SUSY particles decay via off-shell W and Z bosons, where the mass of the boson is less than its PDG mass, resulting in soft leptons and small p_{T} LSP, $\tilde{\chi}_1^0$. This scenario is referred to as "electroweakino".

The second diagrams shows the direct production of sleptons decaying to two leptons and missing transverse momenta from the LSP. The sleptons are winos and the LSP is a bino. This scenario is referred to as "slepton".

In the electroweakino scenario, the leptons from the decay of the Z boson are closer together due to the boost in the system than in the slepton scenario. In both cases, the leptons form a same-flavor, opposite-charge pair.

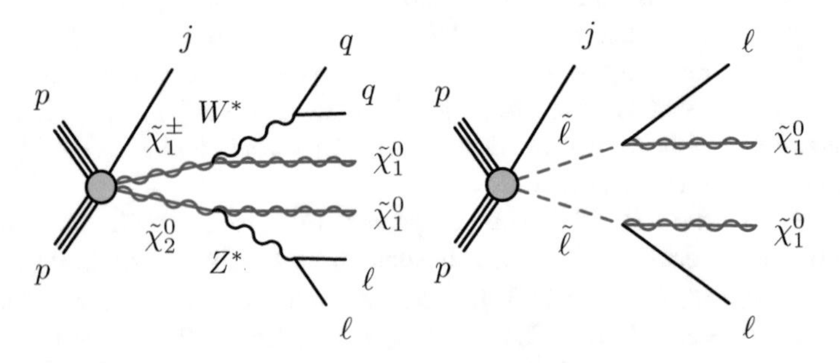

Fig. 10.1 Diagrams for the production of (**a**) $\tilde{\chi}_1^{\pm}\tilde{\chi}_2^0$ decaying via W and Z bosons, (**b**) direct productions of $\tilde{\ell}\tilde{\ell}$, decaying to two leptons and $E_{\mathrm{T}}^{\mathrm{miss}}$ in pp collisions

10.2 Overview of Backgrounds

The backgrounds in this search can be separated into two categories: irreducible and reducible. The irreducible backgrounds come from backgrounds with two prompt leptons, missing transverse energy, and jets. They include diboson (WW and WZ), top ($t\bar{t}$ and tW), and $Z/\gamma \rightarrow \tau\tau$, where the taus decay leptonically. The reducible backgrounds comes from events with a fake or non-prompt lepton such as diboson processes decaying to 0 or 1 lepton, $W + \gamma$, W+jets, $Z(\rightarrow \nu\nu) + \gamma$, and $Z(\rightarrow \nu\nu)$+jets processes.

10.3 Data Set and MC Samples

The proton proton collision data corresponds to an integrated luminosity 36.1 fb^{-1} collected at a center-of-mass energy of 13 TeV in 2015 and 2016. The samples include an ATLAS detector simulation [32], based on Geant4 [33], or a fast simulation [32] that uses a parametrization of the calorimeter response [34] and Geant4 for the other parts of the detector. The simulated events are reconstructed in the same manner as the data.

Table 10.1 summarizes the Monte Carlo (MC) used specifying the generator used to simulate both background and signal events.

Table 10.1 Summary of the signal and background processes with the generator used for the simulation and the order at which the cross section is calculated

Process	Matrix element	Parton shower	PDF set	Cross-section
$Z^{(*)}/\gamma^*$ + jets	SHERPA2.2.1		NNPDF 3.0 NNLO	NNLO
Diboson	SHERPA2.1.1 / 2.2.1 / 2.2.2		NNPDF 3.0 NNLO	Generator NLO
Triboson	SHERPA2.2.1		NNPDF 3.0 NNLO	Generator LO, NLO
$t\bar{t}$	POWHEG- BOX v2	PYTHIA6.428	NLO CT10	NNLO+NNLL
t (s-channel)	POWHEG- BOX v1	PYTHIA6.428	NLO CT10	NNLO+NNLL
t (t-channel)	POWHEG- BOX v1	PYTHIA6.428	NLO CT10f4	NNLO+NNLL
$t + W$	POWHEG- BOX v1	PYTHIA6.428	NLO CT10	NNLO+NNLL
$h(\rightarrow \ell\ell, WW)$	POWHEG- BOX v2	PYTHIA8.186	NLO CTEQ6L1	NLO
$h + W/Z$	MADGRAPH5_aMC@NLO 2.2.2	PYTHIA8.186	NNPDF 2.3 LO	NLO
$t\bar{t} + W/Z/\gamma^*$	MADGRAPH5_aMC@NLO 2.3.3	PYTHIA8.186	NNPDF 3.0 LO	NLO
$t\bar{t} + WW/t\bar{t}$	MADGRAPH5_aMC@NLO 2.2.2	PYTHIA8.186	NNPDF 2.3 LO	NLO
$t + Z$	MADGRAPH5_aMC@NLO 2.2.1	PYTHIA6.428	NNPDF 2.3 LO	LO
$t + WZ$	MADGRAPH5_aMC@NLO 2.3.2	PYTHIA8.186	NNPDF 2.3 LO	NLO
$t + t\bar{t}$	MADGRAPH5_aMC@NLO 2.2.2	PYTHIA8.186	NNPDF 2.3 LO	LO

SHERPA 2.1.1, 2.2.1, and 2.2.2 [35, 36] were used to generate $Z^{(*)}/\gamma^*$ +jets, diboson, and triboson events with matrix elements calculated for up to two partons at NLO and up to four partons at LO. The $Z^{(*)}/\gamma^*$ + jets and diboson samples are simulated down to dilepton invariant masses down to 0.5 GeV for $Z^{(*)}/\gamma^* \rightarrow e^+e^-/\mu^+\mu^-$, and 3.8 GeV for $Z^{(*)}/\gamma^* \rightarrow \tau^+\tau^-$.

POWHEG- BOX v1 and v2 [37–39], and using PYTHIA 6.428 for hadronization were used to simulate $t\bar{t}$ and single-top production at NLO.

POWHEG- BOX v2 was used with PYTHIA 8.186 to simulate Higgs boson production. MADGRAPH5_aMC@NLO v2.2.2 with PYTHIA was used to simulate production of a Higgs boson in association with a W or Z boson, as well as events containing $t\bar{t}$ and one or more electroweak bosons. These processes were generated at NLO in the matrix element except for $t\bar{t} + WW/t\bar{t}, t + t\bar{t}$, and $t + Z$, which were generated at LO.

The Higgsino simplified model includes the production of $\tilde{\chi}_2^0\tilde{\chi}_1^{\pm}$, $\tilde{\chi}_2^0\tilde{\chi}_1^0$ and $\tilde{\chi}_1^+\tilde{\chi}_1^-$; however, the $\tilde{\chi}_2^0\tilde{\chi}_1^0$ and $\tilde{\chi}_1^+\tilde{\chi}_1^-$ process contribute little to the sensitivity. The $\tilde{\chi}_1^0$ and $\tilde{\chi}_2^0$ masses were varied, while the $\tilde{\chi}_1^{\pm}$ masses were set to $m(\tilde{\chi}_1^{\pm}) = \frac{1}{2}[m(\tilde{\chi}_1^0) + m(\tilde{\chi}_2^0)]$. The calculated cross-sections assume pure higgsino states. The $Z^* \rightarrow \ell^+\ell^-$ and $W^* \rightarrow \ell\nu$ branching ratios depend on the mass splittings and were computed using SUSY- HIT v1.5b [40], which accounts for finite b-quark and τ-lepton masses. The SUSY signal processes were generated from LO matrix elements with up to two extra partons, using the MADGRAPH v2.2.3 generator interfaced to PYTHIA8.186.

The slepton simplified model considers direct pair production of the selectron $\tilde{e}_{L,R}$ and smuon $\tilde{\mu}_{L,R}$, where the subscripts L, R denote the left- or right-handed chirality of the partner electron or muon. The four sleptons are assumed to be mass degenerate, i.e. $m(\tilde{e}_L) = m(\tilde{e}_R) = m(\tilde{\mu}_L) = m(\tilde{\mu}_R)$. The sleptons decay with a 100% branching ratio into the corresponding SM partner lepton and the $\tilde{\chi}_1^0$ neutralino. Events were generated at tree level using MADGRAPH5_aMC@NLO v2.2.3 and the NNPDF23LO PDF set with up to two additional partons in the matrix element, and interfaced with PYTHIA v8.186.

10.4 Object Selection

Electrons and muons are identified using identification, isolation, and tracking criteria. Two levels of object selection are used for electrons and muons, described in Table 10.2. Each level "baseline" and "signal" applies the selection of the previous levels along with additional criteria. The baseline leptons use the looser identification criteria and lower lepton p_T in order to provide a higher efficiency of identifying and removing processes decaying to four prompt leptons. Signal leptons satisfy stricter criteria.

In Run 2, leptons can be reconstructed to lower p_T, 4.5 GeV for electrons and 4 GeV for muons, as shown in Fig. 10.2. Reconstructing leptons at such low p_T helps target the signals very small mass splittings.

Table 10.2 Summary of the baseline and signal levels for electron and muon criteria. Each new level contains the selection of the previous level

Cut	Value/description	
	Baseline electron	Baseline muon
Acceptance	$p_T > 4.5\,\text{GeV}$, $\lvert\eta^{\text{cluster}}\rvert <$ 2.47	$p_T > 4\,\text{GeV}$, $\lvert\eta\rvert < 2.5$
Identification	`VeryLooseLLH`	`Medium`
Impact parameter	$\lvert z_0 \sin\theta\rvert < 0.5$ mm,	$\lvert z_0 \sin\theta\rvert < 0.5$ mm,
Reco algorithm	veto author == 16	
	Signal electron	Signal muon
Identification	`TightLLH`	`Medium`
Isolation	`GradientLoose`	`FixedCutTrackOnly`
Impact parameter	$\lvert z_0 \sin\theta\rvert < 0.5$ mm,	$\lvert z_0 \sin\theta\rvert < 0.5$ mm,
	$\lvert d_0/\sigma_{d_0}\rvert < 5$	$\lvert d_0/\sigma_{d_0}\rvert < 3$

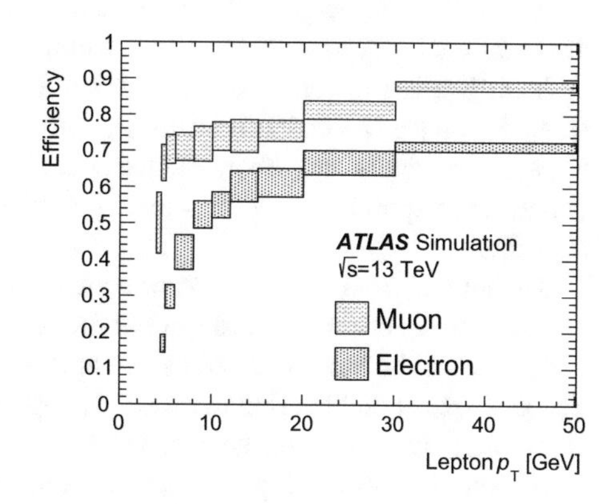

Fig. 10.2 Signal lepton efficiencies for electrons and muons, averaged over all Higgsino and slepton samples. The uncertainty band is derived from the range of efficiencies observed for all signals at a given p_T

Baseline electrons must have $p_T > 4.5\,\text{GeV}$ and fall within the inner detector, with $|\eta| < 2.47$. The electrons must also satisfy the `VeryLooseLLH` quality criteria and $|z_0 \sin\theta| < 0.5$ mm impact parameter. Signal electrons need to pass the impact parameter cuts of and $|d_0/\sigma_{d_0}| < 5$, designed to suppress fake electrons from pileup jets.They also satisfy tighter identification criteria, `TightLLH` and tighter isolation, `GradientLoose`.

Baseline muons must satisfy the `Medium` identification criteria, have $p_T > 4\,\text{GeV}$ and fall within the inner detector, $|\eta| < 2.4$, and $|z_0 \sin\theta| < 0.5$ mm. Signal muons must pass the impact parameter cuts of $|d_0/\sigma_{d_0}| < 3$. They must must fulfill a tighter isolation, `FixedCutTrackOnly`.

Jets are reconstructed from topological clusters using the anti-k_t algorithm with distance parameter $\Delta R = 0.4$. Baseline jets are required to have $p_T > 20$ GeV and fulfill the pseudorapidity requirement of $|\eta| < 4.5$. To suppress jets originating from pileup, jets are further required to pass a JVT cut ($JVT > 0.59$) if the jet p_T

Table 10.3 Summary of the baseline and signal selection for jets and b-jets

Cut	Value/description
	Baseline jet
Collection	AntiKtEMTopo
Acceptance	$p_T > 20$ GeV, $\|\eta\| < 4.5$
	Signal jet
JVT	$\|JVT\| > 0.59$ for jets with $p_T < 60$ GeV and $\|\eta\| < 2.4$
Acceptance	$p_T > 30$ GeV, $\|\eta\| < 2.8$
	Signal b-jet
b-tagger algorithm	MV2c10, 85% efficiency
Acceptance	$p_T > 20$ GeV, $\|\eta\| < 2.5$

is within $20 < p_T < 50$ GeV and it resides within $|\eta| < 2.4$ [41]. Signal jets have the additional requirement of falling within $|\eta| < 2.8$ and $p_T > 30$ GeV.

Identification of jets containing b-hadrons (b-jets), so called b-tagging, is performed with the MV2c10 algorithm, a multivariate discriminant making use of track impact parameters and reconstructed secondary vertices [42, 43]. A requirement is chosen corresponding to a 85% average efficiency obtained for b-jets in simulated $t\bar{t}$ events.

The jet and b-jet selection criteria are summarized in Table 10.3.

Separate algorithms are run in parallel to reconstruct electrons, muons, and jets. A particle can be reconstructed as one or more objects. To resolve these ambiguities, a procedure called "overlap removal" is applied. For electrons, this overlap removal is applied in two steps. At the baseline selection, an electron that shares a track with a muon, and the sub-leading p_T electron from two overlapping electrons are removed. The second step removes electrons if they are within $0.2 < \Delta R < 0.4$ of a jet. For muons, overlap removal is applied to baseline muons to separate prompt muons from those originating from the decay of hadrons in a jet. A baseline muon is removed if it is within $\Delta R < 0.4$ of a jet that at least 3 tracks.

The missing transverse momentum, with magnitude E_T^{miss}, is calculated as the negative vector sum of the transverse momenta of the calibrated selected leptons and jets, and the sum of transverse momenta of additional soft objects in the event, which are reconstructed from tracks in the inner detector or calorimeter cell clusters.

10.5 Event Selection

10.5.1 E_T^{miss} Triggers

Signal electrons require $p_T > 4.5$ GeV and signal muons require $p_T > 4$ GeV, as shown in Table 10.2. Lepton triggers cannot be used to select events because their thresholds are much higher, around 20 GeV, as shown in Table 7.6. Since the events require the presence of at least one jet, the missing energy and leptons system are boosted. As as a result of the boost, the missing energy sums to a large value, which means that that the event can be selected using $E_{\mathrm{T}}^{\mathrm{miss}}$ triggers or jet triggers. Jet triggers require large jet p_T and reduce the signal acceptance too much so $E_{\mathrm{T}}^{\mathrm{miss}}$ triggers are chosen.

Just like other triggers, described in Sect. 3.4, the $E_{\mathrm{T}}^{\mathrm{miss}}$ trigger in Run 2 has two levels, L1 and HLT. $E_{\mathrm{T}}^{\mathrm{miss}}$ at L1 [44, 45] is calculated from the energy deposits in the calorimeter by summing over regions of $\Delta\eta \times \Delta\phi = 0.2 \times 0.2$ for $|\eta| < 2.5$ and over larger regions for larger values of $|\eta|$, called trigger towers. The energy of the trigger towers is calibrated at the electromagnetic energy scale (EM scale). The EM scale correctly reconstructs the energy deposited by particles in an electromagnetic shower in the calorimeter but underestimates the energy deposited by hadrons.

HLT algorithms use more calorimeter information and were designed for the higher pileup conditions of Run 2 [45]. The simplest HLT algorithm is called the "Cell algorithm". This algorithm calculates $E_{\mathrm{T}}^{\mathrm{miss}}$ from all calorimeter cells but reduces the contribution of noise by requiring the components of the missing energy (E_x and E_y) to be larger than twice the size of the energy fluctuation in the cell due to noise and pileup. Energies are kept at the EM scale, just like at L1. Another algorithm is the jet based algorithm, or "MHT", which calculates $E_{\mathrm{T}}^{\mathrm{miss}}$ from trigger level jets. These jets are formed from clusters with transverse momentum of at least 7 GeV, but with correction from pileup and jet calibration, can have transverse momentum much lower. The last algorithm discussed is the pileup fit algorithm, "putfit", which uses a complex approach to mitigate pileup contribution in the calculation of $E_{\mathrm{T}}^{\mathrm{miss}}$. In the pufit algorithm, a transverse energy dependent threshold on the towers is required. Also, events are required to have at least one tower above threshold. A least squares fit is then performed to determine the $E_{\mathrm{T}}^{\mathrm{miss}}$ contribution from pileup. The $E_{\mathrm{T}}^{\mathrm{miss}}$ for the event is then calculated from the towers above threshold and the fit result.

The efficiency of the $E_{\mathrm{T}}^{\mathrm{miss}}$ trigger is limited by the HLT algorithm and not by the L1 threshold, as long as the L1 threshold is below the HLT threshold. In Fig. 10.3, the combined L1 and HLT triggers had the same turn-on curves despite using a different L1 seed.

Since the HLT algorithms limit the trigger turn-on, the different algorithms are compared to select the most efficient one. Figures 10.4, 10.5 show $E_{\mathrm{T}}^{\mathrm{miss}}$ trigger turn-on curves for constant L1 value (L1XE50), varying the cell $E_{\mathrm{T}}^{\mathrm{miss}}$ (xe70, xe75), and varying MHT $E_{\mathrm{T}}^{\mathrm{miss}}$ cuts (xe110_MHT, xe130_MHT) for different mass splittings and for the production different SUSY masses. The trigger with the smallest MHT $E_{\mathrm{T}}^{\mathrm{miss}}$ and no cell $E_{\mathrm{T}}^{\mathrm{miss}}$ requirement has the earliest turn-on for triggers with HLT.

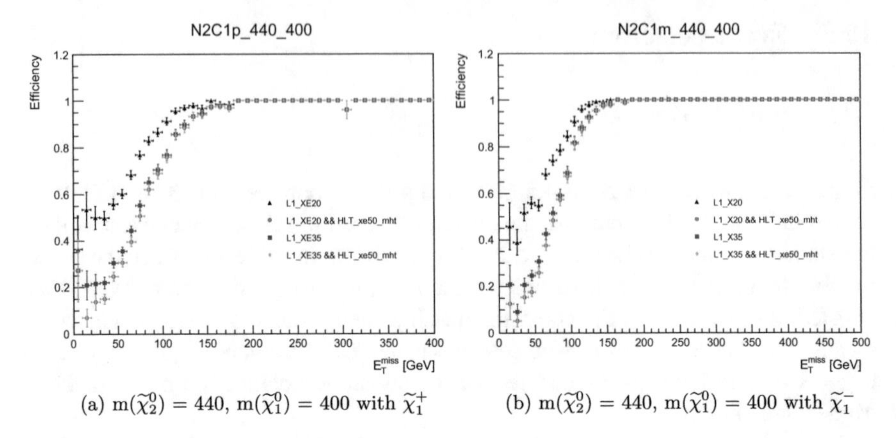

(a) $m(\widetilde{\chi}_2^0) = 440$, $m(\widetilde{\chi}_1^0) = 400$ with $\widetilde{\chi}_1^+$ (b) $m(\widetilde{\chi}_2^0) = 440$, $m(\widetilde{\chi}_1^0) = 400$ with $\widetilde{\chi}_1^-$

Fig. 10.3 Trigger turn-on curves for constant HLT MHT $E_{\mathrm{T}}^{\mathrm{miss}}$ and varying L1 thresholds split by signal with positive and negative chargino

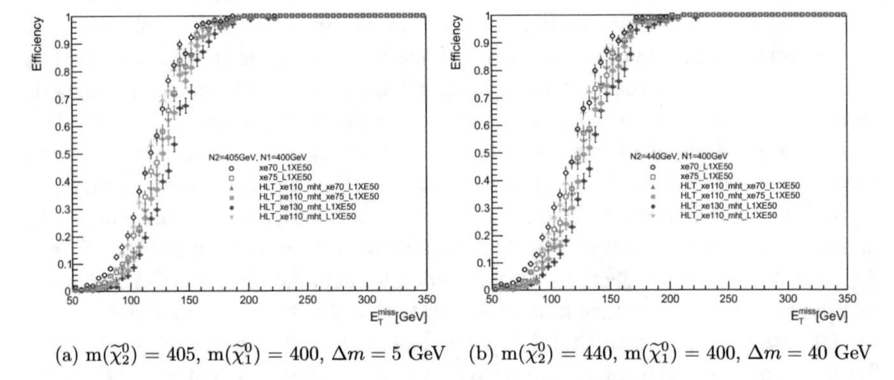

(a) $m(\widetilde{\chi}_2^0) = 405$, $m(\widetilde{\chi}_1^0) = 400$, $\Delta m = 5$ GeV (b) $m(\widetilde{\chi}_2^0) = 440$, $m(\widetilde{\chi}_1^0) = 400$, $\Delta m = 40$ GeV

Fig. 10.4 $E_{\mathrm{T}}^{\mathrm{miss}}$ trigger turn-on curves for constant L1 value (L1XE50), varying the cell $E_{\mathrm{T}}^{\mathrm{miss}}$ (xe70, xe75), and varying MHT $E_{\mathrm{T}}^{\mathrm{miss}}$ cuts (xe110_MHT, xe130_MHT) for different mass splittings, $\Delta m = 5, 40$ GeV

Adding a cell $E_{\mathrm{T}}^{\mathrm{miss}}$ cut is more effective than requiring a larger MHT $E_{\mathrm{T}}^{\mathrm{miss}}$ cut. The triggers reach their maximum efficiency at $E_{\mathrm{T}}^{\mathrm{miss}} > 200$ GeV, regardless of mass splitting or SUSY particle mass.

The final HLT algorithm to check is pufit; therefore, Fig. 10.6 compares MHT, cell, and pufit at constant L1 for three signal masses and mass splittings. For all three signals considered, the least efficient HLT algorithm is the one using cell $E_{\mathrm{T}}^{\mathrm{miss}}$. The pufit and MHT algorithms give similar turn-on curves and are fully efficient for $E_{\mathrm{T}}^{\mathrm{miss}} > 200$GeV; however, the MHT algorithm performs a little better.

Since there is not gain by using HLT triggers calculated with pufit or cell $E_{\mathrm{T}}^{\mathrm{miss}}$, the lowest unprescaled triggers using MHT $E_{\mathrm{T}}^{\mathrm{miss}}$ are chosen, which are 95% efficient for $E_{\mathrm{T}}^{\mathrm{miss}} > 200$ GeV. Candidate events are required the $E_{\mathrm{T}}^{\mathrm{miss}}$ triggers summarized in Table 10.4.

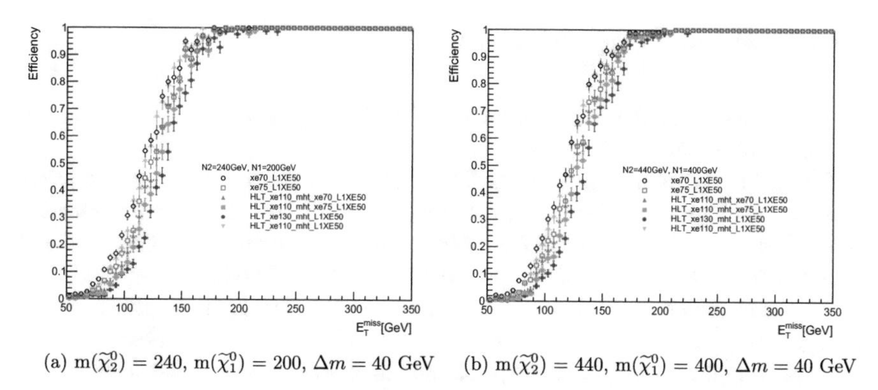

(a) $\mathrm{m}(\widetilde{\chi}_2^0) = 240$, $\mathrm{m}(\widetilde{\chi}_1^0) = 200$, $\Delta m = 40$ GeV (b) $\mathrm{m}(\widetilde{\chi}_2^0) = 440$, $\mathrm{m}(\widetilde{\chi}_1^0) = 400$, $\Delta m = 40$ GeV

Fig. 10.5 $E_{\mathrm{T}}^{\mathrm{miss}}$ trigger turn-on curves for constant L1 value (L1XE50), varying the cell $E_{\mathrm{T}}^{\mathrm{miss}}$ (xe70, xe75), and varying MHT $E_{\mathrm{T}}^{\mathrm{miss}}$ cuts (xe110_MHT, xe130_MHT), the masses of the produced SUSY particles ($\mathrm{m}(\widetilde{\chi}_2^0)$=240, 440)GeV and constant $\Delta m = 40$GeV

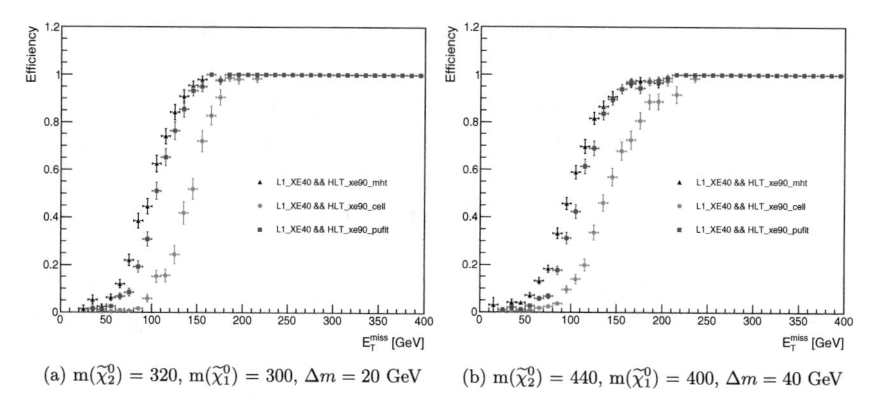

(a) $\mathrm{m}(\widetilde{\chi}_2^0) = 320$, $\mathrm{m}(\widetilde{\chi}_1^0) = 300$, $\Delta m = 20$ GeV (b) $\mathrm{m}(\widetilde{\chi}_2^0) = 440$, $\mathrm{m}(\widetilde{\chi}_1^0) = 400$, $\Delta m = 40$ GeV

Fig. 10.6 Comparing $E_{\mathrm{T}}^{\mathrm{miss}}$ turn-on curves for MHT, cell, and pufit $E_{\mathrm{T}}^{\mathrm{miss}}$ with constant L1 for two different signal masses and mass splittings

Table 10.4 Summary of $E_{\mathrm{T}}^{\mathrm{miss}}$ triggers used in the compressed searches

Year	Period	Trigger
2015	all	HLT_xe70_MHT
2016	A-D3	HLT_xe90_MHT_L1XE50
	D4-F1	HLT_xe100_MHT_L1XE50
	F2-K	HLT_xe110_MHT_L1XE50

10.5.2 $m_{\ell\ell}$ and m_{T2} Shape Fits

Two main discriminating variables are used in this analysis which take advantage of a kinematic endpoint for the signal.

The main discriminating variable for the electroweakino signal is the invariant mass of the dilepton pair. An upper cut of $m_{\ell\ell} < 60$GeV reduces the contamination of on-shell WZ, Z+jets, and $ZZ \rightarrow \ell\ell\nu\nu$ which peak around the mass of the Z boson. The electroweakino signal decays via off-shell W and Z bosons and, as a result, the invariant mass is kinematically bound by the signal mass splitting, $\Delta m = m(\tilde{\chi}_2^0) - m(\tilde{\chi}_1^0)$, as seen in Fig. 10.7. The signal is thus isolated from backgrounds. A shape fit in the invariant mass is performed to take advantage of this feature.

For the slepton signal model, the main discriminating variable is the stransverse mass, m_{T2}, first defined in Sect. 7.8.1.1. In this case, m_{T2} is calculated using an invisible particle mass hypothesis, m_χ. Including m_χ, Eq. (7.5) becomes:

$$m_{T2}^{m_\chi} = \min_{\mathbf{q}_T} \left[\max \left(m_T(\mathbf{p}_T^{\ell 1}, \mathbf{q}_T, m_\chi), m_T(\mathbf{p}_T^{\ell 2}, \mathbf{p}_T^{\text{miss}} - \mathbf{q}_T, m_\chi) \right) \right], \tag{10.1}$$

where $\mathbf{p}_T$ is the transverse momentum vector of lepton 1 or lepton 2, $\mathbf{p}_T^{\text{miss}}$ is the missing energy vector, $\mathbf{q}_T$ is the quantity minimized over, and m_T is defined as,

$$m_T(\mathbf{p}_T, \mathbf{q}_T, m_\chi) = \sqrt{m_\ell^2 + m_\chi^2 + 2(E_T^\ell E_T^q - \mathbf{p}_T \cdot \mathbf{q}_T)} \tag{10.2}$$

Just like the electroweakino signal has a kinematic bound on $m_{\ell\ell}$, the slepton signal has a kinematic bound on $m_{T2}^{m_\chi}$ at $\Delta m = m(\tilde{\ell}) - m(\tilde{\chi}_1^0)$, as shown in Fig. 10.8. $m_{T2}^{m_\chi}$

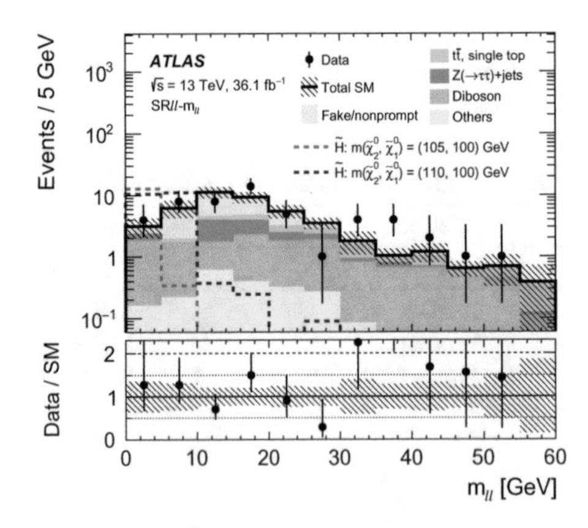

Fig. 10.7 $m_{\ell\ell}$ distribution in the inclusive $m_{\ell\ell}$ SR [1,60]. Background processes containing fewer than two prompt leptons are categorized as 'Fake/non-prompt'. The category 'Others' contains rare backgrounds from triboson, Higgs boson, and the remaining top-quark production processes. The uncertainty bands plotted include all statistical and systematic uncertainties. The dashed lines represent benchmark higgsino signal samples

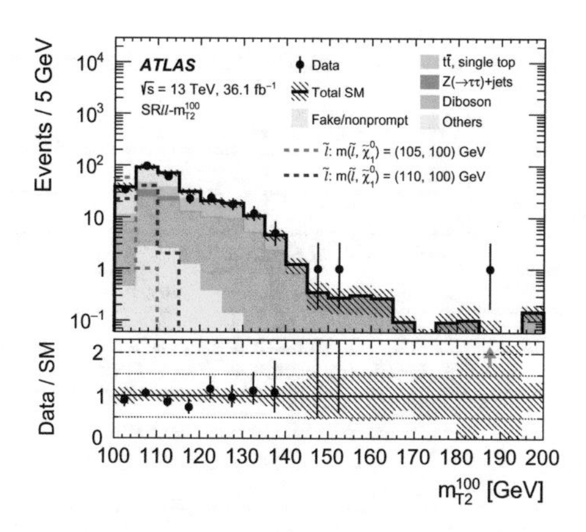

Fig. 10.8 m_{T2} distribution in the inclusive m_{T2}^{100} SR [100,∞]. Background processes containing fewer than two prompt leptons are categorized as 'Fake/non-prompt'. The category 'Others' contains rare backgrounds from triboson, Higgs boson, and the remaining top-quark production processes. The uncertainty bands plotted include all statistical and systematic uncertainties. The dashed lines represent benchmark slepton signal samples

is always less than the lepton mass when the hypothesis invisible mass is set to the neutralino mass, $m_\chi = m(\tilde{\chi}_1^0)$. The background processes, such as WZ do not need this requirement on the hypothesis mass on the invisible particle because neutrinos are effectively massless. The hypothesis mass, m_χ, is set to 100 GeV in this analysis because this assumption gives the most distinct signal kinematic edge. To make use of this feature, a shape fit in m_{T2}^{100} is performed for the slepton signal model.

10.5.3 Definitions of Signal Regions

The electroweakino and slepton signal regions are summarized in Table 10.5. The binning strategy in $m_{\ell\ell}$ for electroweakino signals and m_{T2} for slepton signals is shown in Table 10.6. The electroweakino SRs are divided into seven non-overlapping ranges of $m_{\ell\ell}$, which are further divided by lepton flavor (ee, $\mu\mu$), and referred to as exclusive regions. Seven inclusive regions are also defined, characterized by overlapping ranges of $m_{\ell\ell}$. For the slepton SRs, m_{T2}^{100} is used to define 12 exclusive regions and 6 inclusive regions. When setting model-dependent limits, the exclusive bins are considered while for model independent limits, only the inclusive regions are included.

Events are selected with exactly two baseline leptons that form a same flavor, opposite charge pair. The leading lepton p_{T} for electrons and muons, $p_{\mathrm{T}} > 5$ GeV, is tighter than the signal lepton requirement while the sub-leading lepton has the same p_{T} requirement as signal leptons, defined in Table 10.2, $p_{\mathrm{T}} > 4.5$ GeV for electrons and $p_{\mathrm{T}} > 4$ GeV for muons. Requiring the separation $\Delta R_{\ell\ell} =$

Table 10.5 Summary of event selection criteria. The binning scheme used to define the final signal regions is shown in Table 10.6

Variable	Common requirement	
Number of leptons	=2	
Lepton charge and flavor	e^+e^- or $\mu^+\mu^-$	
Leading lepton $p_T^{\ell_1}$	>5 (5) GeV for electron (muon)	
Sub-leading lepton $p_T^{\ell_2}$	>4.5 (4) GeV for electron (muon)	
$\Delta R_{\ell\ell}$	>0.05	
$m_{\ell\ell}$	$\in [1, 60]$ GeV excluding [3.0, 3.2] GeV	
E_T^{miss}	>200 GeV	
Number of jets	≥1	
Leading jet p_T	>100 GeV	
$\Delta\phi_{j,MET}$	>2.0	
min($\Delta\phi$(any jet, E_T^{miss}))	>0.4	
Number of b-tagged jets	=0	
$m_{\tau\tau}$	< 0 or > 160 GeV	
	Electroweakino SRs	Slepton SRs
$\Delta R_{\ell\ell}$	<2	—
$m_T^{\ell_1}$	<70 GeV	—
E_T^{miss}/H_T^{lep}	$> \max\left(5, 15 - 2\frac{m_{\ell\ell}}{1\ \text{GeV}}\right)$	$> \max\left(3, 15 - 2\left(\frac{m_{T2}^{100}}{1\ \text{GeV}} - 100\right)\right)$
Binned in	$m_{\ell\ell}$	m_{T2}^{100}

Table 10.6 Signal region binning for the electroweakino and slepton SRs. Each SR is defined by the lepton flavor (ee, $\mu\mu$, or $\ell\ell$ for both) and a range of $m_{\ell\ell}$ (for electroweakino SRs) or m_{T2}^{100} (for slepton SRs) in GeV. The inclusive bins are used to set model-independent limits, while the exclusive bins are used to derive exclusion limits on signal models

Electroweakino SRs								
Exclusive	SRee-$m_{\ell\ell}$, SR$\mu\mu$-$m_{\ell\ell}$	[1, 3]	[3.2, 5]	[5, 10]	[10, 20]	[20, 30]	[30, 40]	[40, 60]
Inclusive	SR$\ell\ell$-$m_{\ell\ell}$	[1, 3]	[1, 5]	[1, 10]	[1, 20]	[1, 30]	[1, 40]	[1, 60]
Slepton SRs								
Exclusive	SRee-m_{T2}^{100}, SR$\mu\mu$-m_{T2}^{100}		[100, 102]	[102, 105]	[105, 110]	[110, 120]	[120, 130]	[130, ∞]
Inclusive	SR$\ell\ell$-m_{T2}^{100}		[100, 102]	[100, 105]	[100, 110]	[100, 120]	[100, 130]	[100, ∞]

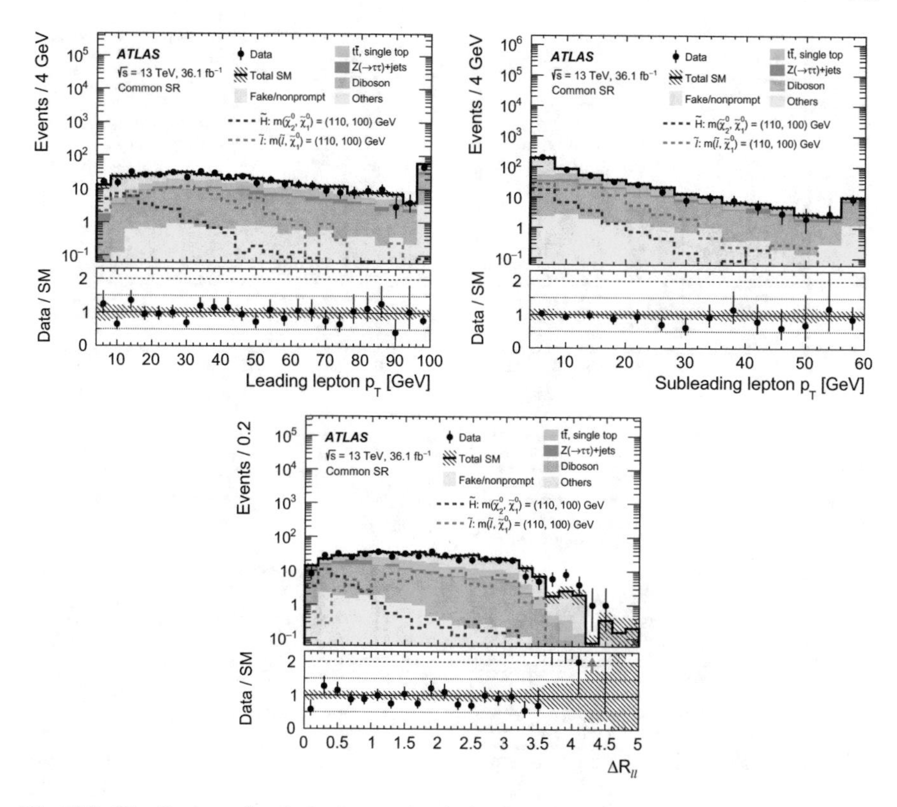

Fig. 10.9 Distributions after the background-only fit of lepton kinematic variables in the common SR in Table 10.5. The category 'Others' contains rare backgrounds from triboson, Higgs boson, and the remaining top-quark production processes. The uncertainty bands plotted include all statistical and systematic uncertainties. The dashed lines represent benchmark higgsino and slepton signal samples

$\sqrt{(\eta_{\ell 1}-\eta_{\ell 2})^2-(\phi_{\ell 1}-\phi_{\ell 2})^2}$ between the two leptons to be greater than 0.05 suppresses nearly collinear lepton pairs originating from photon conversions or muons with spurious pairs of tracks with shared hits. Distribution of lepton p_{T} and $\Delta R_{\ell\ell}$ are shown in Fig. 10.9. At low lepton p_{T}, the dominant background contribution comes from fakes.

The invariant mass of leptons, $m_{\ell\ell}$, is required to be greater than 1 GeV to remove collinear leptons. It is also required to be less than 60 GeV to remove backgrounds that have an on-shell Z boson (WZ, $ZZ \rightarrow \ell\ell\nu\nu$, Z+jets). To suppress contributions from J/ψ decays, a veto on the dilepton mass of [3.0, 3.2] GeV is required.

The reconstructed $E_{\mathrm{T}}^{\mathrm{miss}}$ is required to be greater than 200 GeVto be on plateau for the $E_{\mathrm{T}}^{\mathrm{miss}}$ triggers. For signal events to pass this $E_{\mathrm{T}}^{\mathrm{miss}}$ requirement, the two $\tilde{\chi}_1^0$ momenta must align by recoiling against hadronic initial-state radiation. In order to select this ISR topology, the event has a the leading jet (denoted by j_1) with $p_{\mathrm{T}}^{j_1} >$ 100 GeV and $\Delta\phi_{j,MET} > 2.0$, where $\Delta\phi_{j,MET}$ is the azimuthal separation between

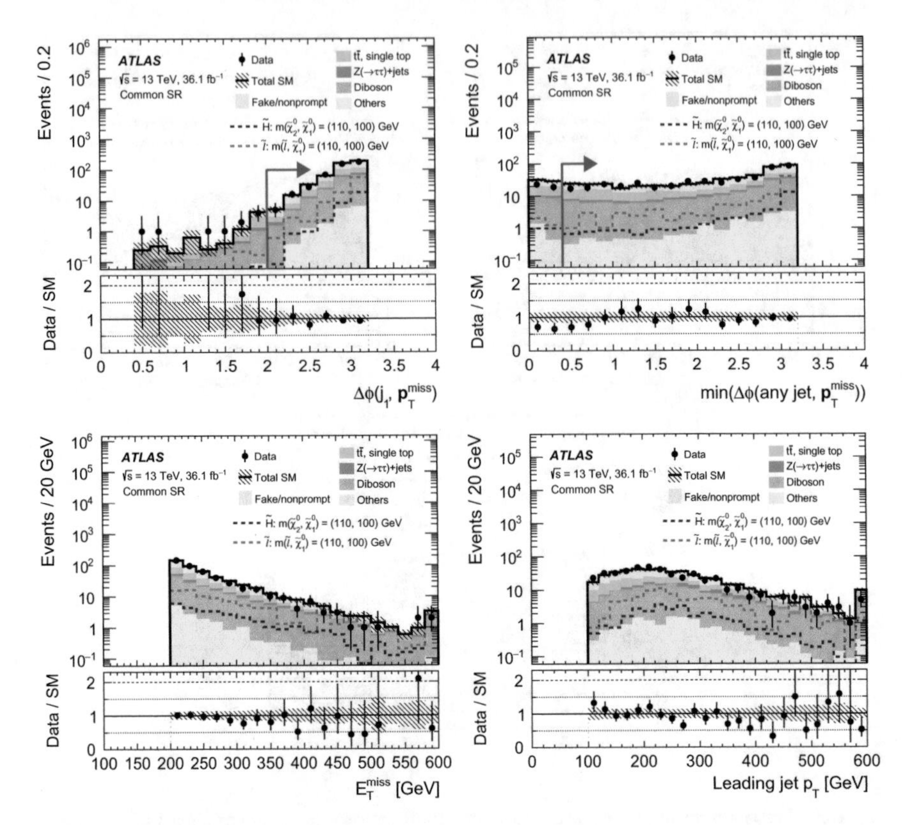

Fig. 10.10 Distributions after the background-only fit of kinematic variables used to select the ISR topology in Table 10.5. Blue arrows in the upper panel denote the final requirement used to define the common SR, otherwise all selections are applied. The category 'Others' contains rare backgrounds from triboson, Higgs boson, and the remaining top-quark production processes. The uncertainty bands plotted include all statistical and systematic uncertainties. The dashed lines represent benchmark higgsino and slepton signal samples

j_1 and $E_{\mathrm{T}}^{\mathrm{miss}}$. Also, a minimum separation of $\min(\Delta\phi(\text{any jet}, E_{\mathrm{T}}^{\mathrm{miss}})) > 0.4$ between any signal jet in the event and $E_{\mathrm{T}}^{\mathrm{miss}}$ reduces the effect of jet-energy mismeasurement on $E_{\mathrm{T}}^{\mathrm{miss}}$. Distributions of the kinematic variables are shown in Fig. 10.10.

To minimize the contribution of events with top quarks, a b-jet veto is applied.

Two ways to calculate $m_{\tau\tau}$ are considered. Both make similar assumptions but calculate $m_{\tau\tau}$ differently. Leptons that arise from the $Z/\gamma^* \to \tau\tau$ decay are energetic and collinear [23]. This assumptions allows for the reconstruction of the four resulting neutrinos. The $E_{\mathrm{T}}^{\mathrm{miss}}$ in the event is due to the four neutrinos with each neutrino's momentum collinear with the momentum of the lepton: $\mathbf{p}_{\nu_i} = \boldsymbol{\xi}_i \mathbf{p}_{\ell_i}$. The i-th tau momenta $\mathbf{p}_{\tau_i} = \mathbf{p}_{\ell_i} + \mathbf{p}_{\nu_i}$. The tau momenta can be calculated as a function of the rescaled lepton p_{T} as:

$$\mathbf{p}_{\tau_i} = (\mathbf{1} + \boldsymbol{\xi}_i)\mathbf{p}_{\ell_i} \quad (10.3)$$

To solve for the two unknown scalars ξ_i, the neutrino masses are constrained using the missing energy vector[1] [30],

$$\mathbf{p}_{\mathrm{T}}^{\mathrm{miss}} = \xi_1 \mathbf{p}_{\mathrm{T}}^{\ell_1} + \xi_2 \mathbf{p}_{\mathrm{T}}^{\ell_2}. \tag{10.4}$$

Thus, the scalars can be calculated by solving,

$$\begin{pmatrix} \xi_1 \\ \xi_2 \end{pmatrix} = \frac{1}{p_x^{\ell_1} p_y^{\ell_2} - p_x^{\ell_2} p_y^{\ell_1}} \begin{pmatrix} p_x^{\mathrm{miss}} p_y^{\ell_2} - p_x^{\ell_2} p_y^{\mathrm{miss}} \\ p_y^{\mathrm{miss}} p_x^{\ell_1} - p_x^{\mathrm{miss}} p_y^{\ell_1} \end{pmatrix}. \tag{10.5}$$

The first method calculates $m_{\tau\tau}$ to be strictly positive [23, 29],

$$m_{\tau\tau}^2 = 2(E^{\ell 1} E^{\ell 2}(1 + |\xi_1|)(1 + |\xi_2|) - \mathbf{p}^{\ell 1} \cdot \mathbf{p}^{\ell 2} (\mathbf{1} + \boldsymbol{\xi}_1)(\mathbf{1} + |\boldsymbol{\xi}_2|)) \tag{10.6}$$

The second method calculates $m_{\tau\tau}$ to be either positive or negative [24, 30].

$$m_{\tau\tau}^2 = \left(p_{\tau_1} + p_{\tau_2}\right)^2 = 2 p_{\ell_1} \cdot p_{\ell_2} (1 + \xi_1)(1 + \xi_2). \tag{10.7}$$

In this case, $m_{\tau\tau}$ being negative represents cases where the assumption of collinearity between the leptons and the $E_{\mathrm{T}}^{\mathrm{miss}}$ fails such as for WW backgrounds or leptons from SUSY processes. Thus, this definition of $m_{\tau\tau}$ is used not to identify events with taus but discriminate signal from background. The positive and negative $m_{\tau\tau}$ variables are defined as,

$$m_{\tau\tau}\left(p_{\ell_1}, p_{\ell_2}, \mathbf{p}_{\mathrm{T}}^{\mathrm{miss}}\right) = \begin{cases} \sqrt{m_{\tau\tau}^2} & m_{\tau\tau}^2 \geq 0, \\ -\sqrt{\left|m_{\tau\tau}^2\right|} & m_{\tau\tau}^2 < 0. \end{cases} \tag{10.8}$$

Using the $m_{\tau\tau}$ calculated with Eq. (10.6), the shape of the reconstructed $m_{\tau\tau}$ is compared with the mass of the truth taus, as shown in Fig. 10.11. Truth $m_{\tau\tau}$ has a peak at the Z-mass and drops sharply outside the Z-mass window while reco $m_{\tau\tau}$ has a wider distribution but still peaks at the Z-mass. This means that in general, $Z(\rightarrow \tau\tau)$+jets events come from an on-shell Z.

Thus, to reduce the tau background, a cut around the Z mass is applied to $m_{\tau\tau}$ calculated with Eq. (10.7), as seen in Fig. 10.12. Equation (10.7) is used to calculate $m_{\tau\tau}$ in the analysis even though it does not match the truth $m_{\tau\tau}$ because more background is suppressed.

The remaining cuts specify the electroweakino regions or the slepton regions. In the electroweakino signal models, the leptons come from the decay of the Z boson and, as a result, have small angular separation. By contrast, the two leptons originate from different legs in the slepton system, so the restrictions on their angular separation are weaker, as shown in Fig. 10.13. Due to the recoil of the SUSY particle system

[1] The other way one could constrain ξ is by assuming the taus recoil against jets as: $-\mathbf{p}_{\mathrm{T}}^{\mathrm{jet}} = (1 + \xi_1)\mathbf{p}_{\mathrm{T}}^{\ell_1} + (1 + \xi_2)\mathbf{p}_{\mathrm{T}}^{\ell_2}$ [24, 46].

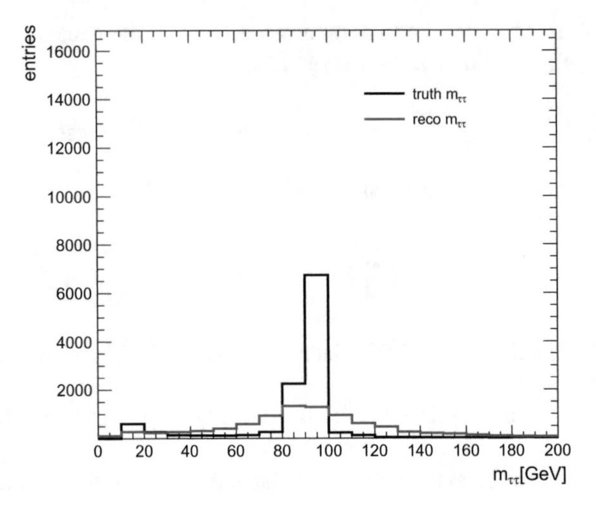

Fig. 10.11 Shapes of reco and truth m_{TT}, reco $m_{\ell\ell}$

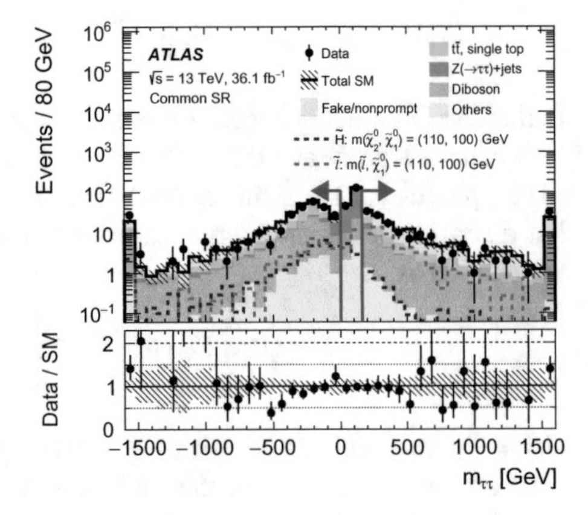

Fig. 10.12 Distributions after the background-only fit of m_{TT} in the common SR in Table 10.5. The category 'Others' contains rare backgrounds from triboson, Higgs boson, and the remaining top-quark production processes. The uncertainty bands plotted include all statistical and systematic uncertainties. The dashed lines represent benchmark higgsino and slepton signal samples

against the ISR jet, the angular separation $\Delta R_{\ell\ell}$ between the two leptons is required to be smaller than 2.0.

The transverse mass, defined in Eq. (5.8), is calculated with the leading lepton and $E_{\mathrm{T}}^{\mathrm{miss}}$ and is required to be smaller than 70 GeV to reduce the background from $t\bar{t}$, Wt, WW/WZ, and W+jets, as shown in Fig. 10.14.

The scalar sum of the lepton transverse momenta, $H_{\mathrm{T}}^{\mathrm{lep}} = p_{\mathrm{T}}^{\ell 1} + p_{\mathrm{T}}^{\ell 2}$, is smaller in the electroweakino and slepton signal regions than for SM backgrounds. The ratio $E_{\mathrm{T}}^{\mathrm{miss}}/H_{\mathrm{T}}^{\mathrm{lep}}$ discriminates between signal and background by selecting topologies where the $E_{\mathrm{T}}^{\mathrm{miss}}$ recoils against the ISR jet and not hard leptons. This variable is sensitive to both electroweakino and slepton signals, as shown in Fig. 10.15.

The minimum value of $E_{\mathrm{T}}^{\mathrm{miss}}/H_{\mathrm{T}}^{\mathrm{lep}}$ is adjusted according to the size of the mass splittings, inferred using $m_{\ell\ell}$ for electroweakinos and m_{T2}^{100} for sleptons. For electroweakinos, $E_{\mathrm{T}}^{\mathrm{miss}}/H_{\mathrm{T}}^{\mathrm{lep}}$ is required to be greater than $\max\left(5, 15 - 2\frac{m_{\ell\ell}}{1\ \mathrm{GeV}}\right)$ while for sleptons, it is greater than $\max\left(3, 15 - 2\left(\frac{m_{\mathrm{T2}}^{100}}{1\ \mathrm{GeV}} - 100\right)\right)$, as shown in Fig. 10.16.

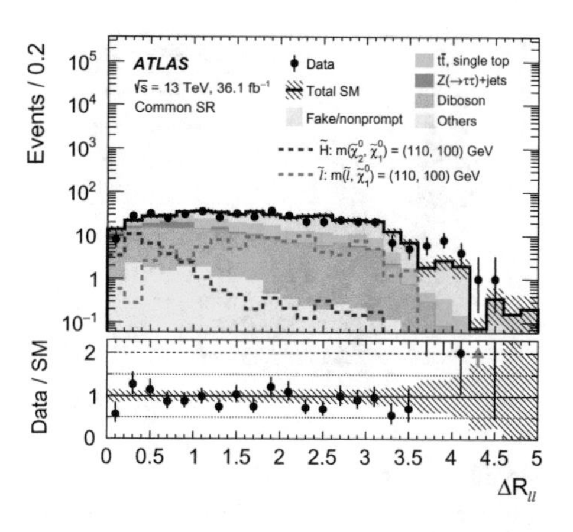

Fig. 10.13 Distributions after the background-only fit of $\Delta R_{\ell\ell}$ in the common SR in Table 10.5. The category 'Others' contains rare backgrounds from triboson, Higgs boson, and the remaining top-quark production processes. The uncertainty bands plotted include all statistical and systematic uncertainties. The dashed lines represent benchmark higgsino and slepton signal samples

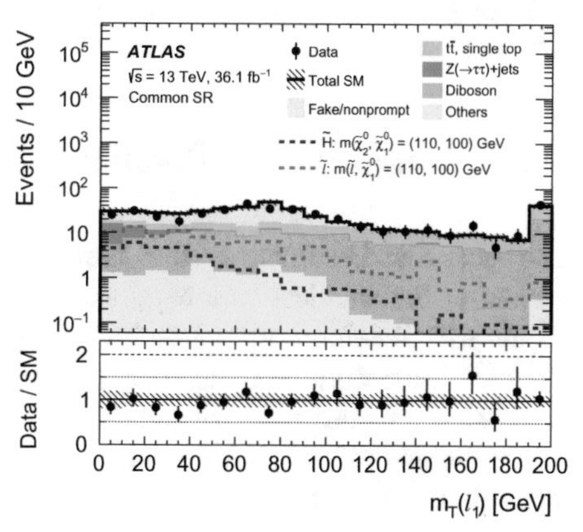

Fig. 10.14 Distributions after the background-only fit of m_{T} in the common SR in Table 10.5. The category 'Others' contains rare backgrounds from triboson, Higgs boson, and the remaining top-quark production processes. The uncertainty bands plotted include all statistical and systematic uncertainties. The dashed lines represent benchmark higgsino and slepton signal samples

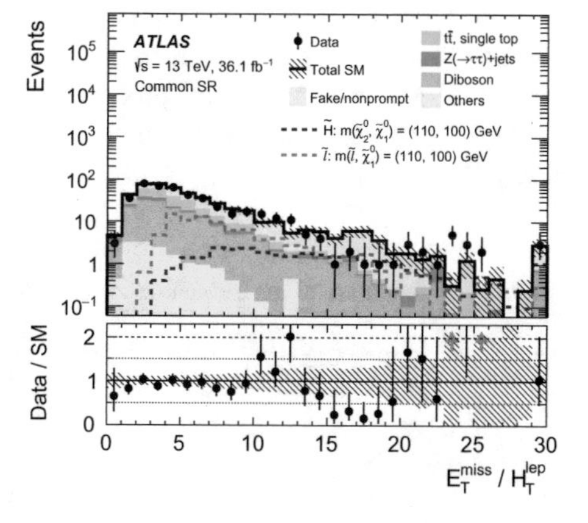

Fig. 10.15 Distributions after the background-only fit of $E_{\mathrm{T}}^{\mathrm{miss}}/H_{\mathrm{T}}^{\mathrm{lep}}$ in the common SR in Table 10.5. The category 'Others' contains rare backgrounds from triboson, Higgs boson, and the remaining top-quark production processes. The uncertainty bands plotted include all statistical and systematic uncertainties. The dashed lines represent benchmark higgsino and slepton signal samples

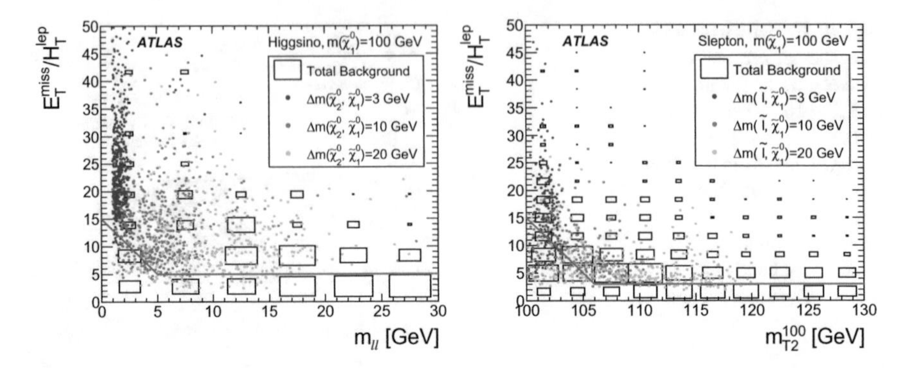

Fig. 10.16 Distributions of $E_{\mathrm{T}}^{\mathrm{miss}}/H_{\mathrm{T}}^{\mathrm{lep}}$ for the electroweakino (left) and slepton (right) SRs, after applying all region selection criteria except those on $E_{\mathrm{T}}^{\mathrm{miss}}/H_{\mathrm{T}}^{\mathrm{lep}}$, $m_{\ell\ell}$, and m_{T2}^{100}. The solid red line indicates the requirement applied in the signal region; events in the region below the red line are rejected

10.6 Background Estimation

The background estimate strategy is summarized in Table 10.7. The dominant background at the smallest lepton p_{T} and smallest bins in $m_{\ell\ell}$ and m_{T2}^{100} comes from events with fake/non-prompt leptons from $t\bar{t}$ and W+jets. This background arises from jets misidentified as leptons, photon conversions, or semileptonic decays of heavy-flavor hadrons.The $t\bar{t}$ and W+jets background is estimated using the Fake Factor method [47, 48], described in Sect. 5.6.1. Diboson processes decaying to one lepton in the final state also have a fake lepton but do not contribute as much in the signal region.

The $Z/\gamma(\rightarrow \tau\tau)$+jets and top ($t\bar{t}$, Wt) backgrounds are normalized in a simultaneous fit to the observed data counts in control regions (CRs).

VV, Higgs, VVV, and $t\bar{t}V$ processes are estimated using the simulation.

Table 10.8 summarizes the control regions and validation regions used in this analysis.

Table 10.7 Summary of the estimation methods for each background process

Background	Estimation
W+jets, $VV(1\ell)$, $t\bar{t}(1\ell)$	FF (under "Fake/non-prompt")
$t\bar{t}+Wt$	NF
$Z/\gamma(\rightarrow \tau\tau)$+jets	NF
$WW/WZ(\geq 2\ell)$	MC
Higgs/ $VVV/t\bar{t}V$	MC (under "Others")

Table 10.8 Definition of control and validation regions. The common selection criteria in Table 10.5 are applied unless otherwise specified

Region	Leptons	$E_{\mathrm{T}}^{\mathrm{miss}}/H_{\mathrm{T}}^{\mathrm{lep}}$	Additional requirements
CR-top	$e^{\pm}e^{\mp}$, $\mu^{\pm}\mu^{\mp}$, $e^{\pm}\mu^{\mp}$, $\mu^{\pm}e^{\mp}$	> 5	≥ 1 b-tagged jet(s)
CR-tau	$e^{\pm}e^{\mp}$, $\mu^{\pm}\mu^{\mp}$, $e^{\pm}\mu^{\mp}$, $\mu^{\pm}e^{\mp}$	$\in [4, 8]$	$m_{\tau\tau} \in$ [60, 120] GeV
VR-VV	$e^{\pm}e^{\mp}$, $\mu^{\pm}\mu^{\mp}$, $e^{\pm}\mu^{\mp}$, $\mu^{\pm}e^{\mp}$	< 3	
VR-SS	$e^{\pm}e^{\pm}$, $\mu^{\pm}\mu^{\pm}$, $e^{\pm}\mu^{\pm}$, $\mu^{\pm}e^{\pm}$	> 5	
VRDF-$m_{\ell\ell}$	$e^{\pm}\mu^{\mp}$, $\mu^{\pm}e^{\mp}$	$> \max\left(5, 15 - 2\frac{m_{\ell\ell}}{1\ \mathrm{GeV}}\right)$	$\Delta R_{\ell\ell} < 2$, $m_{\mathrm{T}}^{\ell_1} < 70$ GeV
VRDF-m_{T2}^{100}	$e^{\pm}\mu^{\mp}$, $\mu^{\pm}e^{\mp}$	$> \max\left(3, 15 - 2\left(\frac{m_{\mathrm{T2}}^{100}}{1\ \mathrm{GeV}} - 100\right)\right)$	

10.6.1 *Top and Tau Backgrounds*

The top and tau backgrounds are fit simultaneously using control regions.

The leptonic top background, $t\bar{t} + Wt$, enters the SR due to a $b-$jet not being identified. In order to estimate the contribution of this background, a sample enriched in top quarks is selected by requiring events with at least one $b-$jet. The common selection criteria in Table 10.5 are applied so that the CR is kinematically similar to the SR. The requirement on $E_{\mathrm{T}}^{\mathrm{miss}}/H_{\mathrm{T}}^{\mathrm{lep}}$ is relaxed to increase the amount of events in the control region, denoted as CR-top. The normalization factor derived in this region is 1.02 ± 0.09.

The tau background comes from $Z/\gamma(\rightarrow \tau\tau)$+jets events. The tau CR is constructed by requiring events to have $m_{\tau\tau}$ between 60 and 120 GeV. $E_{\mathrm{T}}^{\mathrm{miss}}/H_{\mathrm{T}}^{\mathrm{lep}}$ is required to have a value between 4 and 8 to reduce potential contamination from signal events. To increase the statistics of this region, different flavour, opposite charge lepton pairs are allowed. The normalization factor derived in this region, denoted CR-tau, is 0.72 ± 0.14.

Distributions after the performing the simultaneous fit in CR-tau and CR-top are shown in Fig. 10.17. There is good agreement between the observed data and the expected background events.

10.6.2 *Diboson Background*

It is difficult to select a sample of diboson events pure enough to be used to constrain their contribution to the SRs while remaining kinematically close to the SR. The

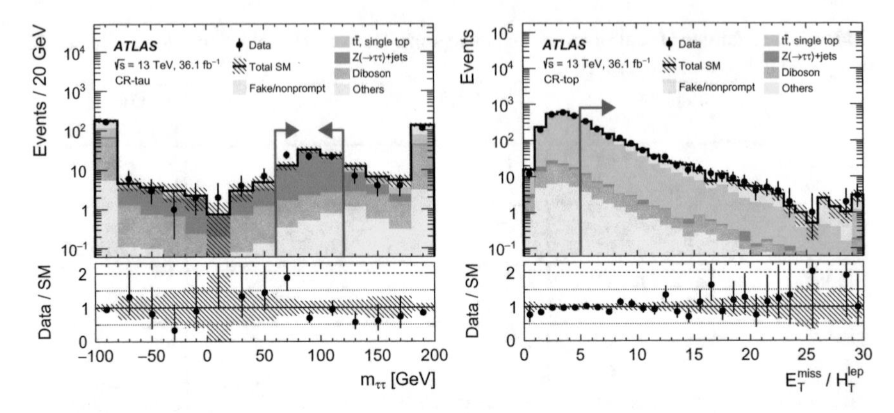

Fig. 10.17 Distributions after the background-only fit of $m_{\tau\tau}$ in CR-tau (left) and $E_{\mathrm{T}}^{\mathrm{miss}}/H_{\mathrm{T}}^{\mathrm{lep}}$ in CR-top (right). The full event selection of the corresponding regions is applied, except for the requirement that is imposed on the variable being plotted. The category 'Others' contains rare backgrounds from triboson, Higgs boson, and the remaining top-quark production processes. The uncertainty bands plotted include all statistical and systematic uncertainties

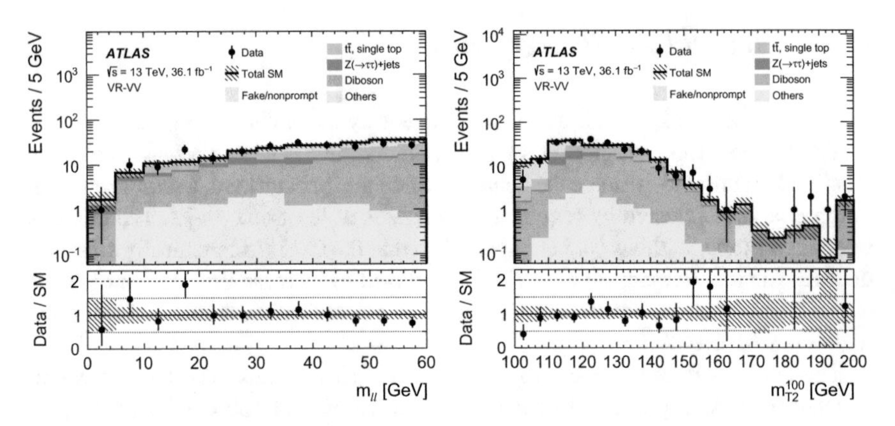

Fig. 10.18 Distributions after the background-only fit of $m_{\ell\ell}$ (left) and m_{T2}^{100} (right) in VR-VV. The full event selection of the corresponding regions is applied, except for the requirement that is imposed on the variable being plotted. The category 'Others' contains rare backgrounds from triboson, Higgs boson, and the remaining top-quark production processes. The uncertainty bands plotted include all statistical and systematic uncertainties

diboson background is therefore estimated with MC simulation. A diboson VR, denoted by VR-VV, is constructed by requiring $E_{\mathrm{T}}^{\mathrm{miss}}/H_{\mathrm{T}}^{\mathrm{lep}} < 3.0$ to be orthogonal to the electroweakino and slepton SRs. This region consists of approximately 40% diboson events, 25% fake/non-prompt lepton events, 23% top events, and smaller contributions from tau and other processes. This region is used to test the modeling of the diboson background and the associated systematic uncertainties. Figure 10.18 shows $m_{\ell\ell}$ and m_{T2}^{100} distributions in VR-VV which have good agreement between the data and the expected background.

10.6.3 Reducible Background

The primary reducible background is W+jets, where a jet fakes a lepton. This background contribution is estimated using the Fake Factor Method described in Sect. 5.6.1. In Chaps. 5, 7, and 8, the Fake Factor measurement directly targets the source of the fake background, Z+jets, by selecting a Z+jets enriched region. In this case, a region targeting directly the production of W+jets would not be pure in these events due to contamination of $t\bar{t}$ events, thus a dijet selection is used to measure fake factors. Because of composition differences between dijet events and W+jets events, composition differences need to be taken into account as a systematic.

The Fake Factor method identifies "ID" leptons whose criteria are identical to signal leptons, described in Table 10.2 and an an "anti-ID" criteria, defined in Table 10.9. The anti-ID criteria is enriched in fake leptons by inverting or relaxing identification and isolation criteria. Dijet events are selected using single lepton triggers and are required to have exactly one ID or anti-ID lepton.

The Fake Factor is the ratio of ID to anti-ID leptons and is binned in lepton p_T for electrons and muons and additionally computed for events with a $b-$jet veto and events with at least one $b-$jet for muons. The derivation of the Fake Factor in the two lepton case is described in Sect. 5.6.1. In order to properly calculate the Fake Factor, the contribution from backgrounds with two prompt leptons must be subtracted from the data.

The Fake Factor is derived in a region orthogonal to the signal selection selection by requiring the leading jet p_T to be greater than 100 GeV. Events are also required to have $m_T < 40$ GeV. The selection for the Fake Factor measurement and validation regions is summarized in Table 10.10.

Table 10.9 Definition of the anti-ID criteria for the Fake Factor measurement region

Electrons	Muons
$p_T > 4.5$ GeV	$p_T > 4$ GeV
$\lvert\eta\rvert < 2.47$	$\lvert\eta\rvert < 2.5$
$\lvert\Delta z_0 \sin\theta\rvert < 0.5$ mm	$\lvert\Delta z_0 \sin\theta\rvert < 0.5$ mm
Pass `LooseAndBLayer` identification	Pass `Medium` identification
Pass OR requirements	No OR requirements
(!`Tight` identification	($\lvert d_0$significance$\rvert > 3$
|| $\lvert d_0$significance$\rvert > 5$	|| !`FixedCutTightTrackOnly` isolation)
|| !`GradientLoose` isolation)	

Table 10.10 Summary of measurement and validation regions used for the dijet Fake Factor estimate

	m_T [GeV]	lead jet p_T (GeV)
FF measurement region	< 40	> 100

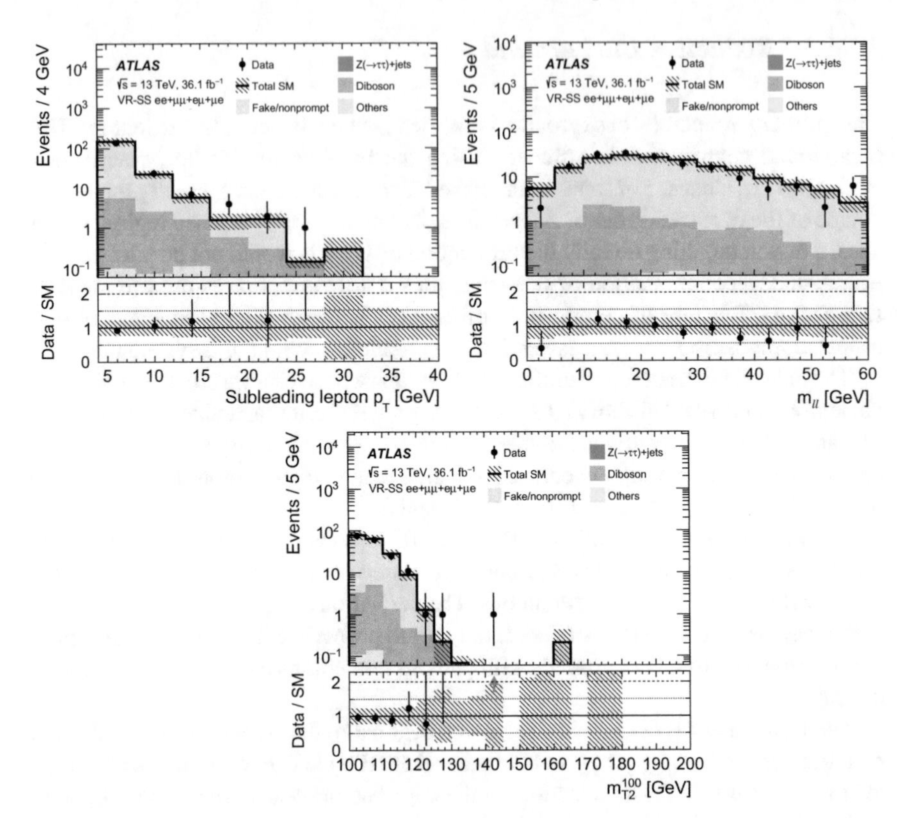

Fig. 10.19 Distributions after the background-only fit of the sub-leading lepton p_{T}, $m_{\ell\ell}$, and m_{T2}^{100} in VR-SS . The full event selection of the corresponding regions is applied, except for the requirement that is imposed on the variable being plotted. The category 'Others' contains rare backgrounds from triboson, Higgs boson, and the remaining top-quark production processes. The uncertainty bands plotted include all statistical and systematic uncertainties

To validate the fake estimate, a fake VR, VR-SS is defined with same charged leptons ($e^{\pm}e^{\pm}$, $\mu^{\pm}\mu^{\pm}$, $e^{\pm}\mu^{\pm}$, $\mu^{\pm}e^{\pm}$) because the electric charge of fake leptons originating from hadrons or photon conversions are uncorrelated to the charge of the W boson, while leptons from the electroweakino and sleptons processes have opposite charge leptons. $E_{\mathrm{T}}^{\mathrm{miss}}/H_{\mathrm{T}}^{\mathrm{lep}}$ is required to be greater than 5 GeV to validate fakes used for both electroweakino and slepton signal regions and keep enough statistics. The selection is summarized in Table 10.8. Figure 10.19 shows the good agreement between data and background in the sub-leading lepton p_{T}, $m_{\ell\ell}$, and m_{T2}^{100} distributions in this validation region.

The Fake Factors are applied in the signal by regions by requiring that events satisfy the signal region requirements defined in Table 10.5 except that one signal lepton is replaced by an anti-ID lepton. The appropriate Fake Factor derived is applied to that event. The estimate for the number of two lepton events containing at least one fake lepton is shown in Eq. (5.21).

There are several sources of uncertainties for the Fake Factor method. First is the statistical uncertainty on the Fake Factor.

Second, as the MC samples are used to subtract the diboson contribution from the data, the uncertainty associated to this subtraction must be evaluated. To do so, the prompt MC yield is scaled up and down by 20%, and the Fake Factor is recalculated. The largest difference with respect to the nominal Fake Factor is then used as the Fake Factor's uncertainty on the prompt subtraction and assigned as a symmetric uncertainty.

Third, a kinematic dependence is assigned by measuring the the Fake Factors as a function of other variables such as lepton η, $\Delta\phi$(lead jet, $E_{\mathrm{T}}^{\mathrm{miss}}$), and lead jet p_{T}. The resulting uncertainty is calculated to be 25% for both electrons and muons, driven by the variation of the fake factors as a function of η.

Finally, a closure systematic is assigned to cover kinematic and composition differences between the Fake Factor measurement region and the signal region. To do so, the root mean square of the difference between the fake estimate and the data in VR-SS, where no cut on $E_{\mathrm{T}}^{\mathrm{miss}}/H_{\mathrm{T}}^{\mathrm{lep}}$ is applied, is calculated as a function of lepton p_{T}. This uncertainty is determined to be 38% for electrons with $p_{\mathrm{T}} < 7$ GeV, 97% for muons with 7 GeV $< p_{\mathrm{T}} <$10 GeV,and 0% everywhere else. This 0% is assigned because the fake lepton estimate and the data agree within their uncertainties in VR-SS for the other p_{T} bins considered.

These systematic uncertainties are then added in quadrature to determine a total Fake Factor systematic uncertainty.

A second source of reducible background comes from background processes which can satisfy the $E_{\mathrm{T}}^{\mathrm{miss}} > 200$ GeV requirement due to the mismeasurement of the momenta of leptons or jets by the detector such as Drell-Yan dilepton production. This background was found to be negligible and its contribution is estimated using MC.

10.6.4 Different Flavor VR

Different flavor opposite charge VR are used to check the modeling of the background sources which are symmetric in $ee + \mu\mu$ and $e\mu + \mu e$ events (from Z$\rightarrow \tau\tau$, and top backgrounds), as shown in Table 10.8. VRDF-iMLL[1,60] has the same selection as SR$\ell\ell$-$m_{\ell\ell}$ [1,60] (summarized in Tables 10.5, 10.6), except for requiring different flavor leptons instead of same flavor leptons. Similarly, VRDF-iMT2[100,∞] has the same selection as SR$\ell\ell$-m_{T2}^{100}[100,∞], except for the requirement on lepton flavor. Figure 10.20 shows the good agreement between data and background in the sub-leading lepton p_{T} and $m_{\ell\ell}$ distributions in VRDF-iMLLg and Fig. 10.21 shows the good agreement between data and the expected background in the m_{T2}^{100} and $E_{\mathrm{T}}^{\mathrm{miss}}/H_{\mathrm{T}}^{\mathrm{lep}}$ distributions in VRDF-iMT2f.

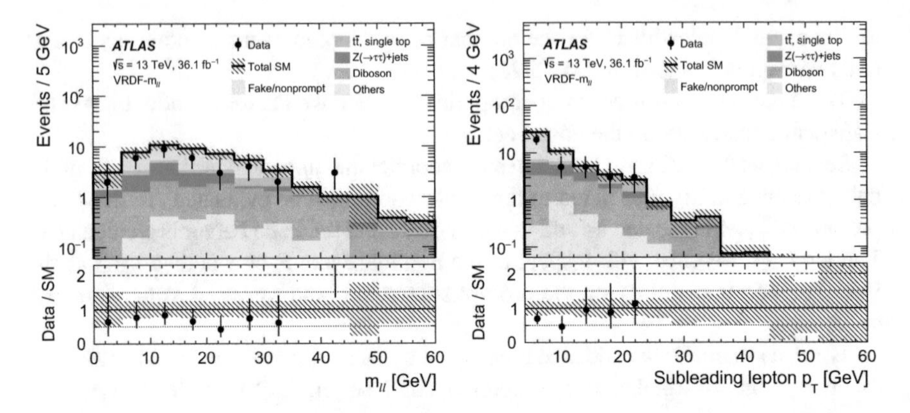

Fig. 10.20 Distributions after the background-only fit of the sub-leading lepton p_{T} and $m_{\ell\ell}$ distributions in VRDF-iMLL[1,60]. The full event selection of the corresponding regions is applied, except for the requirement that is imposed on the variable being plotted. The category 'Others' contains rare backgrounds from triboson, Higgs boson, and the remaining top-quark production processes. The uncertainty bands plotted include all statistical and systematic uncertainties

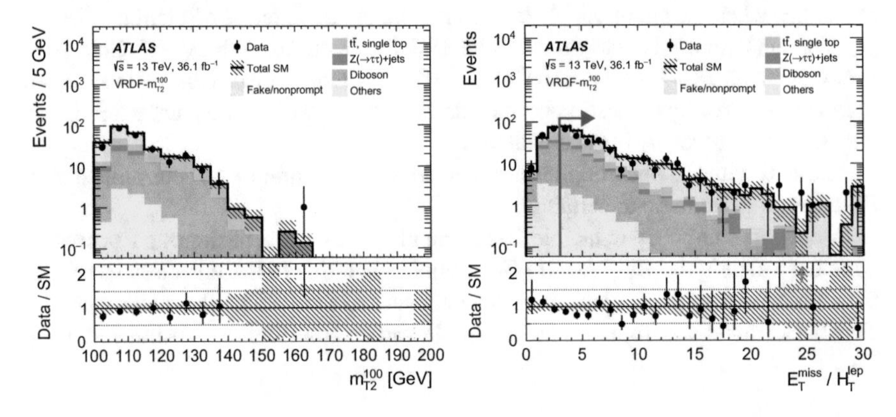

Fig. 10.21 Distributions after the background-only fit of m_{T2}^{100} and $E_{\mathrm{T}}^{\mathrm{miss}}/H_{\mathrm{T}}^{\mathrm{lep}}$ distributions in VRDF-iMT2[100,∞]. The full event selection of the corresponding regions is applied, except for the requirement that is imposed on the variable being plotted. The category 'Others' contains rare backgrounds from triboson, Higgs boson, and the remaining top-quark production processes. The uncertainty bands plotted include all statistical and systematic uncertainties

10.6.5 Summary of Background Estimation

To determine the background prediction, a simultaneous fit to data in CR-tau and CR-top is performed and applied to $Z/\gamma(\rightarrow\tau\tau)$+jets and top ($t\bar{t}$ and Wt) MC.

All validation regions with the normalization factors applied are shown in Fig. 10.22. There is good background modeling in all the VRs.

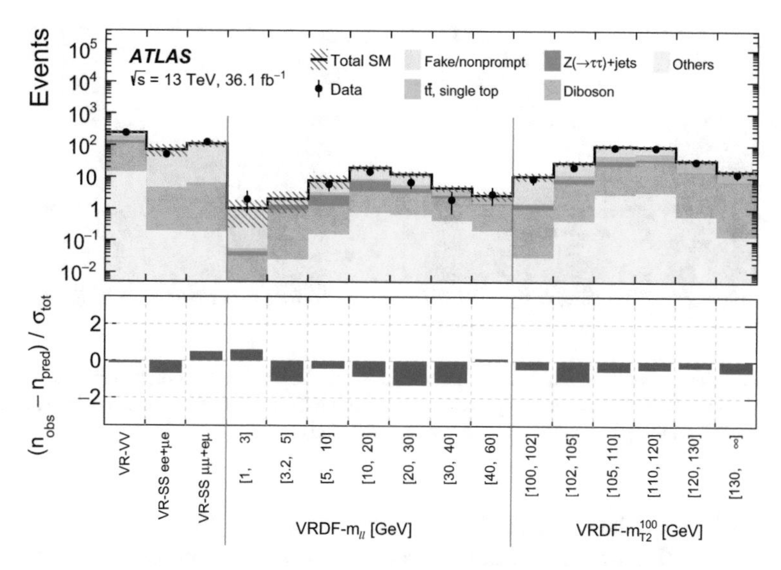

Fig. 10.22 Comparison of observed and expected event yields in the validation regions after the background-only fit. The full event selection of the corresponding regions is applied, except for the requirement that is imposed on the variable being plotted. The category 'Others' contains rare backgrounds from triboson, Higgs boson, and the remaining top-quark production processes. The uncertainty bands plotted include all statistical and systematic uncertainties

10.7 Uncertainties

There are several sources of experimental and theoretical uncertainties. The largest sources of systematic come from the uncertainty associated with the Fake Factor which were discussed in Sect. 10.6.3.

Other large sources of systematic uncertainties come from jet energy scale (JES) and resolution (JER). The JES and JER uncertainties are derived as a function of jet p_{T} and η, as well as the jet flavor composition. They are derived using data and simulation using dijet, Z+jet, and γ+jet samples [49, 50].

The systematic uncertainties related to the $E_{\mathrm{T}}^{\mathrm{miss}}$ modeling in the simulation are estimated by propagating the uncertainties in the energy or momentum scale of each of the physics objects, as well as the uncertainties in the soft term's resolution and scale [51].

Other systematics are derived on the muon (electron) momentum (energy) resolution, momentum (energy) scale, reconstruction, and isolation efficiencies. Uncertainties due to the trigger efficiency, and b-tagging efficiency were also calculated. These uncertainties were found to be negligible.

Theoretical uncertainties on the $t\bar{t}$, Wt, $Z/\gamma(\rightarrow \tau\tau)$+jets, and diboson backgrounds are the choice of PDF set, QCD renormalization (μ_R) and factorization (μ_F) scales, and the choice of the strong coupling constant (α_s). Further discussion about the calculation of these systematics can be found in Sect. 5.8.

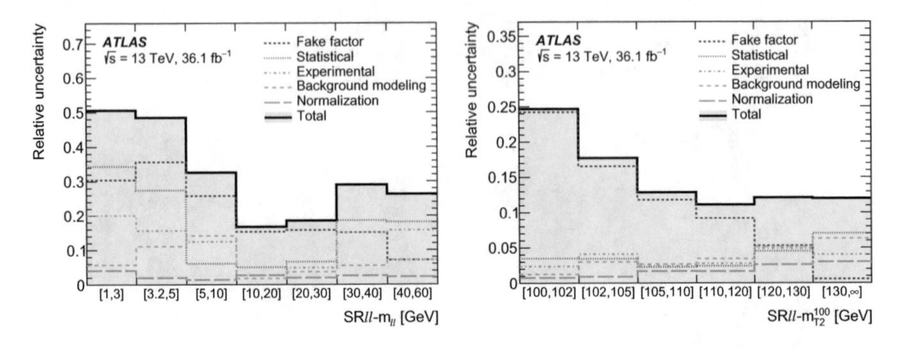

Fig. 10.23 The relative systematic uncertainties in the background prediction in the exclusive electroweakino (left) and slepton (right) SRs. The individual uncertainties can be correlated and do not necessarily add up in quadrature to the total uncertainty

A 2.1% uncertainty is applied to the integrated luminosity.

Figure 10.23 shows the size of the various uncertainties in the background predictions in the exclusive electroweakino and slepton SRs. The uncertainties related to the Fake Factor method are displayed separately from the remaining experimental uncertainties due to their relatively large contribution. The breakdown also includes the uncertainties in the normalization factors obtained from CR-tau and CR-top.

Theoretical uncertainties are also calculated for the SUSY signal models by varying by a factor of two the parameters corresponding to the renormalization, factorization, CKKW-L matching scales, and the PYTHIA tune parameters. The overall uncertainties in the signal acceptance range from about 20 to 40% and depend on the SUSY particle mass splitting and the production process. Uncertainties in the signal acceptance due to PDF uncertainties are evaluated following the PDF4LHC15 recommendations [52] and amount to 15% at most for large $\tilde{\chi}_2^0$ or $\tilde{\ell}$ masses. Uncertainties in the shape of the $m_{\ell\ell}$ or m_{T2}^{100} signal distributions due to the sources above are found to be small, and are neglected.

10.8 Results

The HistFitter package [53] is used to implement the statistical interpretation based on a profile likelihood method [54]. Systematic uncertainties are treated as nuisance parameters in the likelihood. To determine the background contribution in the SRs independent of the signal, a background-only fit is performed using CR-tau and CR-top to constrain the fit parameters.

Since no significant excess of data above the expected background is observed, two types of exclusion limits for new physics are calculated using the CLs technique: exclusion limits and discovery limits first discussed in Sects. 7.9.2–7.9.3.

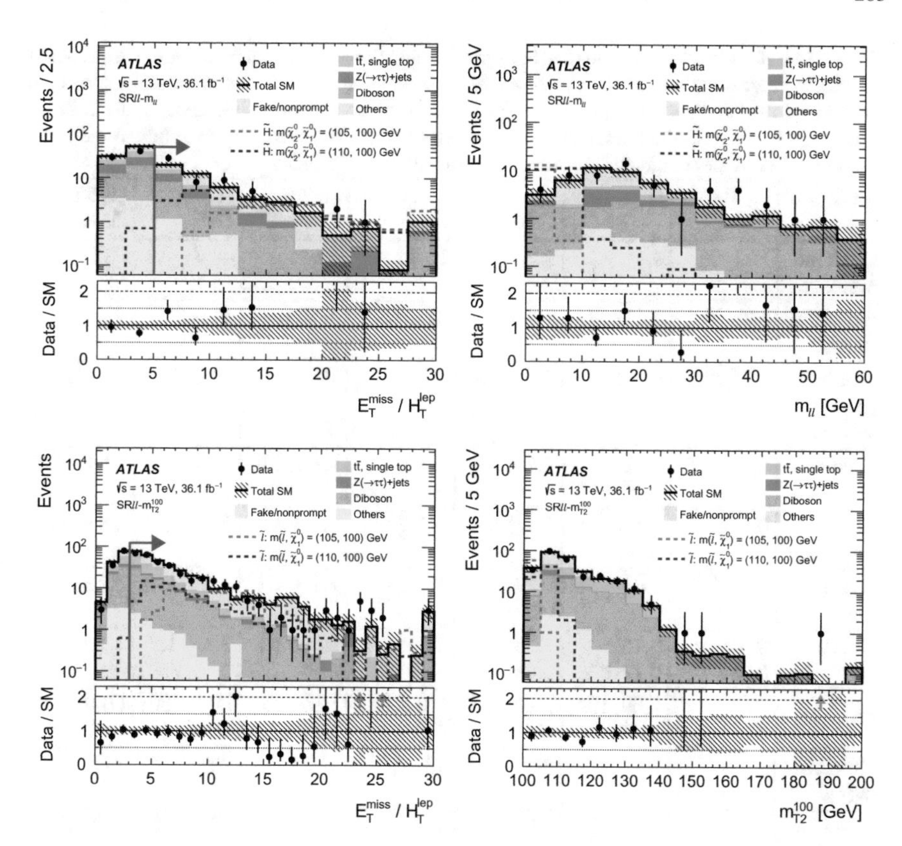

Fig. 10.24 Distributions after the background-only fit in SR$\ell\ell$-m_{T2}^{100}(top) and SR$\ell\ell$-m_{T2}^{100}(bottom). The full event selection of the corresponding regions is applied, except for the requirement that is imposed on the variable being plotted. The category 'Others' contains rare backgrounds from triboson, Higgs boson, and the remaining top-quark production processes. The uncertainty bands plotted include all statistical and systematic uncertainties

10.8.1 Background-Only Fit

Figure 10.24 shows kinematic distributions of the data and the expected backgrounds for the inclusive SRs and Table 10.11 shows the yields in the inclusive SRs. No significant excesses of the data above the expected background are observed.

10.8.2 Discovery Fit

Model independent upper limits are set on the visible cross-section $\langle\epsilon\sigma_{obs}^{95}\rangle$ as well as on the observed (S_{obs}^{95}) and expected (S_{exp}^{95}) number of events from new physics

Table 10.11 Observed event yields and background-only fit results for the inclusive electroweakino and slepton signal regions. Background processes containing fewer than two prompt leptons are categorized as 'Fake/non-prompt'. The category 'Others' contains rare backgrounds from triboson, Higgs boson, and the remaining top-quark production processes. Uncertainties in the fitted background estimates combine statistical and systematic uncertainties

SR$\ell\ell$-$m_{\ell\ell}$	[1, 3] GeV	[1, 5] GeV	[1, 10] GeV	[1, 20] GeV	[1, 30] GeV	[1, 40] GeV	[1, 60] GeV
Observed events	1	4	12	34	40	48	52
Fitted SM events	1.7 ± 0.9	3.1 ± 1.2	8.9 ± 2.5	29 ± 6	38 ± 6	41 ± 7	43 ± 7
Fake/non-prompt leptons	$0.4^{+0.6}_{-0.4}$	$0.7^{+0.9}_{-0.7}$	5.0 ± 1.9	16 ± 4	19 ± 5	19 ± 5	19 ± 5
Diboson	0.9 ± 0.5	1.7 ± 0.7	3.2 ± 1.0	6.3 ± 1.5	9.4 ± 1.9	10.7 ± 2.2	12.4 ± 2.5
$Z(\rightarrow \tau\tau)$+jets	$0.4^{+0.4}_{-0.4}$	0.5 ± 0.4	$0.0^{+0.7}_{-0.0}$	$3.2^{+3.2}_{-3.2}$	4.3 ± 3.3	4.6 ± 3.3	4.6 ± 3.3
$t\bar{t}$, single top	$0.01^{+0.10}_{-0.01}$	$0.01^{+0.10}_{-0.01}$	0.29 ± 0.20	1.8 ± 0.7	3.1 ± 1.1	4.0 ± 1.2	5.0 ± 1.4
Others	0.049 ± 0.032	0.16 ± 0.12	0.38 ± 0.25	1.4 ± 0.8	2.0 ± 1.1	2.1 ± 1.2	2.1 ± 1.2
SR$\ell\ell$-m_{T2}^{100}		[100, 102] GeV	[100, 105] GeV	[100, 110] GeV	[100, 120] GeV	[100, 130] GeV	[100, ∞] GeV
Observed events		8	34	131	215	257	277
Fitted SM events		12.4 ± 3.1	38 ± 7	129 ± 18	232 ± 29	271 ± 32	289 ± 33
Fake/non-prompt leptons		9.9 ± 3.0	26 ± 7	71 ± 16	111 ± 24	115 ± 25	114 ± 25
Diboson		1.30 ± 0.34	7.6 ± 1.6	30 ± 5	60 ± 10	79 ± 14	89 ± 15
$Z(\rightarrow \tau\tau)$+jets		0.6 ± 0.4	$0.1^{+2.2}_{-0.1}$	9 ± 4	14 ± 5	14 ± 5	14 ± 5
$t\bar{t}$, single top		0.59 ± 0.28	3.2 ± 1.0	17 ± 4	41 ± 10	56 ± 14	64 ± 15
Others		0.035 ± 0.028	0.45 ± 0.26	3.2 ± 1.7	6.9 ± 3.6	7 ± 4	8 ± 4

Table 10.12 Left to right: The first two columns present observed (N_{obs}) and expected (N_{exp}) event yields in the inclusive signal regions. The next two columns show the observed 95% CL upper limits on the visible cross-section $\left(\langle\epsilon\sigma\rangle^{95}_{\text{obs}}\right)$ and on the number of signal events $\left(S^{95}_{\text{obs}}\right)$. The fifth column $\left(S^{95}_{\text{exp}}\right)$ shows what the 95% CL upper limit on the number of signal events would be, given an observed number of events equal to the expected number (and $\pm 1\ \sigma$ deviations from the expectation) of background events. The last two columns indicate the CL_B value, i.e. the confidence level observed for the background-only hypothesis, and the discovery p-value ($p(s=0)$) which is capped at 0.5

Signal region	N_{obs}	N_{exp}	$\langle\epsilon\sigma\rangle^{95}_{\text{obs}}$ [fb]	S^{95}_{obs}	S^{95}_{exp}	CL_B	$p(s=0)$
SR$\ell\ell$-$m_{\ell\ell}$ [1, 3]	2	2.0 ± 1.0	0.16	5.6	$5.4^{+1.0}_{-1.2}$	0.55	0.44
SR$\ell\ell$-$m_{\ell\ell}$ [1, 5]	5	3.4 ± 1.4	0.22	7.8	$6.2^{+2.1}_{-1.1}$	0.76	0.25
SR$\ell\ell$-$m_{\ell\ell}$ [1, 10]	14	10.5 ± 2.7	0.35	12.6	$9.5^{+3.4}_{-1.6}$	0.80	0.21
SR$\ell\ell$-$m_{\ell\ell}$ [1, 20]	37	33 ± 6	0.60	22	18^{+8}_{-5}	0.66	0.31
SR$\ell\ell$-$m_{\ell\ell}$ [1, 30]	44	42 ± 7	0.62	22	22^{+9}_{-5}	0.56	0.37
SR$\ell\ell$-$m_{\ell\ell}$ [1, 40]	52	45 ± 7	0.74	27	21^{+9}_{-5}	0.73	0.24
SR$\ell\ell$-$m_{\ell\ell}$ [1, 60]	57	48 ± 7	0.80	29	21^{+9}_{-6}	0.80	0.19
SR$\ell\ell$-m^{100}_{T2} [100, 102]	9	13.5 ± 3.2	0.20	7.2	10^{+4}_{-3}	0.17	0.50
SR$\ell\ell$-m^{100}_{T2} [100, 105]	39	42 ± 8	0.61	22	25^{+7}_{-9}	0.40	0.50
SR$\ell\ell$-m^{100}_{T2} [100, 110]	146	146 ± 21	1.4	50	50^{+20}_{-16}	0.49	0.50
SR$\ell\ell$-m^{100}_{T2} [100, 120]	244	254 ± 32	1.8	66	74^{+24}_{-20}	0.39	0.50
SR$\ell\ell$-m^{100}_{T2} [100, 130]	298	293 ± 36	2.2	81	79^{+29}_{-22}	0.56	0.31
SR$\ell\ell$-m^{100}_{T2} [100, ∞]	324	310 ± 40	2.7	98	90^{+32}_{-23}	0.62	0.27

processes. The p-value and the corresponding significance for the background-only hypothesis is evaluated. Table 10.12 summarizes these results for the inclusive electroweakino and slepton signal regions.

Table 10.13 Observed event yields and exclusion fit results with the signal strength parameter set to zero for the exclusive electroweakino and slepton signal regions. Background processes containing fewer than two prompt leptons are categorized as 'Fake/non-prompt'. The category 'Others' contains rare backgrounds from triboson, Higgs boson, and the remaining top-quark production processes. Uncertainties in the fitted background estimates combine statistical and systematic uncertainties

SRee-$m_{\ell\ell}$	[1, 3] GeV	[3.2, 5] GeV	[5, 10] GeV	[10, 20] GeV	[20, 30] GeV	[30, 40] GeV	[40, 60] GeV
Observed events	0	1	1	10	4	6	2
Fitted SM events	$0.01^{+0.11}_{-0.01}$	$0.6^{+0.7}_{-0.6}$	2.4 ± 1.0	8.3 ± 1.6	4.0 ± 1.0	2.4 ± 0.6	1.4 ± 0.5
Fake/non-prompt leptons	$0.00^{+0.08}_{-0.00}$	$0.02^{+0.12}_{-0.02}$	1.4 ± 0.9	4.0 ± 1.5	1.6 ± 0.9	0.7 ± 0.6	$0.02^{+0.11}_{-0.02}$
Diboson	$0.007^{+0.014}_{-0.007}$	$0.28^{+0.29}_{-0.28}$	0.51 ± 0.28	1.9 ± 0.6	1.36 ± 0.31	0.72 ± 0.22	0.80 ± 0.28
$Z(\rightarrow \tau\tau)$+jets	$0.000^{+0.007}_{-0.000}$	$0.3^{+0.8}_{-0.3}$	$0.3^{+0.5}_{-0.3}$	1.7 ± 0.7	$0.25^{+0.26}_{-0.25}$	0.20 ± 0.18	$0.04^{+0.28}_{-0.04}$
$t\bar{t}$, single top	$0.00^{+0.08}_{-0.00}$	$0.02^{+0.12}_{-0.02}$	$0.11^{+0.14}_{-0.11}$	0.44 ± 0.29	0.63 ± 0.35	0.7 ± 0.4	0.6 ± 0.4
Others	$0.002^{+0.015}_{-0.002}$	$0.012^{+0.013}_{-0.012}$	0.12 ± 0.11	0.25 ± 0.16	0.21 ± 0.12	$0.05^{+0.06}_{-0.05}$	$0.0018^{+0.0033}_{-0.0018}$
SR$\mu\mu$-$m_{\ell\ell}$	[1, 3] GeV	[3.2, 5] GeV	[5, 10] GeV	[10, 20] GeV	[20, 30] GeV	[30, 40] GeV	[40, 60] GeV
Observed events	1	2	7	12	2	2	2
Fitted SM events	1.1 ± 0.6	1.3 ± 0.6	4.9 ± 1.3	13.1 ± 2.2	4.2 ± 1.0	1.4 ± 0.6	1.6 ± 0.6
Fake/non-prompt leptons	$0.00^{+0.33}_{-0.00}$	$0.4^{+0.5}_{-0.4}$	3.0 ± 1.3	7.3 ± 2.1	$0.4^{+0.8}_{-0.4}$	$0.03^{+0.19}_{-0.03}$	$0.0^{+0.5}_{-0.0}$
Diboson	0.9 ± 0.5	0.7 ± 0.4	1.3 ± 0.6	1.4 ± 0.5	1.9 ± 0.4	0.9 ± 0.5	0.97 ± 0.28
$Z(\rightarrow \tau\tau)$+jets	$0.18^{+0.25}_{-0.18}$	0.13 ± 0.12	$0.3^{+0.5}_{-0.3}$	2.4 ± 0.8	0.7 ± 0.4	$0.001^{+0.011}_{-0.001}$	$0.05^{+0.06}_{-0.05}$
$t\bar{t}$, single top	$0.01^{+0.10}_{-0.01}$	$0.02^{+0.12}_{-0.02}$	0.19 ± 0.13	1.4 ± 0.6	0.8 ± 0.4	0.37 ± 0.21	0.51 ± 0.33
Others	0.047 ± 0.030	$0.07^{+0.09}_{-0.07}$	0.13 ± 0.12	0.7 ± 0.5	0.35 ± 0.20	0.09 ± 0.07	0.020 ± 0.020
SRee-m_{T2}^{100}		[100, 102] GeV	[102, 105] GeV	[105, 110] GeV	[110, 120] GeV	[120, 130] GeV	[130, ∞] GeV
Observed events		3	10	37	42	10	7
Fitted SM events		3.5 ± 1.2	11.0 ± 2.0	33 ± 4	42 ± 4	15.7 ± 2.0	7.5 ± 1.1

(continued)

Table 10.13 (continued)

SRee-$m_{\ell\ell}$	[1, 3] GeV	[3.2, 5] GeV	[5, 10] GeV	[10, 20] GeV	[20, 30] GeV	[30, 40] GeV	[40, 60] GeV
Fake/non-prompt leptons		2.9 ± 1.2	6.8 ± 2.0	13 ± 4	14 ± 4	1.9 ± 1.2	$0.01^{+0.10}_{-0.01}$
Diboson		0.33 ± 0.12	2.3 ± 0.6	8.5 ± 1.6	12.7 ± 2.4	7.4 ± 1.4	4.3 ± 0.9
$Z(\rightarrow \tau\tau)$+jets		$0.13^{+0.23}_{-0.13}$	0.6 ± 0.4	4.1 ± 1.8	2.9 ± 1.0	$0.00^{+0.08}_{-0.00}$	$0.00^{+0.20}_{-0.00}$
$t\bar{t}$, single top		0.08 ± 0.08	1.2 ± 0.5	6.5 ± 1.6	10.7 ± 2.4	6.3 ± 1.4	3.2 ± 0.9
Others		$0.011^{+0.012}_{-0.011}$	0.17 ± 0.11	0.8 ± 0.4	1.3 ± 0.7	0.14 ± 0.09	0.06 ± 0.04
SR$\mu\mu$-m_{T2}^{100}		[100, 102] GeV	[102, 105] GeV	[105, 110] GeV	[110, 120] GeV	[120, 130] GeV	[130, ∞] GeV
Observed events		5	16	60	42	32	13
Fitted SM events		6.8 ± 1.5	15.0 ± 2.1	57 ± 5	53 ± 4	24.9 ± 2.9	11.0 ± 1.4
Fake/non-prompt leptons		5.1 ± 1.5	8.2 ± 2.1	26 ± 5	18 ± 4	1.2 ± 0.8	$0.02^{+0.17}_{-0.02}$
Diboson		0.89 ± 0.22	4.1 ± 0.9	14.3 ± 2.2	18.0 ± 2.7	12.9 ± 2.2	5.9 ± 1.1
$Z(\rightarrow \tau\tau)$+jets		0.31 ± 0.23	$1.0^{+1.3}_{-1.0}$	6.6 ± 1.7	$1.6^{+1.8}_{-1.6}$	$0.03^{+0.25}_{-0.03}$	$0.02^{+0.24}_{-0.02}$
$t\bar{t}$, single top		0.43 ± 0.22	1.4 ± 0.5	8.3 ± 2.2	12.4 ± 2.9	10.5 ± 2.6	5.0 ± 1.3
Others		$0.020^{+0.024}_{-0.020}$	0.24 ± 0.15	1.8 ± 1.0	2.4 ± 1.3	0.35 ± 0.23	0.11 ± 0.07

10.8.3 Exclusion Fit

The results are interpreted as constraints on the SUSY models shown in Fig. 10.1 using the exclusive electroweakino and slepton SRs. The background-only fit now allows for a signal model with a corresponding signal strength parameter in a simultaneous fit of all CRs and relevant SRs, referred to as the exclusion fit. When an electroweakino signal is assumed, the 14 exclusive SRee-$m_{\ell\ell}$ and SR$\mu\mu$-$m_{\ell\ell}$ regions binned in $m_{\ell\ell}$ are considered. By statistically combining these SRs, the signal shape of the $m_{\ell\ell}$ spectrum can be exploited to improve the sensitivity. When a slepton signal is assumed, the 12 exclusive SRee-m_{T2}^{100} and SR$\mu\mu$-m_{T2}^{100} regions binned in m_{T2}^{100} are used for the fit.

Table 10.13 summarizes the fitted and observed event yields in the exclusive electroweakino and slepton SRs using an exclusion fit configuration where the signal strength parameter is fixed to zero. Figure 10.25 illustrates the compatibility of the fitted and observed event yields in these regions. No significant differences between the fitted background and the observed event yields are found in the exclusive SRs.

The exclusion limits are projected into the next-to-lightest neutralino mass $\Delta m(\tilde{\chi}_2^0, \tilde{\chi}_1^0)$ versus $m(\tilde{\chi}_2^0)$ plane, where $\tilde{\chi}_2^0$ are excluded up to masses of $\sim$130 GeV for $\Delta m(\tilde{\chi}_2^0, \tilde{\chi}_1^0)$ between 5 GeV and 10 GeV, and down to $\Delta m(\tilde{\chi}_2^0, \tilde{\chi}_1^0) \sim 3$ GeV for $m(\tilde{\chi}_2^0) \sim 100$ GeV, as shown in Fig. 10.26.

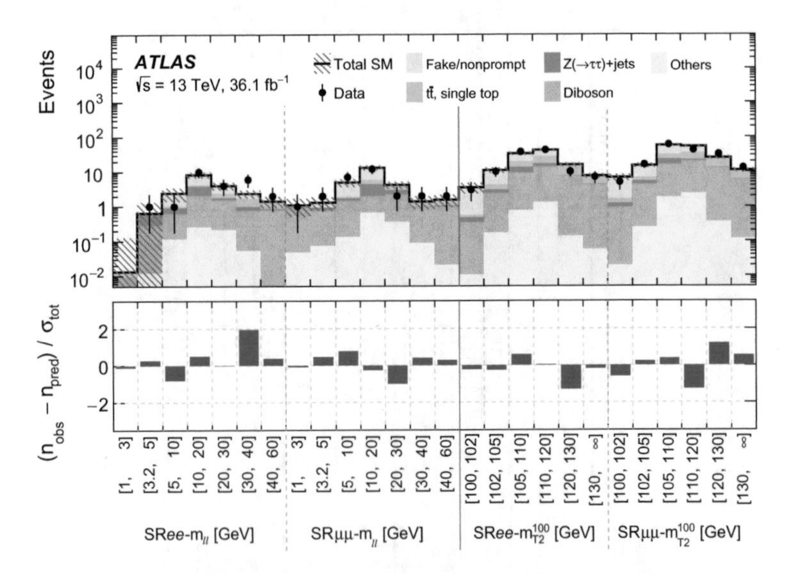

Fig. 10.25 Comparison of observed and expected event yields after the exclusion fit with the signal strength parameter set to zero in the exclusive signal regions. The full event selection of the corresponding regions is applied, except for the requirement that is imposed on the variable being plotted. The category 'Others' contains rare backgrounds from triboson, Higgs boson, and the remaining top-quark production processes. The uncertainty bands plotted include all statistical and systematic uncertainties

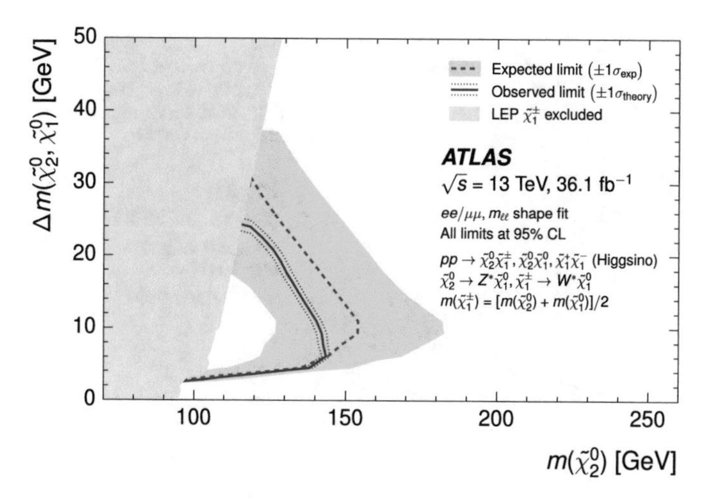

Fig. 10.26 Expected 95% CL exclusion sensitivity (blue dashed line) with $\pm 1\sigma_{\text{exp}}$ (yellow band) from experimental systematic uncertainties and observed limits (red solid line) with $\pm 1\sigma_{\text{theory}}$ (dotted red line) from signal cross-section uncertainties for simplified models of direct Higgsino production. A fit of signals to the $m_{\ell\ell}$ spectrum is used to derive the limit, projected onto the $\Delta m(\tilde{\chi}_2^0, \tilde{\chi}_1^0)$ vs. $m(\tilde{\chi}_2^0)$ plane. The chargino $\tilde{\chi}_1^{\pm}$ mass is assumed to be halfway between the two lightest neutralino masses. The gray regions denote the lower chargino mass limit from LEP [5]

Figure 10.27 shows the 95% CL limits on the slepton simplified model, based on an exclusion fit that exploits the shape of the m_{T2}^{100} spectrum using the exclusive slepton SRs. Here, $\tilde{\ell}$ with masses of up to ~180 GeV are excluded for $\Delta m(\tilde{\ell}, \tilde{\chi}_1^0) \sim 5$ GeV, and down to mass splittings $\Delta m(\tilde{\ell}, \tilde{\chi}_1^0)$ of approximately 1 GeV for $m(\tilde{\ell}) \sim 70$ GeV. A fourfold degeneracy is assumed in selectron and smuon masses.

10.9 Wino-Bino Reinterpretation

The electroweakino search assumes that the SUSY particles are Higgsinos; however, as shown in Chap. 7, the diagram in Fig. 10.1 can have a wino-bino interpretation. The Higgsino cross-section must be scaled by a factor of 4, as shown in Fig. 6.1. The main difference comes from the $m_{\ell\ell}$ shape, as shown in Fig. 10.28. The $m_{\ell\ell}$ spectrum can be calculated using the following equation:

$$\frac{\mathrm{d}\Gamma}{2m\mathrm{d}m} \propto \frac{\sqrt{m^4 - m^2(m_{\tilde{\chi}_1^0}^2 + m_{\tilde{\chi}_2^0}^2) + (m_{\tilde{\chi}_2^0}^2 - m_{\tilde{\chi}_1^0}^2)^2}}{(m^2 - m_Z^2)^2}\left[-2m^4 + m^2(m_{\tilde{\chi}_1^0}^2 \pm 6m_{\tilde{\chi}_1^0}m_{\tilde{\chi}_2^0} + m_{\tilde{\chi}_2^0}^2) + (m_{\tilde{\chi}_1^0}^2 - m_{\tilde{\chi}_2^0}^2)^2\right], \tag{10.9}$$

where the ± depends on the assumption of the mixture of the eigenstates: + for wino-bino and − for higgsino. There is good agreement between the simulation

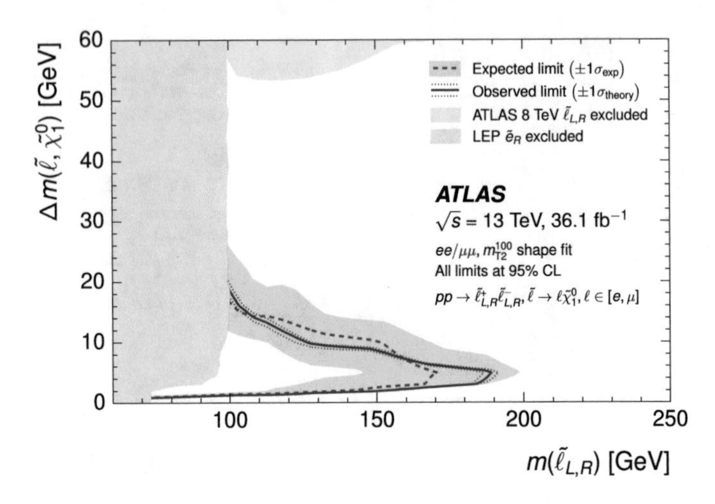

Fig. 10.27 Expected 95% CL exclusion sensitivity (blue dashed line) with $\pm 1\sigma_{\text{exp}}$ (yellow band) from experimental systematic uncertainties and observed limits (red solid line) with $\pm 1\sigma_{\text{theory}}$ (dotted red line) from signal cross-section uncertainties for simplified models of direct slepton production. The gray region is the $\widetilde{e}_R$ limit from LEP [5, 9], while the blue region is the fourfold mass degenerate slepton limit from ATLAS Run 1 [16]

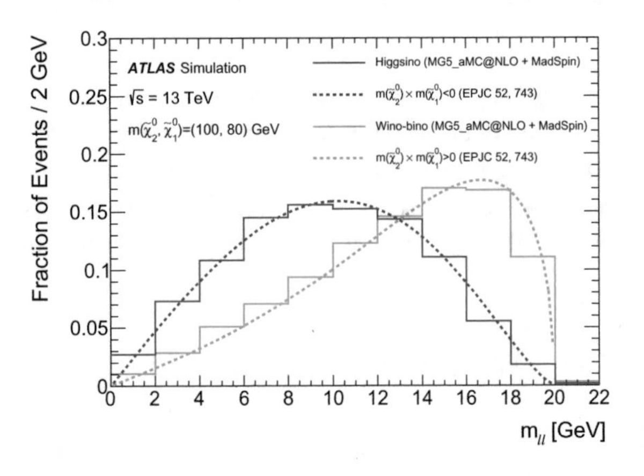

Fig. 10.28 $m_{\ell\ell}$ for direct Higgsino (blue) and direct wino (red) productions. Simulation (red) is compared with the theoretical calculation (dashed)

and the theoretical calculation for both wino-bino and higgsino models, as shown in Fig. 10.28.

Another difference between the two models is the assumption on the mass of the chargino. The higgsino model assumes that $m(\widetilde{\chi}^0_2)$ is halfway between $m(\widetilde{\chi}^{\pm}_1)$ and $m(\tilde{\chi}^0_1)$ while in wino-bino models, the assumption is that $m(\widetilde{\chi}^0_2) = m(\widetilde{\chi}^{\pm}_1)$.

This difference in assumption causes the acceptances to be different between reweighting Higgsino samples (correcting for cross-section normalized and $m_{\ell\ell}$ shape) and wino-sample simulated samples, as shown in Fig. 10.29. As a result, to reinterpret the Higgsino SR using a wino-bino interpretation, the direct wino production was simulated instead of using reweighted Higgsino samples.

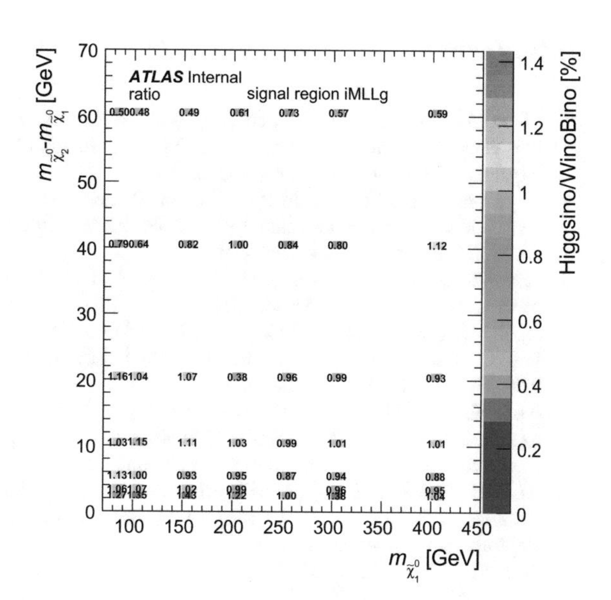

Fig. 10.29 Ratio of wino-bino acceptances to $m_{\ell\ell}$ reweighted Higgsino acceptances

The 95% CL limits of the wino–bino simplified model are shown in Fig. 10.30 (bottom), where $\widetilde{\chi}_2^0$ neutralino is excluded up to masses of $\sim$170 GeV for $\Delta m(\widetilde{\chi}_2^0, \widetilde{\chi}_1^0)$ $\sim$ 10 GeV, and down $\Delta m(\widetilde{\chi}_2^0, \widetilde{\chi}_1^0) \sim 2.5$ GeV for $m(\widetilde{\chi}_2^0) \sim 100$ GeV.

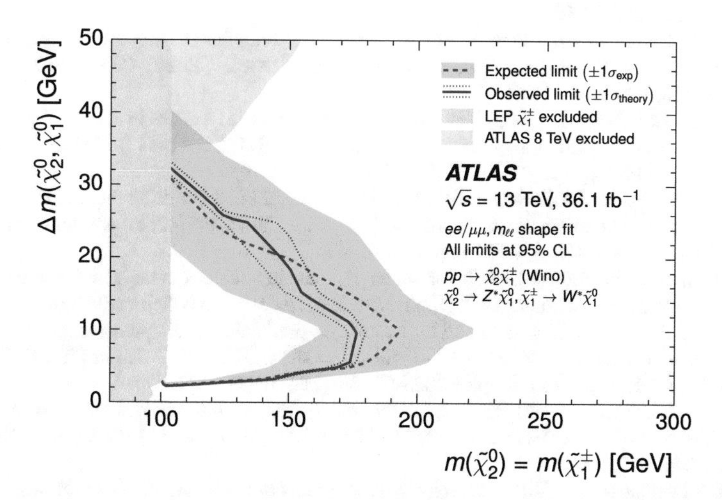

Fig. 10.30 Expected 95% CL exclusion sensitivity (blue dashed line) with $\pm 1\sigma_{\text{exp}}$ (yellow band) from experimental systematic uncertainties and observed limits (red solid line) with $\pm 1\sigma_{\text{theory}}$ (dotted red line) from signal cross-section uncertainties for simplified models of direct wino production. A fit of signals to the $m_{\ell\ell}$ spectrum is used to derive the limit, which is projected into the $\Delta m(\widetilde{\chi}_2^0, \widetilde{\chi}_1^0)$ vs. $m(\widetilde{\chi}_2^0)$ plane, $m(\widetilde{\chi}_2^0) = m(\widetilde{\chi}_1^\pm)$ is assumed. The gray regions denote the lower chargino mass limit from LEP [5]. The blue region in the lower plot indicates the limit from the $2\ell + 3\ell$ combination of ATLAS Run 1 [16, 17]

References

1. Muon g-2 Collaboration, Bennett GW, et al (2006) Final report of the Muon E821 anomalous magnetic moment measurement at BNL. Phys Rev D 73:072003. https://doi.org/10.1103/PhysRevD.73.072003, arXiv:hep-ex/0602035
2. Hagiwara K, Liao R, Martin AD, Nomura D, Teubner T (2011) $(g-2)_\mu$ and $\alpha(M_Z^2)$ re-evaluated using new precise data. J Phys G 38:085003. https://doi.org/10.1088/0954-3899/38/8/085003, arXiv:1105.3149 [hep-ph]
3. Endo M, Hamaguchi K, Iwamoto S, Yoshinaga T (2014) Muon g-2 vs LHC in Supersymmetric Models. JHEP 01:123. https://doi.org/10.1007/JHEP01(2014)123, arXiv:1303.4256 [hep-ph]
4. Ajaib MA, Dutta B, Ghosh T, Gogoladze I, Shafi Q (2015) Neutralinos and sleptons at the LHC in light of muon $(g-2)_\mu$. Phys Rev D 92(7):075033. https://doi.org/10.1103/PhysRevD.92.075033, arXiv:1505.05896 [hep-ph]
5. ALEPH, DELPHI, L3, OPAL Experiments (2002) Combined LEP Chargino Results, up to 208 GeV for low DM, LEPSUSYWG/02-04.1. http://lepsusy.web.cern.ch/lepsusy/www/inoslowdmsummer02/charginolowdm_pub.html
6. ALEPH, DELPHI, L3, OPAL Experiments (2004) Combined LEP Selectron/Smuon/Stau Results, 183–208 GeV, Lepsusywg/04-01.1. http://lepsusy.web.cern.ch/lepsusy/www/sleptons_summer04/slep_final.html
7. ALEPH Collaboration (2002) Search for scalar leptons in e^+e^- collisions at center-of-mass energies up to 209 GeV. Phys Lett B 526:206–220. https://doi.org/10.1016/S0370-2693(01)01494-0, arXiv:hep-ex/0112011
8. ALEPH Collaboration (2002) Search for charginos nearly mass degenerate with the lightest neutralino in e^+e^- collisions at center-of-mass energies up to 209 GeV. Phys Lett B 533:223–236. https://doi.org/10.1016/S0370-2693(02)01584-8, arXiv:hep-ex/0203020
9. ALEPH Collaboration (2002) Absolute lower limits on the masses of selectrons and sneutrinos in the MSSM. Phys Lett B 544:73–88. https://doi.org/10.1016/S0370-2693(02)02471-1, arXiv:hep-ex/0207056
10. ALEPH Collaboration (2004) Absolute mass lower limit for the lightest neutralino of the MSSM from e^+e^- data at $\sqrt{s}$ up to 209 GeV. Phys Lett B 583:247–263.https://doi.org/10.1016/j.physletb.2003.12.066
11. DELPHI Collaboration (2003) Searches for supersymmetric particles in e^+e^- collisions up to 208 GeV and interpretation of the results within the MSSM. Eur Phys J C 31:421–479. https://doi.org/10.1140/epjc/s2003-01355-5, arXiv:hep-ex/0311019
12. L3 Collaboration (2000) Search for charginos with a small mass difference with the lightest supersymmetric particle at $\sqrt{s} = 189$ GeV. Phys Lett B 482:31–42. https://doi.org/10.1016/S0370-2693(00)00488-3, arXiv:hep-ex/0002043
13. L3 Collaboration (2004) Search for scalar leptons and scalar quarks at LEP. Phys Lett B 580:37–49. https://doi.org/10.1016/j.physletb.2003.10.010, arXiv:hep-ex/0310007
14. OPAL Collaboration (2004) Search for anomalous production of dilepton events with missing transverse momentum in e^+e^- collisions at $\sqrt{s} = 183$ GeV to 209 GeV. Eur Phys J C 32:453–473. https://doi.org/10.1140/epjc/s2003-01466-y, arXiv:hep-ex/0309014
15. OPAL Collaboration (2003) Search for nearly mass degenerate charginos and neutralinos at LEP. Eur Phys J C 29:479–489. https://doi.org/10.1140/epjc/s2003-01237-x, arXiv:hep-ex/0210043
16. ATLAS Collaboration (2014) Search for direct production of charginos, neutralinos and sleptons in final states with two leptons and missing transverse momentum in pp collisions at $\sqrt{s} = 8$ TeV with the ATLAS detector. JHEP 05:071. https://doi.org/10.1007/JHEP05(2014)071, arXiv:1403.5294 [hep-ex]
17. ATLAS Collaboration (2014) Search for direct production of charginos and neutralinos in events with three leptons and missing transverse momentum in $\sqrt{s} = 8$ TeV pp collisions with the ATLAS detector. JHEP 04:169. https://doi.org/10.1007/JHEP04(2014)169, arXiv:1402.7029 [hep-ex]

18. ATLAS Collaboration (2016) Search for the electroweak production of supersymmetric particles in $\sqrt{s} = 8$ TeV pp collisions with the ATLAS detector. Phys Rev D 93:052002. https://doi.org/10.1103/PhysRevD.93.052002, arXiv:1509.07152 [hep-ex]
19. ATLAS Collaboration (2016) Dark matter interpretations of ATLAS searches for the electroweak production of supersymmetric particles in $\sqrt{s} = 8$ TeV proton–proton collisions. JHEP 09:175. https://doi.org/10.1007/JHEP09(2016)175, arXiv:1608.00872 [hep-ex]
20. CMS Collaboration (2014) Searches for electroweak production of charginos, neutralinos, and sleptons decaying to leptons and W, Z, and Higgs bosons in pp collisions at 8 TeV. Eur Phys J C 74:3036. https://doi.org/10.1140/epjc/s10052-014-3036-7, arXiv:1405.7570 [hep-ex]
21. CMS Collaboration (2016) Search for supersymmetry in events with soft leptons, low jet multiplicity, and missing transverse energy in proton–proton collisions at $\sqrt{s} = 8$ TeV. Phys Lett B 759:9. https://doi.org/10.1016/j.physletb.2016.05.033, arXiv:1512.08002 [hep-ex]
22. CMS Collaboration, Search for electroweak production of charginos and neutralinos in multilepton final states in proton–proton collisions at $\sqrt{s} = 13$ TeV. arXiv:1709.05406 [hep-ex]
23. Han Z, Kribs GD, Martin A, Menon A (2014) Hunting quasidegenerate Higgsinos. Phys Rev D 89(7):075007. https://doi.org/10.1103/PhysRevD.89.075007, arXiv:1401.1235 [hep-ph]
24. Baer H, Mustafayev A, Tata X (2014) Monojet plus soft dilepton signal from light higgsino pair production at LHC14. Phys Rev D 90(11):115007. https://doi.org/10.1103/PhysRevD.90.115007, arXiv:1409.7058 [hep-ph]
25. Giudice GF, Han T, Wang K, Wang L-T (2010) Nearly degenerate gauginos and dark matter at the LHC. Phys Rev D 81:115011. https://doi.org/10.1103/PhysRevD.81.115011, arXiv:1004.4902 [hep-ph]
26. Schwaller P, Zurita J (2014) Compressed electroweakino spectra at the LHC. JHEP 03:060. https://doi.org/10.1007/JHEP03(2014)060, arXiv:1312.7350 [hep-ph]
27. Gori S, Jung S, Wang L-T (2013) Cornering electroweakinos at the LHC. JHEP 10:191. https://doi.org/10.1007/JHEP10(2013)191, arXiv:1307.5952 [hep-ph]
28. Dutta B, Ghosh T, Gurrola A, Johns W, Kamon T, Sheldon P, Sinha K, Wang K, Wu S (2015) Probing compressed sleptons at the LHC using vector boson fusion processes. Phys Rev D 91(5):055025. https://doi.org/10.1103/PhysRevD.91.055025, arXiv:1411.6043 [hep-ph]
29. Han Z, Liu Y (2015) MT2 to the Rescue—searching for Sleptons in compressed spectra at the LHC. Phys Rev D 92(1):015010. https://doi.org/10.1103/PhysRevD.92.015010, arXiv:1412.0618 [hep-ph]
30. Barr A, Scoville J (2015) A boost for the EW SUSY hunt: monojet-like search for compressed sleptons at LHC14 with 100 fb $^{-1}$. JHEP 04:147. https://doi.org/10.1007/JHEP04(2015)147, arXiv:1501.02511 [hep-ph]
31. ATLAS Collaboration (2018) Search for electroweak production of supersymmetric states in scenarios with compressed mass spectra at $\sqrt{s} = 13$ TeV with the ATLAS detector. Phys Rev D97(5):052010. https://doi.org/10.1103/PhysRevD.97.052010, arXiv:1712.08119 [hep-ex]
32. ATLAS Collaboration (2010) The ATLAS simulation infrastructure. Eur Phys J C 70:823. https://doi.org/10.1140/epjc/s10052-010-1429-9, arXiv:1005.4568 [physics.ins-det]
33. GEANT4 Collaboration, Agostinelli S, et al (2003) GEANT4—a simulation toolkit. Nucl Instrum Meth A 506:250. https://doi.org/10.1016/S0168-9002(03)01368-8
34. ATLAS Collaboration (2010) The simulation principle and performance of the ATLAS fast calorimeter simulation FastCaloSim, ATL-PHYS-PUB-2010-013. https://cds.cern.ch/record/1300517
35. Gleisberg T, Höche S, Krauss F, Schönherr M, Schumann S, Siegert F, Winter J (2009) Event generation with SHERPA 1.1. JHEP 02:007. https://doi.org/10.1088/1126-6708/2009/02/007, arXiv:0811.4622 [hep-ph]
36. Bothmann E, et al (2019) Event generation with Sherpa 2.2. SciPost Phys 7(3):034. https://doi.org/10.21468/SciPostPhys.7.3.034, arXiv:1905.09127 [hep-ph]
37. Alioli S, Nason P, Oleari C, Re E (2010) A general framework for implementing NLO calculations in shower Monte Carlo programs: the POWHEG BOX. JHEP 06:043. https://doi.org/10.1007/JHEP06(2010)043, arXiv:1002.2581 [hep-ph]

38. Alioli S, Nason P, Oleari C, Re E (2009) NLO Higgs boson production via gluon fusion matched with shower in POWHEG. JHEP 04:002. https://doi.org/10.1088/1126-6708/2009/04/002, arXiv:0812.0578 [hep-ph]
39. Nason P, Oleari C (2010) NLO Higgs boson production via vector-boson fusion matched with shower in POWHEG. JHEP 02:037. https://doi.org/10.1007/JHEP02(2010)037, arXiv:0911.5299 [hep-ph]
40. Djouadi A, Muhlleitner MM, Spira M (2007) Decays of supersymmetric particles: the program SUSY-HIT (SUspect-SdecaY-Hdecay-InTerface). Acta Phys Polon B 38:635–644 arXiv:hep-ph/0609292
41. ATLAS Collaboration (2016) Performance of pile-up mitigation techniques for jets in pp collisions at $\sqrt{s} = 8$ TeV using the ATLAS detector. Eur Phys J C76(11):581. https://doi.org/10.1140/epjc/s10052-016-4395-z, arXiv:1510.03823 [hep-ex]
42. ATLAS Collaboration (2016) Optimisation of the ATLAS b-tagging performance for the 2016 LHC Run, ATL-PHYS-PUB-2016-012. https://cds.cern.ch/record/2160731
43. ATLAS Collaboration (2016) Performance of b-jet identification in the ATLAS experiment. JINST 11:P04008. https://doi.org/10.1088/1748-0221/11/04/P04008, arXiv:1512.01094 [hep-ex]
44. The ATLAS transverse-momentum trigger performance at the LHC in 2011, Tech Rep ATLAS-CONF-2014-002, CERN, Geneva, Feb, 2014. https://cds.cern.ch/record/1647616
45. ATLAS Collaboration (2017) Performance of the ATLAS trigger system in 2015. Eur Phys J C77(5):317. https://doi.org/10.1140/epjc/s10052-017-4852-3, arXiv:1611.09661 [hep-ex]
46. CMS Collaboration Collaboration (2016) Search for new physics in the compressed mass spectra scenario using events with two soft opposite-sign leptons and missing transverse momentum at $\sqrt{s} = 13$ TeV. Tech Rep CMS-PAS-SUS-16-025, CERN, Geneva. https://cds.cern.ch/record/2205866
47. ATLAS Collaboration (2012) Measurement of the WW cross section in $\sqrt{s} = 7$ TeV pp collisions with the ATLAS detector and limits on anomalous gauge couplings. Phys Lett B 712:289. https://doi.org/10.1016/j.physletb.2012.05.003, arXiv:1203.6232 [hep-ex]
48. Prospects for Higgs Boson searches using the $H \to WW^{(*)} \to \ell\nu\ell\nu$ decay mode with the ATLAS detector for 10 TeV. Tech Rep ATL-PHYS-PUB-2010-005, CERN, Geneva, Jun, 2010. https://cds.cern.ch/record/1270568
49. Jet calibration and systematic uncertainties for jets reconstructed in the ATLAS detector at $\sqrt{s} = 13$ TeV. Tech Rep ATL-PHYS-PUB-2015-015, CERN, Geneva, July, 2015. https://cds.cern.ch/record/2037613
50. ATLAS Collaboration (2017) Jet energy scale measurements and their systematic uncertainties in proton-proton collisions at $\sqrt{s} = 13$ TeV with the ATLAS detector. Phys Rev D96(7):072002. https://doi.org/10.1103/PhysRevD.96.072002, arXiv:1703.09665 [hep-ex]
51. Expected performance of missing transverse momentum reconstruction for the ATLAS detector at $\sqrt{s} = 13$ TeV. Tech Rep ATL-PHYS-PUB-2015-023, CERN, Geneva, July, 2015. https://cds.cern.ch/record/2037700
52. Butterworth J, et al (2016) PDF4LHC recommendations for LHC Run II. J Phys G 43:023001. https://doi.org/10.1088/0954-3899/43/2/023001, arXiv:1510.03865 [hep-ph]
53. Baak M, Besjes GJ, Cte D, Koutsman A, Lorenz J, Short D (2015) HistFitter software framework for statistical data analysis. Eur Phys J C 75:153. https://doi.org/10.1140/epjc/s10052-015-3327-7, arXiv:1410.1280 [hep-ex]
54. Cowan G, Cranmer K, Gross E, Vitells O (2011) Asymptotic formulae for likelihood-based tests of new physics. Eur Phys J C 71:1554. https://doi.org/10.1140/epjc/s10052-011-1554-0, arXiv:1007.1727 [physics.data-an]

Chapter 11
Future Work: Search for Compressed Wino-Bino Production in the Three Lepton Final State

This search benefits from the techniques developed in both the wino-bino search, in Chap. 7 and the compressed searches in Chap. 10 because the SUSY particles decay via off-shell bosons just like in the compressed searches but the final state is three leptons and $E_{\mathrm{T}}^{\mathrm{miss}}$ just as in the wino-bino search. In the summary limit shown in Fig. 11.1, the compressed search can be found along the diagonal where $m(\tilde{\chi}_2^0) = m(\tilde{\chi}_1^0)$. The wino-bino and the RJR searches limits can be found start at $m(\tilde{\chi}_1^+, \tilde{\chi}_1^-) = 200$ GeV and $m(\tilde{\chi}_1^0) = 100$ GeV.

This search is complementary to these searches by targeting an area that has not been probed since Run 1 and that has an excess for $m(\tilde{chi}_1^p m/\chi_2^0) < 200$ GeV. The mass splittings targeted by this search are $\Delta m = m(\tilde{\chi}_2^0) - m(\tilde{\chi}_1^0) = [0, 60]$ GeV.

11.1 Signal Signature

Figure 11.2 shows the diagrams for the production of $\tilde{\chi}_1^{\pm}\tilde{\chi}_2^0$ decaying via W and Z bosons in proton-proton collisions. In this search, referred to as the off-shell 3ℓ search, the W and Z bosons are off-shell, meaning that $m(W^*) < m(W)$ and $m(Z^*) < m(Z)$.

For these gauge-boson-mediated decays, the final state considered is the three-lepton (where lepton refers to an electron or muon) final state where both the W and Z bosons decay leptonically. Leptonic decays of taus are indistinguishable from promptly produced electrons and muons and therefore contribute to the signal regions. The final state, in addition to the leptons, has missing energy from the LSP, $\tilde{\chi}_1^0$, and a neutrino since the W decays leptonically.

In this model, the $\tilde{\chi}_1^{\pm}$ and $\tilde{\chi}_2^0$ are the Next-to-Lightest SUSY particles (NLSP) and mass degenerate winos and the LSP, $\tilde{\chi}_1^0$, is a bino. The mass splitting, Δm, refers

© The Editor(s) (if applicable) and The Author(s), under exclusive license to Springer Nature Switzerland AG 2020

E. Resseguie, *Electroweak Physics at the Large Hadron Collider with the ATLAS Detector*, Springer Theses,
https://doi.org/10.1007/978-3-030-57016-3_11

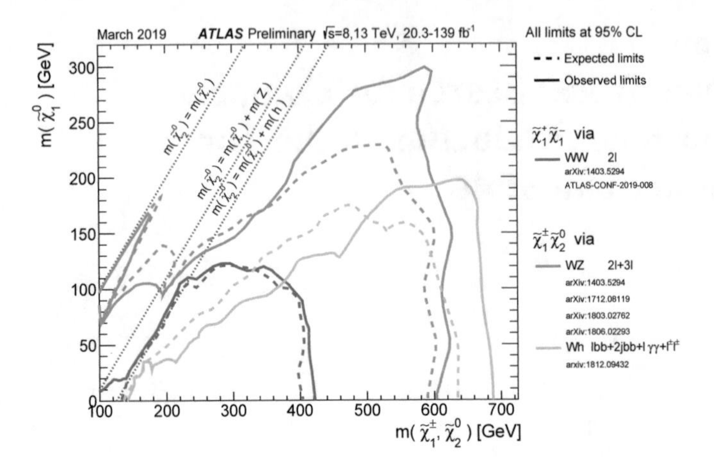

Fig. 11.1 The 95% CL exclusion limits on $\tilde{\chi}_1^+\tilde{\chi}_1^-$ and $\tilde{\chi}_1^\pm\tilde{\chi}_2^0$ production with SM-boson-mediated decays, as a function of the $\tilde{\chi}_1^\pm$, $\tilde{\chi}_2^0$ and $\tilde{\chi}_1^0$ masses. The production cross-section is for pure wino $\tilde{\chi}_1^+\tilde{\chi}_1^-$ and $\tilde{\chi}_1^\pm\tilde{\chi}_2^0$. Each individual exclusion contour represents a union of the excluded regions of one or more analyses [1]

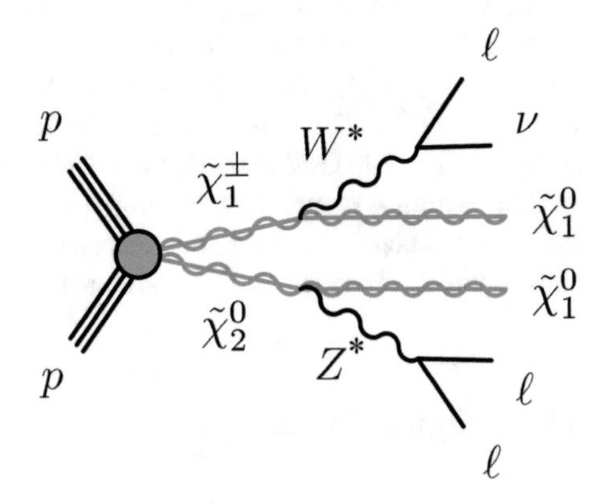

Fig. 11.2 Diagrams for the production of $\tilde{\chi}_1^\pm\tilde{\chi}_2^0$ decaying via off-shell W and Z bosons in pp collisions

to the difference in mass between the wino and the bino. Since the W and Z are off-shell, the mass splitting between the NLSP and the LSP is less than the mass of the Z boson.

Optimization and background estimate will use techniques from both the compressed searches and the on-shell wino-bino search discussed in Chaps. 10 and 7, respectively.

11.2 Overview of Backgrounds

The backgrounds can be separated into two categories: reducible and irreducible.

Reducible backgrounds have at least one fake lepton originating from Z+jets/$Z + \gamma$ processes, shown in Fig. 5.2, where a jet is mis-identified as a lepton or the photon converts to a lepton. This background is especially important at the lowest mass splittings and at the lowest lepton p_T. Other reducible backgrounds come from top backgrounds such as $t\bar{t}$, Wt, and WW where a jet is mis-identified as a lepton or a $b-$jet decays semi-leptonically.

Irreducible backgrounds come from processes with at least three prompt leptons. The largest of these backgrounds is WZ and should be normalized just like in the on-shell wino-bino searches. Other irreducible backgrounds are ZZ, $t\bar{t}V$, and Higgs production.

11.3 Data Set and MC Samples

The samples include an ATLAS detector simulation [2], based on Geant4 [3], or a fast simulation [2] that uses a parametrization of the calorimeter response [4] and Geant4 for the other parts of the detector. The simulated events are reconstructed in the same manner as the data.

Table 11.1 summarizes the Monte Carlo (MC) used specifying the generator used to simulate both background and signal events.

Diboson and triboson processes were simulated with SHERPA 2.2.1 [5, 6] and the cross sections were calculated at NLO.

The $t\bar{t}$ and single top quarks samples in the Wt channel were simulated using POWHEG [7, 8] generator. The $t\bar{t}$ events were normalized using the NNLO+next-to-next-to-leading-logarithm (NNLL) QCD [9] cross-section, while the cross-section for single-top-quark events was calculated at NLO+NNLL [10].

The Z+jets background simulation is simulated using the SHERPA generator and the cross section is calculated at NNLO [11].

Table 11.1 Summary of the signal and background processes with the generator used for the simulation and the order at which the cross section is calculated

Process	Event generator	Parton shower, hadronization	UE tune	PDF	Order α_s
$\tilde{\chi}_1^{\pm}\tilde{\chi}_2^0$	MADGRAPH 2.6.1	PYTHIA 8.230	A14	NNPDF2.3lo	NLO+NLL
WW,WZ,ZZ	SHERPA 2.2.2	SHERPA	Default	NNPDF3.0nlo	NLO
Triboson	SHERPA 2.2.2	SHERPA	Default	NNPDF3.0nlo	(NLO)
Z/W+jets	SHERPA 2.2.1	SHERPA	Default	NNPDF3.0nlo	NNLO
$t\bar{t}$	POWHEG- BOX v2	PYTHIA 8.230	A14	NNPDF2.3lo	NNLO+NNLL
Single top	POWHEG- BOX v2	PYTHIA 8.230	A14	NNPDF2.3lo	(NNLO+NNLL)
Other top	MADGRAPH5_aMC@NLO 2.3.3	PYTHIA 8.212	A14	NNPDF2.3lo	NLO
Higgs	POWHEG	PYTHIA	A14	NNPDF2.3	NNLO + NNLL

Higgs boson processes include gluon-gluon fusion, associated VH production, and vector-boson fusion. They were generated using POWHEG [12] and PYTHIA. The cross sections are calculated at NNLO with NNLL accuracy.

The SUSY signal processes were generated from LO matrix elements with up to two extra partons, using the MADGRAPH v2.2.3 generator interfaced to PYTHIA8.186. Signal cross-sections were calculated at NLO with NLL accuracy [13–17]. The nominal cross-section and its uncertainty were taken from an envelope of cross-section predictions using different PDF sets and factorization and renormalization scales, as described in Ref. [18].

No data is used in these studies. The simulation is normalized an integrated luminosity 139 fb^{-1}, which corresponds to the luminosity of the data collected during Run 2.

11.4 Object and Event Selection

11.4.1 Object Selection

Electrons and muons are identified using identification, isolation, and tracking criteria. Two levels of object selection are used for electrons and muons, described in Table 11.2. Each level "baseline" and "signal" applies the selection of the previous levels along with additional criteria. The baseline leptons use the looser identification criteria and lower lepton p_{T} in order to provide a higher efficiency of identifying and removing processes decaying to four prompt leptons. Signal leptons satisfy stricter criteria.

Baseline electrons must have $p_{\mathrm{T}} > 4.5$ GeV and fall within the inner detector, $|\eta| < 2.47$. The electrons must also satisfy the `VeryLooseLLH` quality criteria and $|z_0 \sin\theta| < 0.5$ mm impact parameter. Signal electrons need to pass the impact parameter cuts of and $|d_0/\sigma_{d_0}| < 5$, designed to suppress fake electrons from pileup jets. They also satisfy tighter identification criteria, `MediumLLH` and tighter isolation, `Gradient`. The electron criteria for the compressed searches, summarized in Table 10.2 are a bit different for the off-shell 3ℓ search. The electron ID is loosened from `TightLLH` to `MediumLLH` to increase the acceptance for the three lepton final state.

There was an additional improvement from the compressed searches: muons can now be reconstructed down to 3 GeV instead of 4 GeV. Baseline muons must satisfy the `Medium` identification criteria, have $p_{\mathrm{T}} > 3$ GeV and fall within the inner detector, $|\eta| < 2.4$, and $|z_0 \sin\theta| < 0.5$ mm. Signal muons must pass the impact parameter cuts of $|d_0/\sigma_{d_0}| < 3$. They must fulfill a tighter isolation, `FCLoose_FixedRad`.

Jets are reconstructed from topological clusters using the anti-k$_t$ algorithm with distance parameter $\Delta R = 0.4$. Baseline jets are required to have $p_{\mathrm{T}} > 20$ GeV and fulfill the pseudorapidity requirement of $|\eta| < 4.5$. To suppress jets originating from pileup, jets are further required to pass a JVT cut ($JVT > 0.59$) if the jet p_{T} is

Table 11.2 Summary of the baseline and signal levels for electron and muon criteria. Each new level contains the selection of the previous level

Cut	Value/description	
	Baseline electron	Baseline muon
Acceptance	$p_T > 4.5$ GeV, $\|\eta^{cluster}\| < 2.47$	$p_T > 3$ GeV, $\|\eta\| < 2.5$
Identification	`VeryLooseLLH`	`Medium`
Impact parameter	$\|z_0 \sin\theta\| < 0.5$ mm	$\|z_0 \sin\theta\| < 0.5$ mm
	Signal electron	Signal muon
Identification	`MediumLLH`	`Medium`
Isolation	`Gradient`	`FCLoose_FixedRad`
Impact parameter	$\|z_0 \sin\theta\| < 0.5$ mm $\|d_0/\sigma_{d_0}\| < 5$	$\|z_0 \sin\theta\| < 0.5$ mm $\|d_0/\sigma_{d_0}\| < 3$

Table 11.3 Summary of the baseline and signal selection for jets and b-jets

Cut	Value/description
	Baseline jet
Collection	`AntiKtEMTopo`
Acceptance	$p_T > 20$ GeV, $\|\eta\| < 4.5$
	Signal jet
JVT	$\|JVT\| > 0.59$ for jets with $p_T < 60$ GeV and $\|\eta\| < 2.4$
Acceptance	$p_T > 30$ GeV, $\|\eta\| < 2.8$
	Signal b-jet
b-tagger algorithm	MV2c10, 85% efficiency
Acceptance	$p_T > 20$ GeV, $\|\eta\| < 2.5$

within $20 < p_T < 50$ GeV and it resides within $|\eta| < 2.4$ [19]. Signal jets have the additional requirement of falling within $|\eta| < 2.8$ and have $p_T > 30$ GeV.

Identification of jets containing b-hadrons (b-jets), so called b-tagging, is performed with the MV2c10 algorithm, a multivariate discriminant making use of track impact parameters and reconstructed secondary vertices [20, 21]. A requirement is chosen corresponding to a 85% average efficiency obtained for b-jets in simulated $t\bar{t}$ events.

The jet and b-jet selection criteria are summarized in Table 11.3.

Separate algorithms are run in parallel to reconstruct electrons, muons, and jets. A particle can be reconstructed as one or more objects. To resolve these ambiguities, a procedure called "overlap removal" is applied. For electrons, this overlap removal is applied in two steps. At the baseline selection, an electron that shares a track with a muon, and the sub-leading p_T electron from two overlapping electrons are removed. The second step removes electrons if they are within $0.2 < \Delta R < 0.4$ of a jet. For muons, overlap removal is applied to baseline muons to separate prompt muons from those originating from the decay of hadrons in a jet. A baseline muon is removed if it is within $\Delta R < 0.4$ of a jet that at least 3 tracks.

The missing transverse momentum, with magnitude $E_{\mathrm{T}}^{\mathrm{miss}}$, is calculated as the negative vector sum of the transverse momenta of the calibrated selected leptons and jets, and the sum of transverse momenta of additional soft objects in the event, which are reconstructed from tracks in the inner detector or calorimeter cell clusters.

11.4.2 Overall Strategy and Trigger

Two signal regions are considered: jet veto region and the ISR region where at least one jet is required. In the jet veto region, the $E_{\mathrm{T}}^{\mathrm{miss}}$ is softer since it recoils against the WZ system while in the ISR region, the $E_{\mathrm{T}}^{\mathrm{miss}}$ recoils against the jet, resulting in a harder $E_{\mathrm{T}}^{\mathrm{miss}}$. These topologies can be used to target different mass splittings. The ISR topology, described in Sect. 7.5.3.1, is used to target small mass splittings, where the missing energy summing to a large value due to the recoil against the jet.

The jet veto region targets signals with larger mass splitting, which have leptons with larger transverse momenta than in the ISR region. As a result, lepton triggers are used to target this region. To maximize the signal acceptance, single, di-, and tri-lepton triggers are used and the lepton p_{T} is applied based on the trigger's plateau, as shown in Fig. 11.3. A $E_{\mathrm{T}}^{\mathrm{miss}}$ cut of $E_{\mathrm{T}}^{\mathrm{miss}} > 50\,\mathrm{GeV}$ is applied to reduce the Z+jets contributions.

In the ISR jet region, the $E_{\mathrm{T}}^{\mathrm{miss}}$ cut is found to be optimal at 200 GeV, as shown in Fig. 11.4. The lead jet p_{T} is correlated with $E_{\mathrm{T}}^{\mathrm{miss}}$, and found that at $E_{\mathrm{T}}^{\mathrm{miss}} > 200\,\mathrm{GeV}$, there is no additional gain in significance when making a lead jet p_{T} cut, as seen in the correlation plots for signals with different mass splittings in Fig. 11.5. The $E_{\mathrm{T}}^{\mathrm{miss}}$ triggers also reach full efficiency at 200 GeV, as seen in Fig. 10.6; therefore, the ISR region uses $E_{\mathrm{T}}^{\mathrm{miss}}$ triggers to select events. Using $E_{\mathrm{T}}^{\mathrm{miss}}$ triggers instead of lepton triggers targets the smallest mass splittings which have soft lepton.

In both the jet veto and ISR regions, a $b-$jet veto is applied to reduce $t\bar{t}$ contamination. Events are selected with three leptons which form at least one same flavor opposite charge pair.

11.4.3 Lepton Assignment for the Off-Shell Three Lepton Search

To understand the impact of minimizing and maximizing $m_{\ell\ell}$ to determine the choice of assignment of leptons, only events where there is ambiguity in assignment, eee and $\mu\mu\mu$ events, are used. Events are selected with $E_{\mathrm{T}}^{\mathrm{miss}} > 50\,\mathrm{GeV}$ (to reduce low-$E_{\mathrm{T}}^{\mathrm{miss}}$ Z+jets contamination), at least a pair of leptons forming a same flavor opposite

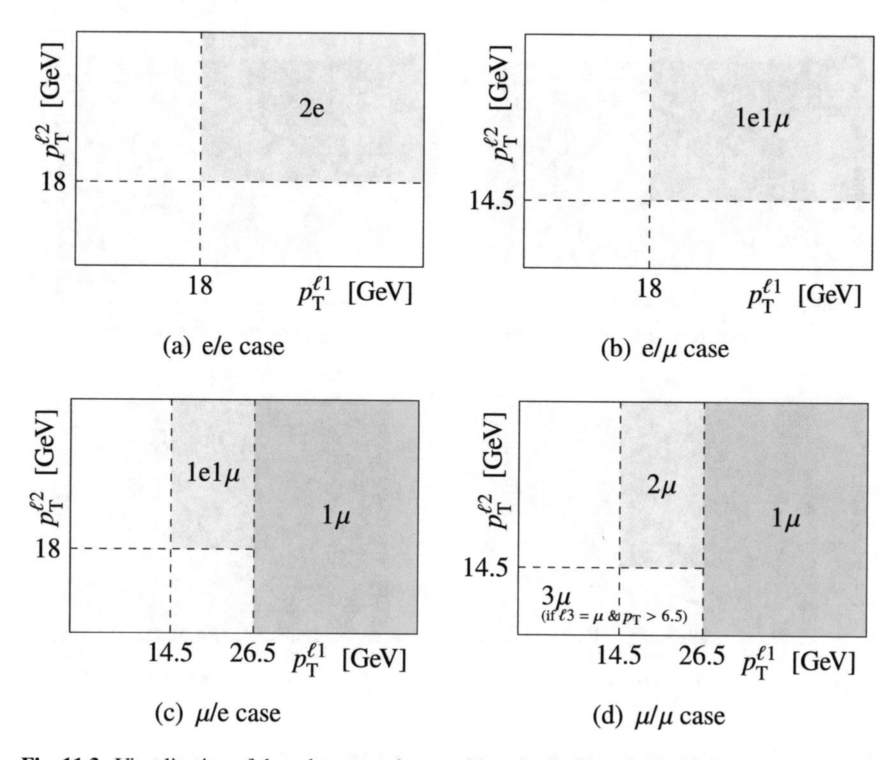

Fig. 11.3 Visualization of the scheme used to combine single, di- and trilepton triggers

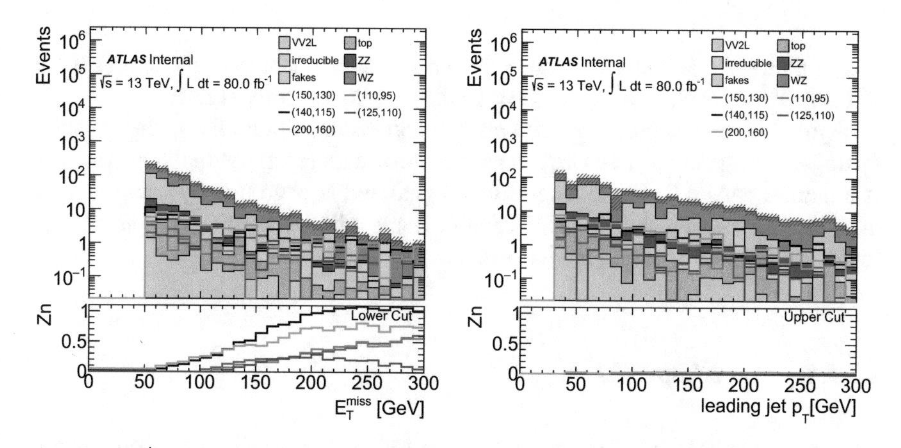

Fig. 11.4 E_T^{miss} and lead jet p_T distributions in a selection with at least 1 jet

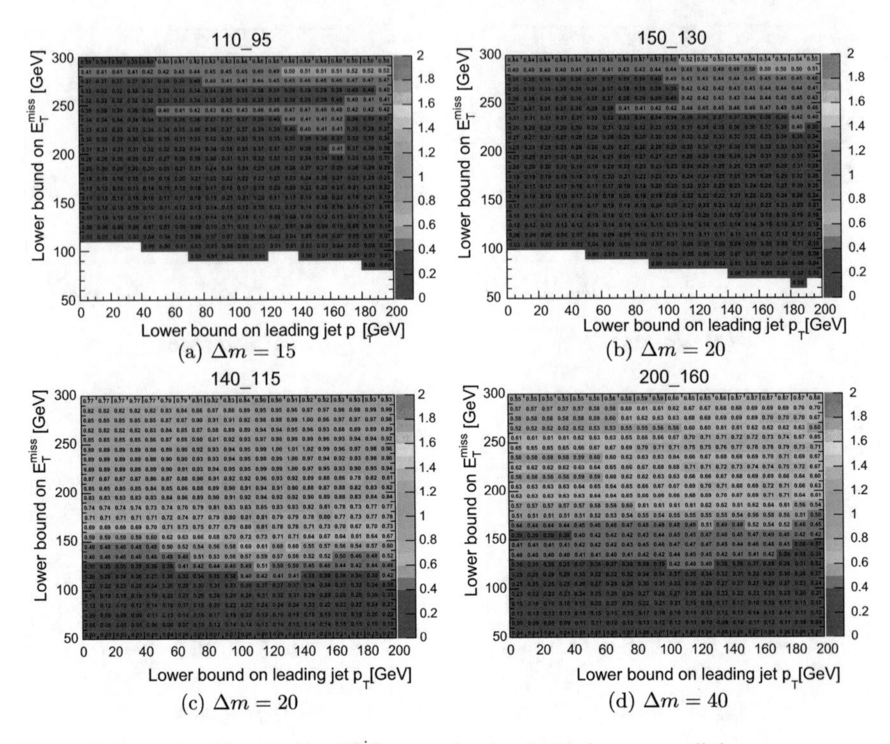

(a) $\Delta m = 15$ (b) $\Delta m = 20$

(c) $\Delta m = 20$ (d) $\Delta m = 40$

Fig. 11.5 Lower bound on lead jet $E_{\mathrm{T}}^{\mathrm{miss}}$ vs. p_T for signal with four mass splittings

charge pair, a $b-$jet veto (to reduce $t\bar{t}$ contamination), lepton triggers, and $p_{\mathrm{T}}^{\ell 1} > 25\,\mathrm{GeV}$, $p_{\mathrm{T}}^{\ell 2} > 15\,\mathrm{GeV}$, $p_{\mathrm{T}}^{\ell 3} > 10\,\mathrm{GeV}$ so the triggers are fully efficient.

Figure 11.6 shows both $m_{\ell\ell}^{\mathrm{min}}$ and $m_{\ell\ell}^{\mathrm{max}}$ calculated by either minimizing or maximizing $m_{\ell\ell}$, respectively. The $m_{\ell\ell}^{\mathrm{min}}$ variable shows a kinematic endpoint at Δm. This is a natural candidate for distinguishing signal and background by using a shape fit to target each signal. The $m_{\ell\ell}^{\mathrm{max}}$ variable helps with background suppression by removing on-shell WZ and Z+jets events.

11.4.4 $m_{\ell\ell}$ Shape Fit

Thus, same flavor opposite charge leptons are assigned to the Z boson by minimizing $m_{\ell\ell}$ and the remaining third lepton is assigned to the W boson. The $m_{\ell\ell}$ resulting from this assignment is labeled as $m_{\ell\ell}^{\mathrm{min}}$. When assigning labels by maximizing $m_{\ell\ell}$, the resulting variable is labeled as $m_{\ell\ell}^{\mathrm{max}}$. The signal has a kinematic edge at $m_{\ell\ell} = \Delta m(m(\tilde{\chi}_1^{\pm}) - m(\tilde{\chi}_1^0))$ as seen in Fig. 11.7. To make use of this signal feature, a shape fit in $m_{\ell\ell}$ will therefore be used in the optimization with each bin of the fit targeting

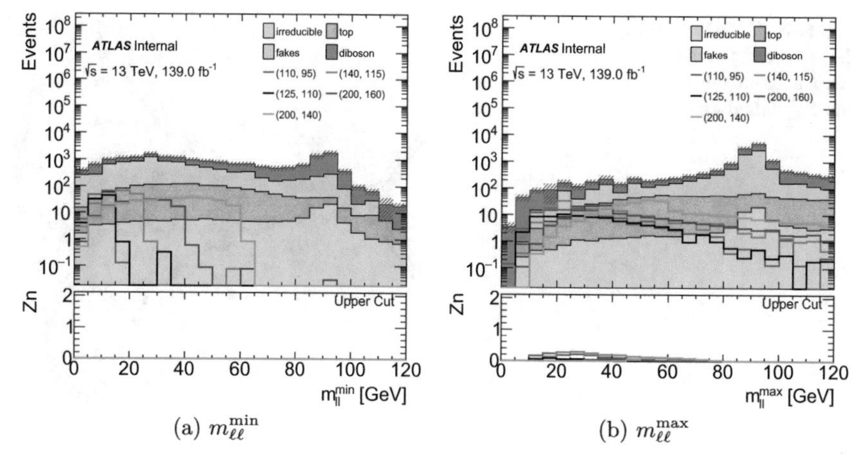

(a) $m_{\ell\ell}^{\min}$ (b) $m_{\ell\ell}^{\max}$

Fig. 11.6 $m_{\ell\ell}^{\min}$ and $m_{\ell\ell}^{\max}$ for eee and $\mu\mu\mu$ events

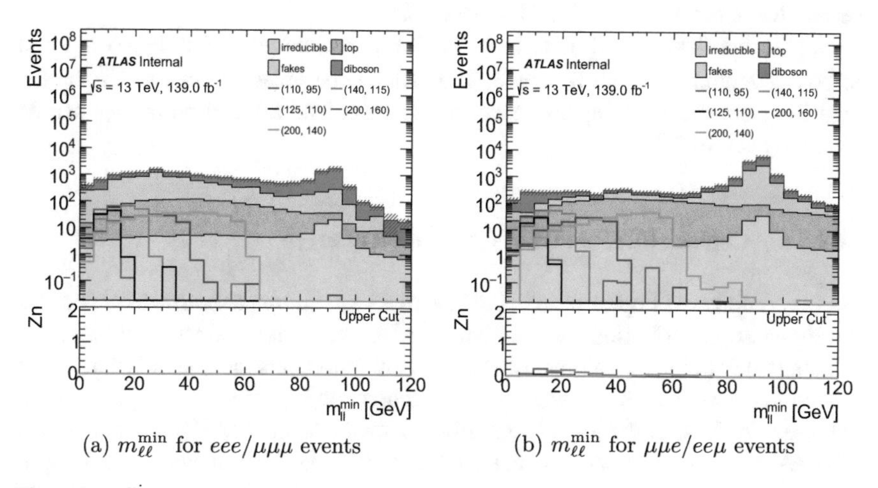

(a) $m_{\ell\ell}^{\min}$ for $eee/\mu\mu\mu$ events (b) $m_{\ell\ell}^{\min}$ for $\mu\mu e/ee\mu$ events

Fig. 11.7 $m_{\ell\ell}^{\min}$ for $eee/\mu\mu\mu$ events in **a**, $\mu\mu e/ee\mu$ events in **b**

different mass splittings, just like in the compressed electroweakino search, discussed in Chap. 10.

11.4.5 Optimizing with m_{T2}

The variable $m_{\mathrm{T2}}^{m_\chi}$ is usually used in analyses where each leg decays to an invisible particle and a visible particle. In the compressed slepton search, discussed in Chap. 10, the m_{T2} variable was found to have a kinematic edge at the mass splitting

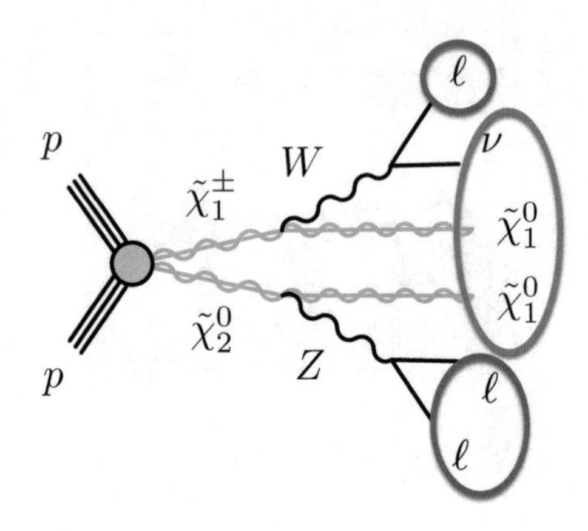

Fig. 11.8 Cartoon showing the calculating of m_{T2} variable in the three lepton and $E_{\mathrm{T}}^{\mathrm{miss}}$ final state

between the slepton and the LSP. Thus, a shape fit in m_{T2} was performed for this search. More details can be found in Sect. 10.5.2.

For the three lepton final state, the two visible particles are the Z system comprised of both Z leptons, and the W lepton. The invisible particles are the two $\tilde{\chi}_1^0$ and the neutrino from the W decay, as shown in Fig. 11.8. The calculation of m_{T2} can be found in Sect. 10.5.2.

11.4.5.1 Choice of LSP Mass in Calculation of m_{T2}

For the compressed slepton search, the LSP mass was chosen to be 100 GeV because the signal m_{T2} distribution had a kinematic edge at Δm at the slepton mass. In this search, to determine the optimal LSP mass assumption, this parameter is varied and plotted to determine which gives the signal kinematic edge for both the jet veto and ISR regions. Both in the jet veto and ISR regions, when the LSP mass is less than 100 GeV, there is no distinct kinematic edge for the signal, as shown in Fig. 11.9.

Thus, the value of this variable is bounded kinematically by the mass splitting of $m(\tilde{\chi}_1^\pm) - m(\tilde{\chi}_1^0)$, if the LSP mass in the m_{T2} calculation is set to 100 GeV.

The m_{T2} distribution has a kinematic edge with LSP mass set at 100 GeV; however, we could consider an asymmetric m_{T2}. This is because the $\tilde{\chi}_1^\pm$ leg has both a neutrino and the LSP as the invisible part (Y) while the $\tilde{\chi}_2^0$ only has the LSP. We will take the LSP mass of the $\tilde{\chi}_2^0$ leg to be 100 GeV.

Figure 11.10 shows the asymmetric m_{T2} with varying values of Y for fixed LSP mass of 100 GeV. As the mass of Y, the neutrino + LSP, increases, the signal kinematic edge becomes less defined. Thus, m_{T2} calculated with a symmetric mass of the invisible system should be used in this search.

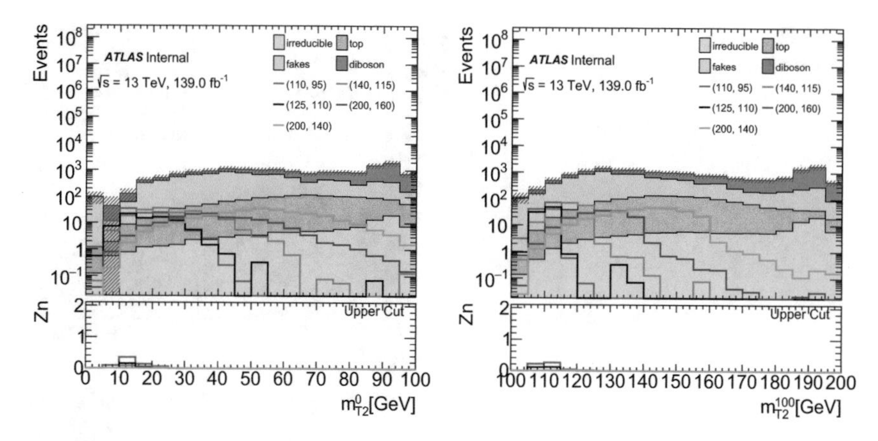

Fig. 11.9 m_{T2} calculated varying the assumption on the LSP mass

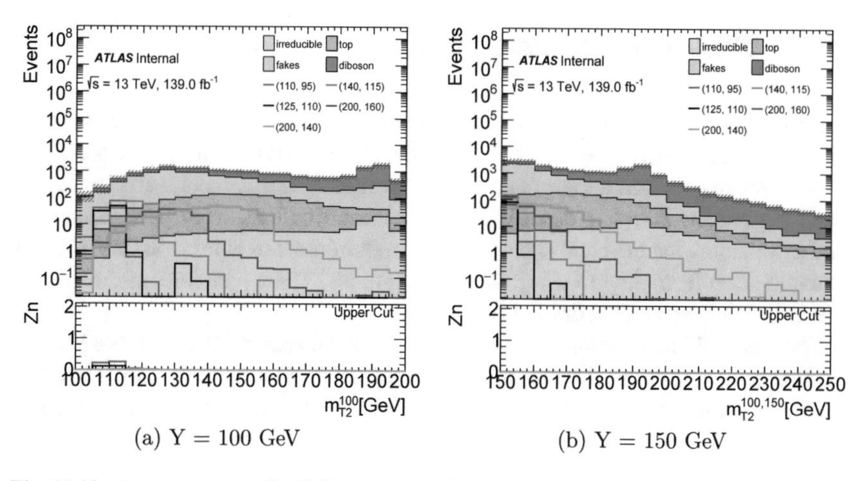

Fig. 11.10 Asymmetric m_{T2} for different values of Y, LSP mass kept at 100 GeV

11.4.5.2 Upper Cut on m_{T2} to Make Use of the Kinematic Signal Edge

The value of m_{T2} is bounded kinematically by the mass splitting of $m(\tilde{\chi}_1^{\pm}) - m(\tilde{\chi}_1^0)$, if the invisible particle is set to the LSP mass of 100 GeV and if the mass is taken as symmetric, as shown in Fig. 11.11. To remove the offset from the LSP mass assumption, a new variable is used that is called $\Delta m_{\mathrm{T2}} = m_{\mathrm{T2}}^{m_{\chi}} - 100$. This second kinematic edge can be used by optimizing the Δm_{T2} cut by placing an upper cut on this variable in each $m_{\ell\ell}^{\mathrm{min}}$ bin.

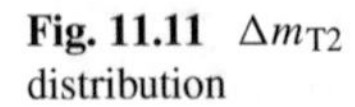

Fig. 11.11 Δm_{T2} distribution

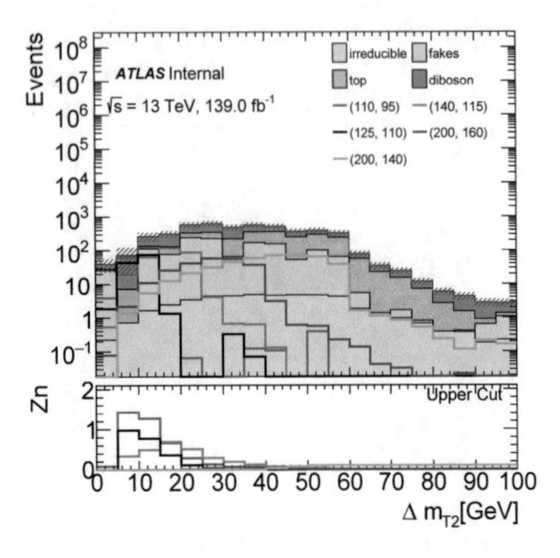

11.4.6 Summary of Signal Regions

The event selection cuts used in both the jet veto and ISR regions are summarized in Table 11.4. A cut on $m_{\ell\ell}^{\min} < 60\,\mathrm{GeV}$ selects the off-shell signal while a cut of $m_{\ell\ell}^{\max} < 60\,\mathrm{GeV}$reduces contamination from the on-shell WZ and Z+jets backgrounds. A matching cut on $\Delta m_{\mathrm{T2}} < 60\,\mathrm{GeV}$ minimizes contributions from backgrounds as well as selecting the off-shell signal. A $E_{\mathrm{T}}^{\mathrm{miss}}$ requirement of at least 50 GeV reduces contamination from the low $E_{\mathrm{T}}^{\mathrm{miss}}$ fake backgrounds. The $b-$jet veto minimizes top events which have at least one $b-$jet.

The cuts used in the signal region optimization are summarized in Table 11.5.

For the jet veto region, the variable that provides the most discrimination is $m_{\mathrm{T}}^{\mathrm{minMll}}$, calculated using the W lepton, after assigning the Z leptons by minimizing $m_{\ell\ell}$

Table 11.4 Summary of event selection criteria. The binning scheme used to define the final signal regions is shown in Table 11.5

Variable	Requirement
Number of leptons	$= 3$
Lepton charge and flavor	$e^+e^-\ell$ or $\mu^+\mu^-\ell$
Leading lepton $p_{\mathrm{T}}^{\ell_1}$	>4.5 (3) GeV for electron (muon)
Sub-leading lepton $p_{\mathrm{T}}^{\ell_2}$	>4.5 (3) GeV for electron (muon)
$m_{\ell\ell}^{\min}$	<60 GeV
$m_{\ell\ell}^{\max}$	<60 GeV
Δm_{T2}	<60 GeV
$E_{\mathrm{T}}^{\mathrm{miss}}$	>50 GeV
Number of b-tagged jets	$= 0$
Binned in	$m_{\ell\ell}^{\min}$

Table 11.5 Summary of cuts in each of the $m_{\ell\ell}^{\min}$ bins

bin	$m_{\ell\ell}^{\min}$	Δm_{T2}	N_{jets}	$E_{\mathrm{T}}^{\mathrm{miss}}$	lepton p_{T}	$m_{\mathrm{T}}^{\mathrm{minMll}}$	$p_{\mathrm{T}}^{\ell\ell\ell}/E_{\mathrm{T}}^{\mathrm{miss}}$
OffZhighMet-0ja	[1,10]	<10	= 0	>50	lep trigger p_{T}	<50	–
OffZhighMet-0jb	[10,15]	<15	= 0	>50	lep trigger p_{T}	<50	–
OffZhighMet-0jc	[15,20]	<20	= 0	>50	25,15,10	<50	–
OffZhighMet-0jd	[20,30]	<30	= 0	>50	25,15,10	<50	–
OffZhighMet-0je	[30,40]	<40	= 0	>50	25,15,10	<70	–
OffZhighMet-0jf	[40,60]	<60	= 0	>50	25,15,10	<70	–
OffZhighMet-1ja	[1,10]	<10	≥1	>200	4.5(e), 3(μ)	–	<0.2
OffZhighMet-1jb	[10,15]	<15	≥1	>200	4.5(e), 3(μ)	–	<0.2
OffZhighMet-1jc	[15,20]	<20	≥1	>200	4.5(e), 3(μ)	–	<0.3
OffZhighMet-1jd	[20,30]	<30	≥1	>200	4.5(e), 3(μ)	–	<0.3
OffZhighMet-1je	[30,40]	<40	≥1	>200	4.5(e), 3(μ)	–	<0.3
OffZhighMet-1jf	[40,60]	<60	≥1	>200	4.5(e), 3(μ)	–	<0.3

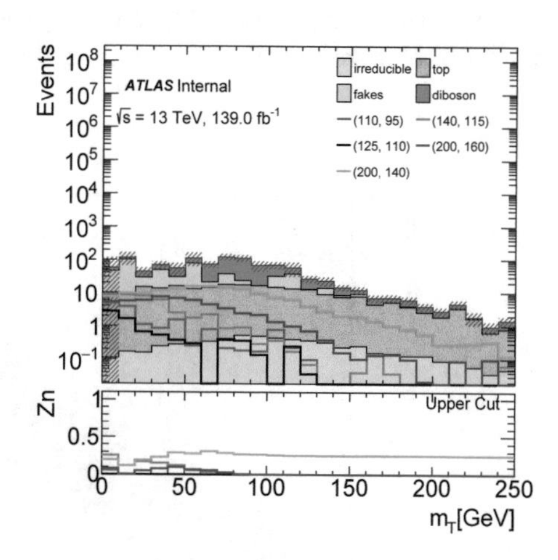

Fig. 11.12 $m_{\mathrm{T}}^{\mathrm{minMll}}$ in the event selection defined in Table 11.4 with a jet veto applied

($m_{\ell\ell}^{\min}$), and $E_{\mathrm{T}}^{\mathrm{miss}}$. This variable is optimized in each $m_{\ell\ell}^{\min}$ bin, as shown in Table 11.5. This variable reduces the WZ background by having an upper cut on this variable, as shown in Fig. 11.12. The cut on m_{T} can be tighter for smaller Δm than for larger Δm.

In the ISR region, a variable that has shown discrimination between signal and background is $p_{\mathrm{T}}^{\ell\ell\ell}/E_{\mathrm{T}}^{\mathrm{miss}}$, where $p_{\mathrm{T}}^{\ell\ell\ell}$ is the p_{T} of the vector sum of all three lepton. The variable $p_{\mathrm{T}}^{\ell\ell\ell}$ was used in the conventional 3ℓ search in the ISR region because it's a measure of the recoil of the lepton and $E_{\mathrm{T}}^{\mathrm{miss}}$ system against the ISR jet. The signal acceptance steeply drops as $p_{\mathrm{T}}^{\ell\ell\ell}/E_{\mathrm{T}}^{\mathrm{miss}}$ increases while the background remains flat. This cut can be further optimized as a function of Δm (Fig. 11.13).

After applying all the cuts in Tables 11.4 and 11.5, the significance for the simplified model with $\tilde{\chi}_1^{\pm}\tilde{\chi}_2^0$ is calculated assuming a flat 30% systematic uncertainty on

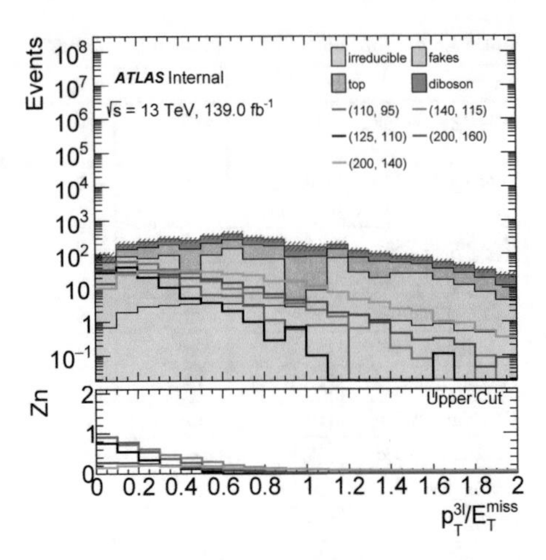

Fig. 11.13 $p_{\mathrm{T}}^{\ell\ell\ell}/E_{\mathrm{T}}^{\mathrm{miss}}$ in the event selection defined in Table 11.4 with at least one jet

the total background. In this binned approach the significance of the 6 jet veto and the 6 ISR bins are statistically combined, providing maximal sensitivity as shown in Fig. 11.14.

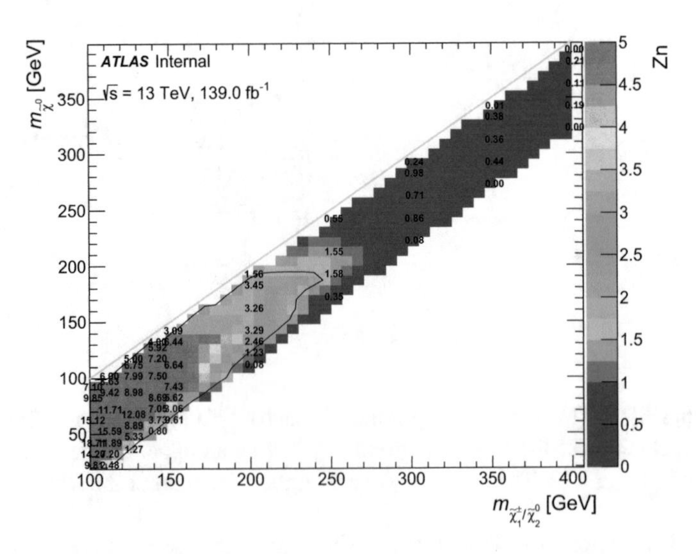

Fig. 11.14 Expected sensitivity assuming a flat 30% uncertainty on the background. The jet veto and ISR bins are statistically combined

11.5 Next Steps

The main remaining step is the background estimation. It should be similar to the previous searches discussed. The fake background will be estimated using the data-driven Fake Factor method. The Fake Factors will be measured in an on-shell region, just as discussed in Sect. 7.6.2.1, and validated in an off-shell selection to be closer to the signal region. The top-like background will be estimated in a control region using different flavor opposite charge leptons just as in Sect. 7.6.2.2. The fake background in the on-shell wino-bino search was found to have a dependence on η; however, due to small size of the fake background and the lack of statistics, the Fake Factors were not binned in both p_{T} and η. This is an improvement that can be made in this search.

The WZ background should be normalized to data in a control region that will have a jet veto and an ISR region just as for the on-shell wino-bino search discussed in Sect. 7.6.1. The challenge is to create a region that has a high WZ purity. An off-shell selection is not pure in WZ; about 60% of the background is WZ; therefore, this background could be normalized in an on-shell region and validated in an off-shell region to check the modeling in a region kinematically close to the SR.

References

1. ATLAS Collaboration (2019) SUSY summary plots. http://atlas.web.cern.ch/Atlas/GROUPS/PHYSICS/CombinedSummaryPlots/SUSY/ATLAS_SUSY_EWSummary_WH/history.html
2. ATLAS Collaboration (2010) The ATLAS simulation infrastructure. Eur Phys J C 70:823. arXiv:1005.4568 [physics.ins-det]
3. GEANT4 Collaboration, Agostinelli S et al (2003) GEANT4—a simulation toolkit. Nucl Instrum Meth A 506:250
4. ATLAS Collaboration (2010) The simulation principle and performance of the ATLAS fast calorimeter simulation FastCaloSim. ATL-PHYS-PUB-2010-013. https://cds.cern.ch/record/1300517
5. Bothmann E et al (2019) Event generation with sherpa 2.2. SciPost Phys 7(3):034. arXiv:1905.09127 [hep-ph]
6. ATLAS Collaboration (2016) Multi-boson simulation for 13 TeV ATLAS analyses. ATL-PHYS-PUB-2016-002. https://cds.cern.ch/record/2119986
7. Re E (2011) Single-top Wt-channel production matched with parton showers using the POWHEG method. Eur Phys J C 71:1547 arXiv:1009.2450 [hep-ph]
8. Frixione S, Nason P, Ridolfi G (2007) A positive-weight next-to-leading-order Monte Carlo for heavy flavour hadroproduction. JHEP 09:126 arXiv:0707.3088 [hep-ph]
9. Czakon M, Mitov A (2014) Top++: a program for the calculation of the top-pair cross-section at hadron colliders. Comput Phys Commun 185:2930 arXiv:1112.5675 [hep-ph]
10. Kidonakis N (2010) Two-loop soft anomalous dimensions for single top quark associated production with a W^- or H^-. Phys Rev D 82:054018 arXiv:1005.4451 [hep-ph]
11. Catani S, Cieri L, Ferrera G, de Florian D, Grazzini M (2009) Vector boson production at hadron colliders: a fully exclusive QCD calculation at NNLO. Phys Rev Lett 103:082001 arXiv:0903.2120 [hep-ph]
12. Alioli S, Nason P, Oleari C, Re E (2010) A general framework for implementing NLO calculations in shower Monte Carlo programs: the POWHEG BOX. JHEP 06:043 arXiv:1002.2581 [hep-ph]

13. Beenakker W, Hopker R, Spira M, Zerwas P (1997) Squark and gluino production at hadron colliders. Nucl Phys B 492:51 arXiv:hep-ph/9610490
14. Kulesza A, Motyka L (2009) Threshold resummation for squark-antisquark and gluino-pair production at the LHC. Phys Rev Lett 102:111802 arXiv:0807.2405
15. Kulesza A, Motyka L (2009) Soft gluon resummation for the production of gluino-gluino and squark-antisquark pairs at the LHC. Phys Rev D 80:095004 arXiv:0905.4749 [hep-ph]
16. Beenakker W, Brensing S, Kramer M, Kulesza A, Laenen E et al (2009) Soft-gluon resummation for squark and gluino hadroproduction. JHEP 12:041 arXiv:0909.4418 [hep-ph]
17. Beenakker W, Brensing S, Kramer M, Kulesza A, Laenen E et al (2011) Squark and gluino hadroproduction. Int J Mod Phys A 26:2637 arXiv:1105.1110 [hep-ph]
18. Borschensky C, Kramer M, Kulesza A, Mangano M, Padhi S, Plehn T, Portell X (2014) Squark and gluino production cross sections in pp collisions at $\sqrt{s} = 13, 14, 33$ and 100 TeV. Eur Phys J C 74:3174 arXiv:1407.5066 [hep-ph]
19. ATLAS Collaboration (2016) Performance of pile-up mitigation techniques for jets in *pp* collisions at $\sqrt{s} = 8$ TeV using the ATLAS detector. Eur Phys J C76(11):581. arXiv:1510.03823 [hep-ex]
20. ATLAS Collaboration (2016) Optimisation of the ATLAS *b*-tagging performance for the 2016 LHC run. ATL-PHYS-PUB-2016-012. https://cds.cern.ch/record/2160731
21. ATLAS Collaboration (2016) Performance of *b*-jet identification in the ATLAS experiment. JINST 11:P04008. arXiv:1512.01094 [hep-ex]

Chapter 12
Conclusion

This thesis presented a measurement and searches using $\sqrt{s} = 13$ TeV data collected taken during Run 2 with the ATLAS detector, as well as tests of electronics for planned upgrades to the ATLAS inner detector.

For the detector upgrade studies, the prototype chip, HCC130, worked as expected. There were a few bugs uncovered that improved the design of the production chips, the HCCStar and the AMAC. The HCC130 was irradiated with gamma rays and was found to not have a large current increase at low total ionizing dose. It also ran without issues for many months during irradiation.

The WZ cross section was measured for the first time at $\sqrt{s} = 13$ TeV with 3.2 fb^{-1} of data collected. This measurement considered the leptonic decays of the gauge bosons to electrons or muons. The total cross section was 50.6 $\pm$ 2.6 (stat.) $\pm$ 2.0 (sys.) $\pm$ 0.9 th. $\pm$ 1.2 (lumi.) fb, which is in good agreement with the new SM NNLO prediction from MATRIX of $48.2^{+1.1}_{-1.0}$(scale) pb. Understanding of the WZ production and signal was important to the electroweak SUSY searches that were discussed.

The SUSY searches presented were searches for the electroweak production of neutralinos, charginos, and sleptons decaying into final states with exactly two or three electrons or muons and missing transverse momentum with 36.1 fb^{-1} of data. The first search used a wino-bino benchmark model with $\tilde{\chi}_1^{\pm}\tilde{\chi}_2^0$ production with decays via gauge bosons to three leptons and missing energy. No significant excess was observed. After combining this search with the search of $\tilde{\chi}_1^{\pm}\tilde{\chi}_2^0$ production where each sparticle decays via an SM gauge boson giving a final state with two leptons consistent with a Z boson and two jets consistent with a W boson, for a massless $\tilde{\chi}_1^0$ neutralino, $\tilde{\chi}_1^{\pm}/\tilde{\chi}_2^0$ masses up to approximately 580 GeV are excluded.

Another search targeting the same benchmark model using the RJR technique saw excesses in two orthogonal bins with local significances of 3.0 σ and 2.1 σ. To understand the tension in the limit and to study the RJR phase space, a new

© The Editor(s) (if applicable) and The Author(s), under exclusive license to Springer Nature Switzerland AG 2020

E. Resseguie, *Electroweak Physics at the Large Hadron Collider with the ATLAS Detector*, Springer Theses,
https://doi.org/10.1007/978-3-030-57016-3_12

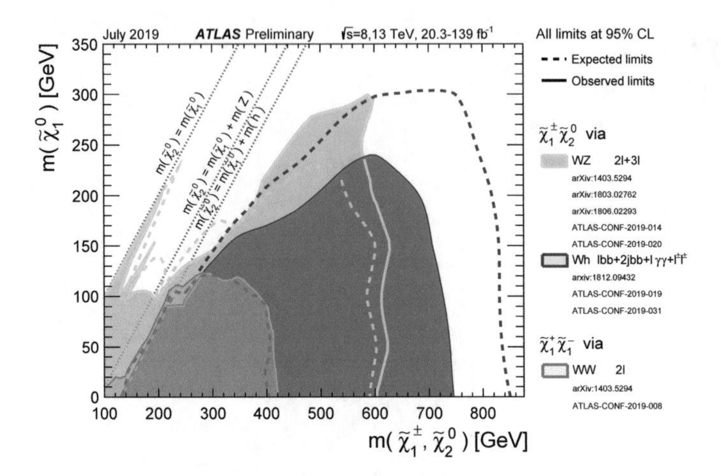

Fig. 12.1 The 95% CL exclusion limits on $\tilde{\chi}_1^+\tilde{\chi}_1^-$ and $\tilde{\chi}_1^\pm\tilde{\chi}_2^0$ production with SM-boson-mediated decays, as a function of the $\tilde{\chi}_1^\pm$, $\tilde{\chi}_2^0$ and $\tilde{\chi}_1^0$ masses. The production cross-section is for pure wino $\tilde{\chi}_1^+\tilde{\chi}_1^-$ and $\tilde{\chi}_1^\pm\tilde{\chi}_2^0$. Each individual exclusion contour represents a union of the excluded regions of one or more analyses [1]

technique, emulated RJR (eRJR) was developed. The RJR excess was reproduced using the eRJR technique and the eRJR search was extended to include the full Run 2 dataset, corresponding to 139 fb^{-1}. In the full dataset the observed event yields are found to be in agreement with Standard Model expectations, with no significant excess seen in either signal region.

Compressed searches in the two leptons and missing energy final state were also discussed with the Higgsino and wino-bino benchmark models for the production of $\tilde{\chi}_1^\pm\tilde{\chi}_2^0$ and the wino-bino model only for the direct slepton production. This search was challenging due to the presence of soft leptons. The dilepton invariant mass and stransverse mass were the main discriminating variables used to construct signal regions. No excess was observed. For the Higgsino simplified model, exclusion limits are set on the $\tilde{\chi}_2^0$ neutralino up to masses of ~130 GeV and down to mass splittings $\Delta m(\tilde{\chi}_2^0, \tilde{\chi}_1^0) \sim 3$ GeV. This Higgsino search was the first conducted on ATLAS and has extended the limits on Higgsino masses set by LEP. In the wino–bino model, these limits on the $\tilde{\chi}_2^0$ extend to masses of up to ~170 GeV and down to mass splittings of approximately 2.5 GeV. The currents ATLAS limits on the wino–bino model are shown in Fig. 12.1.

Direct pair production of sleptons was excluded for slepton masses up to masses of ~180 GeV and down to mass splittings $\Delta m(\tilde{\ell}, \tilde{\chi}_1^0) \sim 1$ GeV. The currents ATLAS limits on slepton production are shown in Fig. 12.2.

Unfortunately, no excesses were observed either in the SM model measurement or the SUSY searches; however, as the ATLAS experiment prepares to take data during Run 3 and then in HL-LHC, if light SUSY particles are accessible at the LHC,

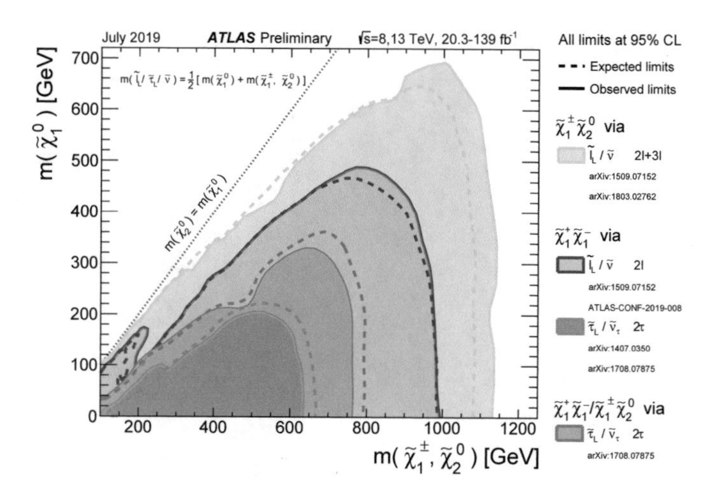

Fig. 12.2 The 95% CL exclusion limits on $\tilde{\chi}_1^+\tilde{\chi}_1^-$ and $\tilde{\chi}_1^\pm\tilde{\chi}_2^0$ production decaying via sleptons, as a function of the $\tilde{\chi}_1^\pm$, $\tilde{\chi}_2^0$ and $\tilde{\chi}_1^0$ masses. The production cross-section is for pure wino $\tilde{\chi}_1^+\tilde{\chi}_1^-$ and $\tilde{\chi}_1^\pm\tilde{\chi}_2^0$. Each individual exclusion contour represents a union of the excluded regions of one or more analyses [1]

there would be an opportunity to not only discover them but be able to measure their properties. If no excesses continue to be observed, exclusions of light SUSY particles would limit some of the most appealing solutions to the Higgs naturalness problem.

Reference

1. ATLAS Collaboration, SUSY July 2019 Summary Plot Update, https://atlas.web.cern.ch/Atlas/GROUPS/PHYSICS/PUBNOTES/ATL-PHYS-PUB-2019-022/

Appendix

A.1 Additional HCC Testing Plots

The number of errors in Registers 3-17 during the irradiation of the HCC130 are shown in Fig. A.1. There were no read-back errors during irradiation.

A.2 eRJR Additional Studies

A.2.1 Overlap of RJR and $m_{\mathrm{T}}^{\mathrm{min}}$ Searches

The kinematics values of the events that do not overlap between the RJR and conventional 3ℓ searches are shown in Tables A.1, A.2 in Appendix A.2.

A.2.1.1 Emulating $p_{\mathrm{T}}^{\mathrm{CM}}$

Because the values of the translated $p_{\mathrm{T}}^{\mathrm{soft}}$ does not match the actual value of $p_{\mathrm{T}}^{\mathrm{CM}}$, different possible definitions of this variable were considered (Table A.3).

The various definitions are shown below:

- $p_{\mathrm{T}}^{\mathrm{soft}}$ v1 = $E_{\mathrm{T}}^{\mathrm{miss}}$ soft term
- $p_{\mathrm{T}}^{\mathrm{soft}}$ v2 = the magnitude of the p_{T} of the vector sum of the four-momenta of the signal jets, leptons, and $\mathbf{p}_{\mathrm{T}}^{\mathrm{miss}}$
- $p_{\mathrm{T}}^{\mathrm{soft}}$v3 = the magnitude of the p_{T} of the vector sum of the four-momenta of the jets (including forward jets), leptons, and $\mathbf{p}_{\mathrm{T}}^{\mathrm{miss}}$
- $p_{\mathrm{T}}^{\mathrm{soft}}$ v4: scalar sum of the p_{T} of the leptons, ISR jets, and $E_{\mathrm{T}}^{\mathrm{miss}}$.

© The Editor(s) (if applicable) and The Author(s), under exclusive license to Springer Nature Switzerland AG 2020

E. Resseguie, *Electroweak Physics at the Large Hadron Collider with the ATLAS Detector*, Springer Theses,
https://doi.org/10.1007/978-3-030-57016-3

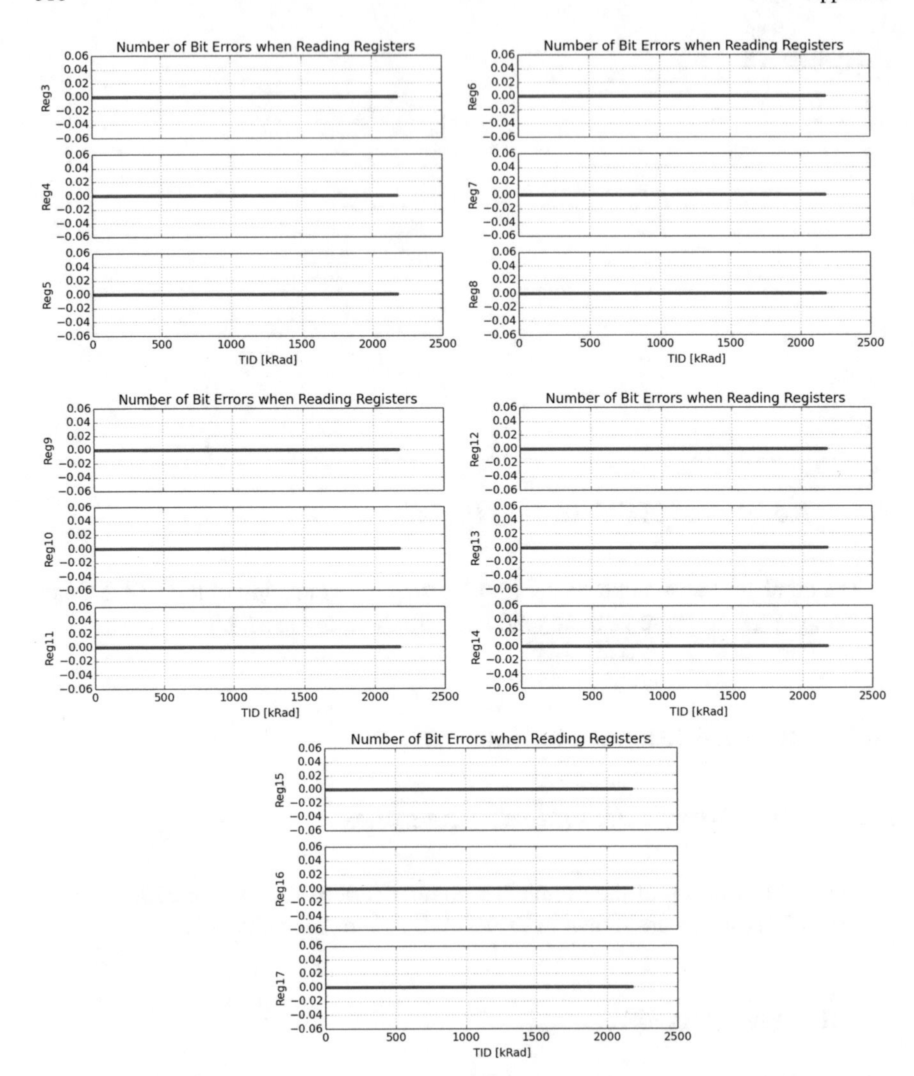

Fig. A.1 Number of error reads for each of the control registers 3-17

The scalar sum of the transverse momenta is the incorrect definition because it does not take into account the direction of the various signal objects and gives a value that is much too high. Most of the signal events do not have forward jets and

Table A.1 Values of kinematic variables for the events which enter the RJR SR-ISR but do not enter the $m_{\mathrm{T}}^{\mathrm{min}}$ SRs

Run number	Event number	$E_{\mathrm{T}}^{\mathrm{miss}}$ (GeV)	$m_{\ell\ell}^{\mathrm{min}}(m_{\ell\ell})$(GeV)	$m_{\mathrm{T}}^{\mathrm{min}}$ (m_{T}) (GeV)	$m_{\mathrm{T}}^{\mathrm{min}}$ SR cut failed
311481	688464706	83.71	84.82 (84.82)	102.42 (102.42)	$E_{\mathrm{T}}^{\mathrm{miss}}$
302347	640107350	97.26	91.63 (91.18)	5.37 (103.41)	$E_{\mathrm{T}}^{\mathrm{miss}}$ and $m_{\mathrm{T}}^{\mathrm{min}}$
300800	2007781950	110.28	89.03 (89.03)	108.08 (108.08)	$E_{\mathrm{T}}^{\mathrm{miss}}$ and $m_{\mathrm{T}}^{\mathrm{min}}$
303638	2322292825	124.924	93.24 (89.46)	87.21 (142.31)	$m_{\mathrm{T}}^{\mathrm{min}}$
311170	481322948	129.79	89.21 (89.21)	100.79 (100.79)	$m_{\mathrm{T}}^{\mathrm{min}}$
311481	1674640922	129.52	93.67 (93.67)	102.15 (102.15)	$m_{\mathrm{T}}^{\mathrm{min}}$
307195	1995121829	160.90	75.54 (103.41)	76.68 (139.34)	$m_{\ell\ell}^{\mathrm{min}}$ and $m_{\mathrm{T}}^{\mathrm{min}}$ cuts
309640	5271389311	106.33	79.40 (88.02)	82.00 (174.42)	$m_{\ell\ell}^{\mathrm{min}}$ and $m_{\mathrm{T}}^{\mathrm{min}}$ cuts
310341	3711343277	89.99	48.38 (87.62)	75.52 (105.66)	$m_{\ell\ell}^{\mathrm{min}}$ and $m_{\mathrm{T}}^{\mathrm{min}}$ cuts

Table A.2 Values of kinematic variables for the events which enter the RJR SR-low but do not enter the conventional 3ℓ SRs

Run number	Event number	$E_{\mathrm{T}}^{\mathrm{miss}}$ (GeV)	$m_{\ell\ell}(m_{\ell\ell}^{\mathrm{min}})$(GeV)	m_{T} $(m_{\mathrm{T}}^{\mathrm{min}})$ (GeV)	$m_{\mathrm{T}}^{\mathrm{min}}$ SR cut failed
310634	3429255513	142.342	81.584 (63.458)	193.354 (148.415)	$m_{\ell\ell}^{\mathrm{min}}$
300415	407813268	54.417	91.110 (160.034)	114.506 (89.356)	$m_{\ell\ell}^{\mathrm{min}}$ and $m_{\mathrm{T}}^{\mathrm{min}}$
300863	1853035891	51.203	95.652 (172.024)	143.784 (73.966)	$m_{\ell\ell}^{\mathrm{min}}$ and $m_{\mathrm{T}}^{\mathrm{min}}$
303846	4572444712	21.218	89.833 (211.731)	103.797 (28.111)	$m_{\ell\ell}^{\mathrm{min}}$ and $m_{\mathrm{T}}^{\mathrm{min}}$
298862	115312201	24.212	90.037 (196.773)	122.310 (81.681)	$m_{\ell\ell}^{\mathrm{min}}$ and $m_{\mathrm{T}}^{\mathrm{min}}$
304128	2922238960	46.057	97.718 (272.621)	165.722 (68.079)	$m_{\ell\ell}^{\mathrm{min}}$ and $m_{\mathrm{T}}^{\mathrm{min}}$
304243	1230198499	46.072	89.790 (170.971)	113.458 (94.390)	$m_{\ell\ell}^{\mathrm{min}}$ and $m_{\mathrm{T}}^{\mathrm{min}}$
306310	4170081299	41.826	91.195 (140.443)	124.745 (56.845)	$m_{\ell\ell}^{\mathrm{min}}$ and $m_{\mathrm{T}}^{\mathrm{min}}$
307732	2241606325	29.334	84.728 (193.333)	104.253 (40.545)	$m_{\ell\ell}^{\mathrm{min}}$ and $m_{\mathrm{T}}^{\mathrm{min}}$
307861	2342711817	21.798	93.929 (178.948)	110.794 (34.109)	$m_{\ell\ell}^{\mathrm{min}}$ and $m_{\mathrm{T}}^{\mathrm{min}}$

(continued)

Table A.2 (continued)

Run number	Event number	$E_{\mathrm{T}}^{\mathrm{miss}}$ (GeV)	$m_{\ell\ell}(m_{\ell\ell}^{\mathrm{min}})$(GeV)	m_{T} ($m_{\mathrm{T}}^{\mathrm{min}}$) (GeV)	$m_{\mathrm{T}}^{\mathrm{min}}$ SR cut failed
309390	3874751426	108.449	95.743 (281.733)	314.252 (10.559)	$m_{\ell\ell}^{\mathrm{min}}$ and $m_{\mathrm{T}}^{\mathrm{min}}$
302300	1124313518	66.029	89.812 (89.812)	105.146 (105.146)	$m_{\mathrm{T}}^{\mathrm{min}}$
304008	789803669	19.493	89.106 (89.106)	113.699 (113.699)	$E_{\mathrm{T}}^{\mathrm{miss}}$
305920	1205419439	49.802	94.912 (94.912)	102.195 (102.195)	$E_{\mathrm{T}}^{\mathrm{miss}}$
309516	2388280800	22.906	91.483 (91.483)	124.142 (124.142)	$E_{\mathrm{T}}^{\mathrm{miss}}$
310809	2270414448	20.106	92.924 (92.924)	113.206 (113.206)	$E_{\mathrm{T}}^{\mathrm{miss}}$
302380	298635234	39.006	90.168 (90.168)	138.919 (138.919)	$E_{\mathrm{T}}^{\mathrm{miss}}$
303304	3387504597	28.264	90.689 (90.689)	160.924 (160.924)	$E_{\mathrm{T}}^{\mathrm{miss}}$
303560	1015290252	25.865	85.297 (85.297)	136.721 (136.721)	$E_{\mathrm{T}}^{\mathrm{miss}}$

Table A.3 Comparing different $p_{\mathrm{T}}^{\mathrm{soft}}$ emulations with the RJR $p_{\mathrm{T}}^{\mathrm{CM}}$ variable

Run number	Event number	$p_{\mathrm{T}}^{\mathrm{soft}}$v1	$p_{\mathrm{T}}^{\mathrm{soft}}$ v2	$p_{\mathrm{T}}^{\mathrm{soft}}$ v3	$p_{\mathrm{T}}^{\mathrm{soft}}$ v4	RJR $p_{\mathrm{T}}^{\mathrm{CM}}$
302347	640107350	9.17	9.17	9.17	447.96	23.00
303638	2322292825	5.45	5.45	5.45	368.27	11.35
303638	2970796970	10.14	10.14	10.14	844.56	19.09
305811	2030981730	5.17	5.17	5.17	607.89	9.92
307195	1995121829	7.70	7.70	7.70	508.15	21.13
309640	5271389311	1.48	1.48	1.48	469.59	16.20
310341	3711343277	16.43	16.43	16.43	306.93	22.56
310809	3633590098	2.86	2.86	2.86	567.00	7.88
311170	481322948	2.65	2.65	2.65	558.53	2.65
311481	688464706	9.06	9.06	9.06	325.15	8.34
311481	1674640922	10.36	13.11	10.36	590.59	23.86
300800	2007781950	2.85	2.85	2.85	350.16	13.76

as a result the definition of $p_{\mathrm{T}}^{\mathrm{soft}}$ which only includes central jets ($p_{\mathrm{T}}^{\mathrm{soft}}$ v2) gives the same value as the soft term.

A.2.2 Comparison of the Values of the eRJR and RJR Kinematic Variables

The RJR and eRJR kinematic variables are compared in SR-ISR data events in Tables A.4, A.5. Most of the values for the kinematic variables agree well between eRJR and RJR except for $p_{\mathrm{T}}^{\mathrm{soft}}$ and $p_{\mathrm{T}}^{\mathrm{CM}}$.

The RJR and eRJR kinematic variables are compared in SR-low data events in Tables A.6, A.7. All the RJR variables are well reproduced by the eRJR technique.

Table A.4 Values of eRJR variables for the 2015–16 data in SR-ISR

Run number	Event number	$E_{\mathrm{T}}^{\mathrm{miss}}$	$\lvert\Delta\phi\left(E_{\mathrm{T}}^{\mathrm{miss}}, \mathrm{jets}\right)\rvert$	$R\left(E_{\mathrm{T}}^{\mathrm{miss}}, \mathrm{jets}\right)$	$p_{\mathrm{T}}^{\mathrm{jets}}$	$p_{\mathrm{T}}^{\mathrm{soft}}$
302347	640107350	97.26	2.97	0.56	172.01	9.17
303638	2322292825	124.92	3.04	0.98	126.49	5.45
303638	2970796970	221.27	2.29	0.98	149.31	10.14
305811	2030981730	201.85	3.00	0.82	242.95	5.17
307195	1995121829	160.90	2.83	0.78	196.64	7.70
309640	5271389311	106.33	2.34	0.54	134.91	1.48
310341	3711343277	89.99	2.78	0.87	97.05	16.43
310809	3633590098	195.35	3.06	0.88	221.98	2.86
311170	481322948	129.79	3.07	0.59	219.06	2.65
311481	688464706	83.71	2.84	0.77	103.79	9.06
311481	1674640922	129.52	2.81	0.73	167.10	13.11
300800	2007781950	110.28	3.13	0.87	127.40	2.85

Table A.5 Values of RJR variables for the 2015–16 data in SR-ISR

Run number	Event number	$p_{\mathrm{T}}^{\mathrm{I}}$	$\Delta\phi_{\mathrm{ISR},E_{\mathrm{T}}^{\mathrm{miss}}}$	R_{ISR}	$p_{\mathrm{T}}^{\mathrm{ISR}}$	$p_{\mathrm{T}}^{\mathrm{CM}}$
302347	640107350	107.81	2.96	0.59	179.68	23.00
303638	2322292825	129.98	3.03	0.98	128.34	11.35
303638	2970796970	223.41	2.31	0.99	152.34	19.09
305811	2030981730	213.50	3.01	0.86	247.15	9.92
307195	1995121829	174.37	2.83	0.81	203.37	21.13
309640	5271389311	118.90	2.46	0.59	155.66	16.20
310341	3711343277	95.65	2.71	0.84	102.98	22.56
310809	3633590098	198.66	3.07	0.88	224.79	7.88
311170	481322948	129.60	3.08	0.59	219.67	2.65
311481	688464706	94.34	2.84	0.85	105.45	8.34
311481	1674640922	136.99	2.80	0.74	176.35	23.86
300800	2007781950	121.07	3.14	0.92	131.97	13.76

Table A.6 Values of eRJR variables for the 2015–16 data in SR-low

Run number	Event number	H^{boost}	$m_{\text{eff}}^{3\ell}/H^{\text{boost}}$	$p_{\text{T}}^{\text{soft}}/(p_{\text{T}}^{\text{soft}}+m_{\text{eff}}^{3\ell})$
284213	3445458672	795.774	0.992	0.044
300415	407813268	261.089	0.985	0.036
300863	1853035891	290.487	0.987	0.024
303846	4572444712	388.091	0.964	0.026
304008	789803669	666.521	0.932	0.021
304128	2922238960	389.783	0.904	0.016
304243	1230198499	278.162	0.942	0.035
298862	115312201	482.600	0.942	0.020
305920	1205419439	291.113	0.929	0.038
307732	2241606325	446.322	0.991	0.003
307861	2342711817	362.653	0.956	0.036
309390	3874751426	545.649	0.974	0.014
309516	2388280800	573.896	0.936	0.044
306310	4170081299	357.530	0.949	0.043
310634	3429255513	329.373	0.913	0.018
310809	2270414448	412.991	0.992	0.040
302300	1124313518	280.395	0.981	0.028
302380	298635234	355.719	0.962	0.033
303304	3387504597	703.934	0.903	0.035
303560	1015290252	404.543	0.989	0.009

Table A.7 Values of RJR variables for the 2015–16 data in SR-low

Run number	Event number	$H_{3,1}^{\text{PP}}$	$H_{\text{T}\,3,1}^{\text{PP}}/H_{3,1}^{\text{PP}}$	$\frac{p_{\text{T}}^{\text{PP}}}{p_{\text{T}}^{\text{PP}}+H_{\text{T}\,3,1}^{\text{PP}}}$
284213	3445458672	640.117	0.991	0.043
300415	407813268	256.0202	0.983	0.038
300863	1853035891	263.238	0.986	0.024
303846	4572444712	292.672	0.963	0.026
304008	789803669	489.965	0.928	0.022
304128	2922238960	388.306	0.903	0.016
304243	1230198499	278.891	0.943	0.035
298862	115312201	480.418	0.942	0.020
305920	1205419439	265.492	0.928	0.038
307732	2241606325	324.887	0.991	0.003
307861	2342711817	340.862	0.953	0.038
309390	3874751426	522.821	0.974	0.014
309516	2388280800	538.585	0.939	0.044
306310	4170081299	278.891	0.943	0.035
310634	3429255513	329.316	0.912	0.019
310809	2270414448	346.907	0.991	0.041
302300	1124313518	281.047	0.981	0.028
302380	298635234	295.679	0.965	0.035
303304	3387504597	535.273	0.906	0.036
303560	1015290252	388.678	0.988	0.009

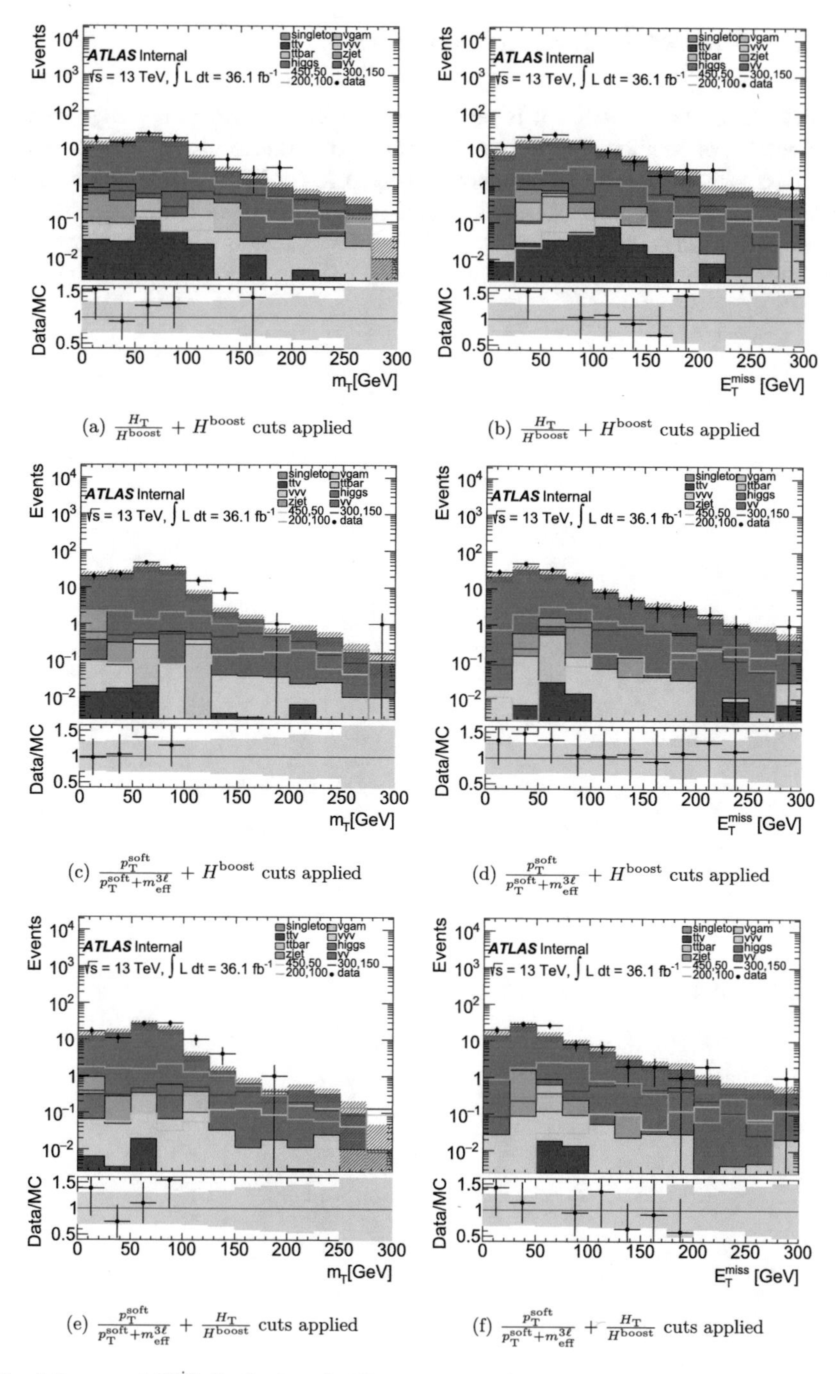

(a) $\frac{H_T}{H^{\text{boost}}} + H^{\text{boost}}$ cuts applied

(b) $\frac{H_T}{H^{\text{boost}}} + H^{\text{boost}}$ cuts applied

(c) $\frac{p_T^{\text{soft}}}{p_T^{\text{soft}} + m_{\text{eff}}^{3\ell}} + H^{\text{boost}}$ cuts applied

(d) $\frac{p_T^{\text{soft}}}{p_T^{\text{soft}} + m_{\text{eff}}^{3\ell}} + H^{\text{boost}}$ cuts applied

(e) $\frac{p_T^{\text{soft}}}{p_T^{\text{soft}} + m_{\text{eff}}^{3\ell}} + \frac{H_T}{H^{\text{boost}}}$ cuts applied

(f) $\frac{p_T^{\text{soft}}}{p_T^{\text{soft}} + m_{\text{eff}}^{3\ell}} + \frac{H_T}{H^{\text{boost}}}$ cuts applied

Fig. A.2 m_T and E_T^{miss} distributions for W muon events after combination of two eRJR cuts. All plots include jet veto, lepton p_T, and $m_{\ell\ell}$ cuts

A.2.3 Further SR-low Excess Studies

To determine which eRJR cut is responsible for the low $E_{\mathrm{T}}^{\mathrm{miss}}$, m_{T} edge excess, distributions of combinations of two RJR mimic cuts after applying jet veto, lepton $p_{\mathrm{T}}, m_{\ell\ell}$ cuts on W muon events are shown in Fig. A.2. The RJR variable combinations that include $p_{\mathrm{T}}^{\mathrm{soft}}/(p_{\mathrm{T}}^{\mathrm{soft}} + m_{\mathrm{eff}}^{3\ell})$ give the largest excess in m_{T} distribution but no excess at low $E_{\mathrm{T}}^{\mathrm{miss}}$. The combination of $m_{\mathrm{eff}}^{3\ell}/H^{\mathrm{boost}} + H^{\mathrm{boost}}$ results in excess at low $E_{\mathrm{T}}^{\mathrm{miss}}$ and some excess at low and at m_{T} edge.

Printed in the United States
by Baker & Taylor Publisher Services